ERGONOMICS
Man in His Working Environment

Ergonomics

MAN IN HIS WORKING ENVIRONMENT

K. F. H. MURRELL

LONDON

CHAPMAN AND HALL

1971

First published 1965 by
Chapman and Hall
11 New Fetter Lane, EC4
Reprinted 1965
Reprinted with corrections 1969
Reprinted 1971
Printed Offset Litho in Great Britain by
Cox & Wyman Ltd, Fakenham, Norfolk
SBN 412 07800 7

Contents

List of Plates

Preface

Until quite recently conditions in industry were often rough. Long hours were worked in insanitary and murky workshops, often with little regard to the effects upon the workpeople who were considered to be expendable. Now, however, these adverse conditions have been recognized and so remedied that there remains little in industrial conditions to disturb the public conscience. This does not mean that conditions of work in office or factory are perfect. The obvious and dramatic abuses of the human frame may have gone, but in their place have arisen stresses and strains which, taking effect only in the long term, are generally undramatic and often unrecognized. They exist none the less.

No organized effort to study the effect of working conditions on man's performance was made until the end of World War I, when the Industrial Fatigue Research Board was set up. For the first time, men trained in the human sciences entered industry to study men at work. They made contributions which set a new standard of scientific investigation into human performance and allowed executive action on the basis of evidence rather than of hunch. The Board's work differed from the contribution of Gilbreth in America in that the principles of Motion Study which he developed were, to a large extent, based on intelligent observation rather than controlled experiment.

During the 1920s the National Institute of Industrial Psychology was founded and there was close collaboration between it and the I.F.R.B. (renamed the Industrial Health Research Board in 1929). Up to 1929 the I.F.R.B. produced 61 reports of industrial studies but in the 1930s the interest in research on performance gradually waned; only 23 studies were published and of these only about half were related to performance. Perhaps the reason for this decline was that during the '30s there was continuous heavy unemployment, manpower was expendable and not worth the research effort to make more efficient. At the same time, with a wide choice of men for the available jobs, psychologists tended to interest themselves mainly in personnel selection. A further contributing factor may have been that machines were still, on the whole, the limiting factor in performance.

With the outbreak of World War II there was rapid development in the military field. Equipment became so complex and operating speeds so high that men were subjected to such stress that they either failed to get the best out of the equipment or suffered complete operational breakdown. It thus became essential for more to be known about the limitations of performance and the capacities of men. Extensive programmes of research were undertaken, for instance, in the United Kingdom at Oxford in what became the Medical Research Climatic Working Efficiency Research Unit, and in the Applied Psychology Research Unit at Cambridge, and in America at the Psychology Branch, Aero-Medical Laboratory, Dayton, Ohio. The Armed Forces themselves carried out both research and application in close collaboration with the laboratories.

As a result of all this work, a quite substantial group of researchers developed an interest in human performance and, when the war ended, many of them continued to work in this field, supported by Government funds and mainly on military problems. Moreover, they continued to work in the separate compartments associated with the discipline to which individually they belonged. The disadvantage of this artificial separation gradually became clear. As a result, a meeting was held in my office in the Admiralty in July 1949, at which an interdisciplinary group was formed for those interested in human work. The Group's first meeting at Oxford in September was attended by interested people in far larger numbers than had been expected. A decision was reached to form a Society which should bring together anatomists, physiologists, psychologists, industrial medical officers, industrial hygienists, design engineers, work study engineers, architects, illuminating engineers; in fact anyone, whatever his background, whose work was related to any aspect of human performance. An immediate need was to find a name for this interdisciplinary field and it was finally decided to coin a new word, ERGONOMICS, from the Greek *ergos*: work; *nomos*: natural laws. It was a simple word from which other parts of speech could be derived; it was capable of translation into languages other than English; and, most important, it did not imply that any one discipline was of more importance than another.

The subsequent growth of the Ergonomics Research Society was quite rapid. Within two years the membership had been widened to include scientists from overseas and they now form a substantial part of the membership. Its annual conferences have provided an opportunity for scientists from all over the World to meet together to discuss problems of common interest and now similar Societies are meeting in several European countries. Development in the international field was sponsored by the European Productivity Agency which sent a team to the U.S.A. in 1956; it reported to an international Technical Seminar in Leyden, in 1957 (Murrell, 1958) and this was followed by a Tripartite Conference in

Zurich in 1959 (Metz, 1960) and a study seminar for Teachers of Engineering in Liège in 1961 (Laner, 1962). One result of this project was the formation of the International Ergonomics Association which held its first meeting in Stockholm in 1961.

This then is the background to ERGONOMICS. The subject is not something scientifically quite new, but rather the logical outcome of many years of gradual development. What is new is its interdisciplinary character breaking down many of the barriers, some largely artificial, between men initially trained in the various basic sciences.

The importance of ergonomics in industry is that technological development can produce effects on operatives which may not always be foreseen. These can have two important consequences. First, stresses to which the human body is unwittingly subjected over a long period may cause loss of efficiency or disability later in life. Secondly, failure to match requirements of a task with the capabilities of the operator may cause reduced output and, in the extreme case, disaster. Unfortunately, the importance of this is only gradually being realized for perhaps three reasons. First, we are all human beings tending to believe that we know all about ourselves; we are inclined, therefore, to work on experience and hunch rather than with the support of experiment. Secondly, we are rather conservative and unwilling to make changes which may have, on the short term, no obvious advantage. Consequences of poor design are often accepted as inevitable. Thirdly, the human body is very adaptable and can take a surprising amount of punishment. As a result, some of the consequences of stress may become evident only after many years.

Yet there is an increasing interest in ergonomics and a realization of the importance of designing man-machine systems as a whole and of using them in the optimum environment. It is this interest which is the reason for this book; in its preparation I have been faced with the question of the audience for whom it is intended. There is no doubt that management will want to know, basically, what part ergonomics could play; designers will want information which will help them to make better machines; some scientists may be interested in learning something of the ergonomic problems of industry, and others something of disciplines with which they are not familiar; the student may want to study the whole subject. In a comprehensive statement on ergonomics all these interests should be met; but to some extent their requirements are conflicting. At one extreme it would be possible to summarize and bring together the relevant research findings without making any attempt at interpretation. This has to some extent already been done in the Tufts University Handbook (1951), and by Hanson and Cornog (1958). The other extreme would be a 'cook book' interpretation of the research results which would permit industry to make use of the data without knowing why the recommendations have been made. This also has been done by several authors (Woodson, 1954; Murrell, 1957; Kellermann, van Wely, and Willems, 1963).

On the other hand, if ergonomics is to make a serious claim for consideration, it must do far more than provide design recommendations which will allow better chairs, dials or controls to be produced, important though these can be. As industry develops more semi-automatic or automatic processes, the emphasis is shifting from the employment of men who were 'doers' to men who are 'controllers'. This produces problems of manpower utilization and organization which the traditional techniques of industry are ill adapted to solve, while even in the current methods of manufacture there are many factors related to industrial performance which are still imperfectly understood. Ergonomic research must be able to suggest the answers to these and other related problems but at the same time engineers and managers should on their part acquire some knowledge of the structure, functions, and capabilities of man; armed with this basic knowledge the reasons for many of the suggestions made as the result of research will be more clearly appreciated.

It will be realized from what has been said already, that the recent growth of interest in research on human performance has been the result of military sponsorship. A substantial part of the current work is being supported by military funds and is carried out on a restricted population of young men. One result is that most of the published material, whether in journals or in books, has had a strong military bias and subjects of importance to industry, such as the nature of 'fatigue', the measurement of work, the mental load on control room operators or the allocation of duties between man and machine, are rarely discussed in an industrial context. This lack of industrial research has made it necessary for me to be more tentative in presentation than I would have wished.

The techniques which are at present used to achieve the most efficient working conditions or to decide how much output can be expected from an individual in a working day have changed, basically, very little in the last fifty years; motion study is still the cornerstone of the study of working methods and time study with a stopwatch and subjective estimates of 'speed and effort' are the basis of work measurement. Fifty years is too long a time for any technique to remain static; with the knowledge now available there is no reason why major advances in these and other fields should not be made; provided always that industry wants an advance and is prepared to make some effort to achieve it.

This is the crux of the matter. Too many people believe that they should be handed a packaged deal on a platter. 'If you can show us how we can use your ergonomic data', they say, 'we will give earnest consideration to its employment'. As a proposition this approach is just not 'on'. There is a large body of data available which needs only a little translation to put it to work. But this translation *must* be done by people who are familiar with the hard facts of industrial life collaborating closely with others who know something of what makes men 'tick'. I have tried to do some of this translation in this book. But I have felt it imperative also to give the reasons for the conclusions drawn and recommendations made, so that those who may have to develop them further will know something of the scientific bases which underlie them. Therefore, the first part of this book will be largely descriptive of the structure, functioning and size of the human body. In the second part research findings will be discussed and their application to design, environmental conditions and organizational problems will be dealt with. The latter will include various aspects of 'fatigue' and the measurement of activity. Wherever possible terms of technical jargon have been avoided, even at the cost of some inexactitude, but when jargon has been inevitable an attempt has been made at definition. Expert readers may find this tedious, but it is hoped that they will accept it as necessary in a text intended for a more general readership.

A great many people have helped me in the preparation of this book, with discussion, advice and information and I would particularly like to thank Otto G. Edholm, Alex Graham, Richard Hellon, John Kalsbeek, Ken Provins, Donald Wallis and Alan Welford.

To my colleagues in Bristol who have for so long had to put up with 'that book' I am particularly grateful; Stephen Griew, Peter Powesland and John Spencer have read and criticized some of the chapters, Bel Forsaith as well as reading and checking the manuscript has helped me in searching the literature, and Graham Bolton has checked the industrial implications of my ideas on work measurement. First the Nuffield Foundation and latterly the Department of Scientific and Industrial Research have provided funds on which to live.

I wish to thank the publishers of the following journals for permission to reproduce illustrations: *The International Journal of Production Research; Ergonomics; Transactions of the Illuminating Engineering Society; Engineering Materials and Design; Engineering; Instrument Practice;* and also the following bodies: The British Productivity Council; Her Majesty's Stationery Office; and the British Standards Institution.

Thanks are also due to Associated Electrical Industries Ltd., for permission to quote results obtained by J. C. Jones and his collaborators (1965) in an experiment carried out while they were members of their staff.

PREFACE

Finally, but by no means least, thanks are due to my secretaries, Miss P. Churches and Miss B. Spencer-Jones who in turn have typed the manuscript, and to D. W. H. Jackson, who checked the proofs.

East Harptree,   K. F. HYWEL MURRELL
February 1963

INTRODUCTION
The Nature of Ergonomics

Ergonomics has been defined as the scientific study of the relationship between man and his working environment. In this sense, the term environment is taken to cover not only the ambient environment in which he may work but also his tools and materials, his methods of work and the organization of his work, either as an individual or within a working group. All these are related to the nature of the man himself; to his abilities, capacities and limitations. Peripheral to ergonomics, but not at present considered to be part of the field, are a man's relations with his fellow workers, his supervisors, his management and his family. These are usually considered to be part of the social sciences but they cannot be ignored since they may play an important part in solving some problems in ergonomics. Another peripheral subject which may in part overlap with ergonomics is industrial hygiene, an important aspect of which is the reduction of toxic and other health hazards. Study of these hazards is not considered to be part of ergonomics but there may well be circumstances under which the ergonomist, the industrial medical officer and industrial hygiene engineer may work together to introduce a safe working method.

A number of scientific disciplines and technologies make a contribution to ergonomics. From anatomy and physiology we learn about the structure and functioning of the human body. Anthropometry gives information on body size. Physiological psychology deals with the functioning of the brain and of the nervous system. Experimental psychology seeks to define the parameters of human behaviour. Industrial medicine can help to define those conditions of work which may prove harmful to the human structure. From physics and to some extent engineering will come knowledge of the conditions with which the worker has to contend. It is in these areas that the principal research efforts are concentrated, the results of which, together with the knowledge accumulated, form the basis of ergonomics. Important though it is, research is only one part of the subject. To have any meaning the results must be applied and tested in the field, and since we are concerned with human work this means in practice that the results should mainly be applied in industry. Their application will be the concern principally of the design engineer, the work study engineer, and the industrial medical officer, and sometimes of the architect, the personnel officer, or the manager. There must be the closest collaboration between the research workers on the one hand and those applying the results on the other, if the maximum effect is to be obtained. Ergonomics can also find applications in the military field. In fact, 'human engineering' in America has been primarily concerned with military problems but, important though these may be, it is not the purpose of this book to deal with them in detail.

Another branch of science which plays an important part in ergonomic procedures is statistics. Traditionally the statistical methods which have been largely used in biological research have been based on agricultural experiments with their emphasis on Latin Squares. These procedures have not, however, always proved satisfactory when dealing with problems of human performance, where the small numbers often involved and the lack of control over the variability of the human subjects raise special problems. Some of these difficulties are now being met with the development of non-parametric statistics.

This then is the field covered by ergonomics. Its object is to increase the efficiency of human activity by providing data which will enable informed decisions to be made. It should enable the cost to the individual to be minimized, in particular by removing those features of design which are likely, in the long term, to cause inefficiency or physical disability. By its activities it should create an awareness in industry of the importance of considering human factors when planning work, thereby making a contribution not only to human welfare but to the national economy as a whole.

Man and His Work

Before going into detailed consideration of the various factors relating man to his work it is necessary to understand the part played by the man himself. In any activity a man receives and processes information, and then acts upon it. The first of these, the receptor function, occurs largely through the sense organs of the eyes and the ears, but information may also be conveyed through the sense of smell, through touch, through sensations of heat or cold, or through kinaesthesia. This information is conveyed through the nervous system to the central mechanism of the brain and spinal cord, where the information is processed to arrive at a decision. This processing may involve the integration of the information being received with information which has already been stored in the brain, and decisions may vary from responses which are automatic to those which involve a high degree of reasoning or logic. Having received the information and processed it, the individual will then take action as a result of the decision and this he will do through his effector mechanism, usually involving muscular activity based on the skeletal framework of the body. Where an individual's activity involves the operation of a piece of equipment he will often form part of a closed loop servo-system displaying many of the feedback characteristics of such a system. Moreover he will usually form that part of the system which makes decisions and it will therefore be appreciated that he has a fundamental part to play in the efficiency of the system. To achieve maximum efficiency a man-machine system must be designed as a whole, with the man being complementary to the machine and the machine being complementary to the abilities of the man.

To understand how these processes function it is desirable to know something of the nervous system, of the functioning and capacity of the central mechanism, of the structure of the body, the bones and the joints, and of the muscles which provide the motive power. Additionally, something needs to be known of the source of the power which drives this mechanism, and of the limits of the output which can be expected from it.

These activities are not of course carried out *in vacuo*. An individual may be working in an environment which is too cold, just right or too hot. He may be subjected to such extremes of heat that his mechanism for

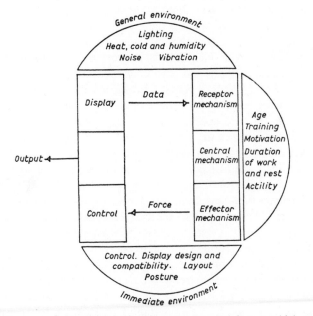

Fig. 1. Man as a component in a closed loop system and factors which may affect his efficiency.

regulating the body temperature may be in danger of breaking down unless he moves to a cooler environment. He may be subjected to noise which may be of such intensity and duration that he may suffer physical impairment of hearing. He has to be in communication with others; this may take the form of drawings, of written instructions or may be verbal. If it is the latter, noise may be an interfering factor. To be able to see he must have light, which should be of a quality and quantity adequate to the needs of the job. His performance may even be affected by the colour of his surroundings. His work must be organized so that he can maintain his maximum efficiency and interest and so that his abilities are being fully utilized. His relationships with other members of his working group should be such that his efficiency is not interfered with. These factors are illustrated in Fig. 1.

These are the important ingredients of successful work design, and anyone who sets out to study a particular work situation should take them all into consideration.

Designing Man-Machine Units

When equipment is intended for human use it should, as far as possible, be designed as a man-machine unit. There are times, however, when this may involve a compromise between the needs of the man and the requirements of the machine. This compromise may in turn require the striking of an economic and operational balance between the two. For instance, if an operating shaft comes to the outside of the machine in a position in which it would be difficult to operate, it will be necessary to balance the first cost of adding additional shafting and gearing in order to bring it out in an accessible position, against the loss of efficiency and possibly of output which would result in the shaft being left in its original position. The decision will not be made any easier when the former course of action will increase the price of the machine, whereas leaving the shaft in the original position will result in a loss of output, which will affect the customer. The decision may ultimately depend on the success of the sales organization in putting across to the customer the idea that a machine designed for efficiency in operation will have a greater output than one designed without consideration for the human operator. This suggests in turn that progressive firms will in time come to use ergonomic data in order to persuade their customers that the machines they are offering, at perhaps higher cost, are going to be worth the extra expenditure in terms of human efficiency and increased output. This approach is beginning to appear in the data sheets of one or two firms. For instance, a firm making gauges has recently been giving the reading distance for gauges of different sizes, together with the accuracy to which each instrument can be read. Another machine tool firm describes the accessibility of controls in its literature.

In most situations it is necessary first to decide whether a particular function can best be carried out by a man or by a machine. In making this decision the functions which man can perform well will be weighed against the prime cost of replacing him by a machine as well as against any supposed operational advantage. There are some things which a man can do very well, better than most machines; conversely there are other functions at which he is rather poor and which should be given to the machine whenever possible. A methodology for an analysis of these problems has been developed and should be used whenever complex systems are under consideration.

Once this role of the man in a system has been decided upon, the equipment must be designed so that he is given the greatest opportunity to function efficiently at all times. He must work in the best conditions and his activity must be organized with due regard to the demands made upon him by the job.

The first step in planning a man-machine unit will usually be to decide what information the operator will need to have in order to carry out his task, and how he best can get this information; he may be able to see things happen or he may need some form of visual, auditory or tactual display. If some form of display is required a type should be chosen which will give the information most quickly and with the least ambiguity. The limits of accuracy in reading the displays must be decided upon and the type of display which can give this information most accurately under the particular circumstances must be chosen. Another aspect of information is that obtained by communication between operators, either on the same equipment or on other pieces of equipment working with it. If communication is verbal, there can be interference from noise.

Some controls will usually be needed, and the degree of speed and accuracy likely to be demanded in their operation, the loads which are likely to be involved, and the possible need for servo assistance will have to be determined. Control types which meet requirements can then be chosen and as far as possible both they and the display should be designed round the operator. But before the areas in which controls or displays are to be placed can be prescribed, it must be decided whether the operator is to sit or stand. It will also be necessary to know if the equipment is likely to be operated most of the time by a woman, when the anthropometric dimensions applicable to a woman should be used. The range of individuals for whom the equipment is to be designed and the space and reach limits of these individuals must also be known.

An appraisal of the probable physical conditions will have to be made. The equipment may emit heat; if so, steps should be taken to reduce the heat load on the operator. If the machine will be noisy, steps should be taken to reduce the noise to a minimum, bearing in mind that the machine may be adding its quota of noise to that already produced by other machines working in the same place. Physical conditions of concern primarily to the users of the equipment are the heating and ventilation of the place where the equipment is going to be used, which should enable the operator to work in comfort; and the lighting, which should be sufficient for the visual demands of the task and located so as to avoid glare or discomfort.

Early consideration will have to be given to the general physical and mental demands of the task of operating the equipment, so as not to over-load the operator. There are many situations which can make excessive demands on an operator and overload can develop unless information reaches him at a speed with which he can cope and in a form which he can easily understand. It should be properly spaced in time and properly located around him, so that he can act upon it without undue mental stress. Machines which force the operator to work at a fixed pace can also be a source of stress. If a study of the demands of the task suggest that overload is possible, steps should be taken to bring the demands within the expected capacity of the operator and so reduce the load.

These requirements, important though they are, do not by any means cover all the situations in which men may be called upon to work and which can be manipulated by the management. Some tasks are, of course, carried out without machines at all or with the minimum of equipment, while others which involve assembly are the responsibility of a methods engineer. These arrangements are largely organizational and similar questions to those which have to be answered by the designers should be checked by those responsible for work organization.

A check list based on the points discussed above should be on the desk of anybody who is concerned, as a designer or an organizer, with human beings at work. Suggested check lists are given at the ends of chapters 7 and 17. The answers to the questions in a check list can give a specification for human performance on a projected task and action needed to meet the specification can be laid down (see also Burger and de Jong, 1962).

In this book is summarized such information as is at present available in a form in which it can be used to meet these human design require-ments. It must be emphasized, however, that there are still a great many gaps in our knowledge of human performance. As a result, much of the information is necessarily somewhat tentative. However, the use of the material which is given in the chapters which follow should enable a better result to be obtained than the use of guess work or intuition. Plainly, among the outstanding characteristics of human beings are their differences as individuals. At the best, recommendations or suggestions can be made only for the hypothetical 'average man'. This must be remembered when applying data to a particular group of people, and the individual characteristics of these people must be taken into account. Suggestions and recommendations must be tempered by the particular circumstances of the case.

PART ONE

The Elements of Ergonomic Practice

CHAPTER ONE

The Physical Basis of Man's Perception of His Environment

Man's knowledge of what goes on around him is obtained primarily through his two main senses of sight and hearing. Take these away from him and he is left in a world which can be defined, with regard to the information that reaches him from his environment, in terms only of touch, smell, taste and sensations of heat or cold. These two primary senses, and in part the sensation of heat, are dependent on transmission by waves, of information from the environment. To avoid repetition, therefore, it seems appropriate at the outset to discuss some of the properties of these waves.

THE NATURE OF WAVE PROPAGATION

The waves which give a sensation of light and radiant heat are a small part of the electromagnetic spectrum, while those which give the sensation of hearing are fluctuations of pressure in an elastic medium. Nevertheless both have the same basic characteristics, being set up by a simple harmonic motion in an emitter; the wave propagated is known as a *sine wave*, whose projection on a plane surface against time is often described as a sinusoidal curve. On such a projection the wave has two symmetrical peaks, one above and one below a mid-line and this complete movement is described as the length of the wave. The wave can thus be completely defined by two variables, its *length* (distance from crest to crest) and the *amplitude* of its displacement above and below the mid-line. Since such waves are moving through a medium at widely varying rates, it is possible also to define a wave in terms of the number of cycles which are completed in unit time, usually one second, and this is known as the *frequency*. As the speeds of both light and sound are constant for specified conditions (approximately 186,000 miles per second for electromagnetic waves, and 760 miles per hour in air at normal temperatures for sound waves) the frequency is equal to the velocity divided by the wave length. It is usual to define sound in terms of its frequency in cycles per second and light in terms of its wavelength, using as units millimicrons or Ångstrom units (Å) (1 mm = 1,000 μ = 1,000,000 mμ = 10,000,000 Å).

Consider a point source emitting waves, which spread out from the source in every direction. If the emitter is placed in the centre of a hollow sphere of one foot radius, the total emission of the source will fall on the whole surface of the sphere; that is on $4\pi r^2$ or 12·57 sq ft. Thus, in the case of light, approximately one twelfth of the total emission from the source will fall on one square foot. If, however, the radius is doubled, the surface of the sphere increases to $4\pi(2 \times r)^2$, and the energy which was distributed over one square foot will now be distributed over four square feet and will be correspondingly less intense. This relationship between distance and area is known as the *inverse square law* which can be expressed as 'the intensity of a stimulus varies inversely with the square of the distance'. The frequency of the waves, which give light its characteristic colour or sound its characteristic tone, will, however, remain unaltered, at any rate over short distances.

LIGHT

Light is that part of the electromagnetic spectrum which will stimulate a response in the receptors of the eye, Fig. 2. Its frequency which is usually

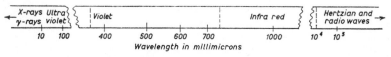

Fig. 2. The electromagnetic spectrum.

expressed as *wavelength*, determines the *colour* of a light and its amplitude determines its *intensity*.

The portion of the spectrum which is seen as light lies between 400 mμ and 700 mμ, the former frequency giving a sensation of violet and the latter a sensation of red. A mixture of most of the wavelengths within this range will give a 'white light', but if any wavelengths predominate the light will be coloured. Thus, a coloured light can be defined in terms of the intensity and wavelength of its components.

The Units of Light

Measurement of the intensity of light is based on the conception of density of *luminous flux* (F) flowing from a standard source of radiant energy which is arbitrary and is called the *international or standard candle*. The unit of luminous flux is the *lumen* (lm) which is defined as the luminous flux produced in unit solid angle at unit distance (1 steradian) by a uniform point source of one standard candle. One standard candle produces 4π lumens and its *intensity* (I) is measured in *candelas* (cd).

We are not usually so interested in the total intensity of light emitted by a source as in the amount of light which falls upon a certain object. This is known as the *illumination* (E). It is usually defined in terms of *lumens per square foot* (lm/ft^2); that is the luminous flux falling upon one square foot of area placed one foot distance from a standard candle, and it is equal to 1/12·57th $\left(\dfrac{I}{4\pi r^2}\right)$ of the total luminous flux which is emitted by a standard candle. This unit was also known as the *foot-candle*, but this term is falling into disuse, lumens per square foot now being preferred. In the c.g.s. system the equivalent unit is the *lux*, that is one lumen per square meter (1 lm/ft^2 = 10·764 lux).

Unhappily, there is confusion in the use of photometric units. Thus the candela and lumen are, at times, used interchangeably, for instance illumination is sometimes expressed in candelas/ft^2 or candelas/in.^{2} rather than in lm/ft^2. Strictly, the lumen measures the rate of flow of the luminous flux from the source, while the candela should be used when describing the intensity of a light source in a given direction, specifically towards the eye, whether the source is an emitter or a surface which is reflecting incident luminous flux.

Except when looking directly at a light source, sensations in the eye are caused by light reflected from the object in view. For this reason, when considering the amount of light required in relation to the task being undertaken, it is the brightness of the object which is of importance. This brightness depends on the illumination and on the proportion of the incident light which the object reflects. Brightness is rather a loose term which is now being replaced by *luminosity* and *luminance*. Luminosity defines the subjective characteristics of the visual sensation whereas luminance (L or B) refers to the quantity of reflected light measured photometrically. These two need not necessarily be the same, since the subjective sensations of brightness can be influenced by various factors not directly related to the light reflected into the eye. The basic unit of luminance is the *lambert* which is defined as the brightness of a perfectly diffusing surface emitting one lumen/cm^2 (or having an intensity of $\dfrac{I}{\pi}$ cd/cm^2). In most biological experiments it is used in the form of the *millilambert*, one thousandth of a lambert. Luminance is also often expressed in the *foot lambert* (ftL) which is defined as, the luminance of a perfectly diffusing surface emitting one lumen (or having an intensity of $\frac{1}{\pi}$ candela) per square foot. One foot lambert is equivalent to 1·076 millilambert. Luminance can also be expressed in terms of the *stilb* (sb) which is 1 cd/cm². It follows that one lambert is equivalent to $\frac{1}{\pi}$ sb.

As has already been mentioned, the luminance of an object depends on the amount of the incident light which is reflected. For non-specular surfaces this is defined in terms of the percentage of the incident flux which is reflected by the surface and is known as the *reflection factor* or *reflectivity* (R). The relationship between reflectivity, luminance and illumination is given by the formula

$$R = \frac{B}{E} \times 100 \qquad (1)$$

where B and E are expressed in lumens per square foot.

The foregoing definitions of the lambert, foot-lambert and reflection factor, it will have been noticed, depend on the light being reflected from a perfectly diffusing surface which has the same luminance from all angles of view. Most of the surfaces which surround us in everyday life are sufficiently non-specular for the foregoing definitions to apply for practical purposes: but, when the surface is specular, conditions are very different and the luminosity will depend on the brightness of the surfaces or light sources which are seen reflected in it as well as the proportion of this light which is reflected.

For example, if the eye looks from a distance of 2 ft at a perfectly diffusing surface, having R = 100 and a luminance of 1 ftL illuminated from a distance of 1 ft by a light source having an intensity of one candela, the area which is seen by foveal vision will be approximately 3 in.²
and therefore the 'light' reaching the fovea will be $\frac{1}{48}$ lm. But if this white surface is replaced by a mirror, the 'light' reaching the fovea will be approximately fifty times greater, but only if the eye is placed in such a position that the light source is seen with foveal vision. Thus direction plays an important part in specular brightness.

All specular surfaces do not, however, reflect all the white light which falls upon them, some wavelengths are absorbed and it is this which gives silver, chromium, brass or copper their characteristic appearance. Thus specular materials have reflection factors in just the same way as have matt surfaces, but clearly any statement on the luminosity of a specular surface, should be expressed in candelas/ft², and must be related to the direction of view and the luminance of the surface reflected. It is usually not related to the incident illumination unless this in turn has some influence on the luminance of the surface reflected.

This distinction is of great importance when considering the light

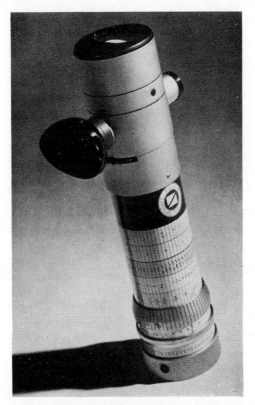

Fig. 3. The S.E.I. Photometer (by courtesy of Ilford Ltd.).

required for work on bright metal, since it is not the light falling on the workplace which may be critical in determining efficiency of seeing but the brightness of the surroundings which may be reflected in the workpiece.

Where two non-specular objects of different reflectivity are adjacent to each other there is said to be a *brightness difference* or *contrast* between them and this depends on the relative amount of incident light reflected by the two objects. Various measures of brightness difference have been used by different workers, none of which appear to be entirely satisfactory. That most commonly used in determining the light required for a task is

$$\frac{R_b - R_d}{R_b} \times 100 = C \qquad (2)$$

where R_b and R_d are the reflection factors of the brighter and darker objects respectively. This measure is useful only under strictly limited conditions, which will be discussed in greater detail in Chapter 14.

Measurement of Light in Industry

Measurement of light, usually known as photometry, can be carried out with a variety of instruments. That most commonly used in the factory is the 'Foot-candle Meter', which is calibrated to read directly in lumens per square foot. Since these instruments have a large angle of acceptance they cannot readily be used for measuring luminance; this is more easily done with a type of photometer which is able to 'look at' a small area of surface and to compare its luminance with that of a standard contained within the instrument itself. A portable instrument of this kind which can be used to obtain direct measurements in foot lamberts is the S. E. I. Photometer, Fig. 3.

The Colour of Light

Measurement of the composition of coloured light is made with a spectro-photometer, which breaks down the incident light into its component wavelengths. The results of this analysis are usually expressed as a curve in which the proportion or quantity of light of each wavelength is plotted against that wavelength. For practical purposes, however, colours are often defined in terms of three *primary colours*, mixtures of which are, subjectively at any rate, supposed to be able to produce the sensation of any other colour. This method of description is known as the *tri-chromatic system*. It was originally conceived by Young and Helmholtz and it is sometimes spoken of as the Young-Helmholtz theory of colour.

There are, in fact, two sets of primary colours, *subtractive* and *additive*. The subtractive colours, since they are the type to be found in coloured pigments, are the commonest in everyday life. These coloured materials when viewed under white light absorb many of the incident wavelengths and reflect only those which are characteristic of the particular colour. Therefore, if two of these pigments are superimposed or mixed they will reflect only those wavelengths which are common to both. The primary subtractive colours are yellow, blue-green and magenta, and when all three are superimposed they should absorb all the incident white light and produce black. Additive colours, on the other hand, are characteristic of incident light, and can be seen by viewing the coloured light sources directly or when the light is projected on a white surface. In either case the wavelengths reaching the eye (or photometer) are substantially those which are emitted by the light sources. Thus if coloured patches of light from

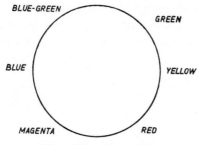

Fig. 4. The colour circle.

two sources are superimposed the result will be the addition of the component wavelengths of one with the component wavelengths of the other, and if the three additive primaries, blue, green and red, are superimposed, the result will be white since the whole of the visible spectrum will be present. This additive function is of great importance, since, as will be seen later, the eyes appear to function in an additive way.

Another way of describing colour is in terms of two pairs of *complementary* colours. Colours are often illustrated by means of the colour circle on which colours are placed according to their wavelength, Fig. 4; on this circle the two pairs of complementaries lie opposite each other. These complementary pairs are green and red, and blue and yellow, and they are characterized by their ability to cancel each other out, as it were, producing, in the case of additive colours, white or near white and, in the case of subtractive colours, black or grey.

Individual colours exhibit three characteristics, *hue, luminosity,* and *saturation.* Hue is the quality which distinguishes a colour from all other colours and depends on its predominant wavelength. Luminosity defines the intensity of the colour and, as with white light, depends on the amplitude of the radiation. Saturation is the measure of purity of the colour and the degree to which its predominant wavelength is mixed with radiation of other wavelengths. If the colour is to retain its hue this additional radiation must be equally distributed on either side of the predominant wavelengths and this will, with additive colours, cause the colour to be de-saturated towards white. With subtractive colours de-saturation will proceed towards black. When brightness is kept constant the saturation of a subtractive colour will lie between de-saturation at grey and saturation at spectral purity. Saturation is thus a relative measure of the extent by which any particular colour differs from a grey having the same brightness. The position of this grey between black and white is a measure of the colour's intensity. For a full description of colour see Murray (1952).

The foregoing description of light is in purely physical terms which can be measured with suitable instruments. The use of colour names to describe a light of a given wavelength is a matter of convenience and does not in this context relate to the subjective sensations of colour which are obtained by man when light falls upon the eye. The structure of the eye and the perceptual activity of vision are described in Chapter 5. The lighting requirements for good vision at work are described in Chapter 14.

HEAT

Heat is the energy which is stored in all matter whether solid, liquid or gaseous, and temperature is a measure of the effect of this stored energy on matter. In biological work heat is usually measured in *calories*, one calorie being the amount of heat required to raise the temperature of one gramme of water from 4° centigrade to 5° centigrade. This unit is however rather small for convenient use and the large calorie or kilo-calorie, normally abbreviated as Kcal, is more commonly used.

So long as a body is in an environment of the same temperature as its own then no heat will be gained or lost by it; but, should it be in an environment of a higher or lower temperature, it can gain or lose heat by *convection, conduction* and *radiation*. Conduction is the transmission of heat energy by direct contact between warmer and cooler solid objects. Convection is the transmission of energy to a gas or liquid which may be circulating round a warmer or colder solid body. Radiation is the exchange of thermal energy between two solid bodies which are not in contact with each other. Radiation, which can take place across a vacuum, is carried by waves of electromagnetic energy which form that part of the electromagnetic spectrum which extends from the extreme end of the visible spectrum at about 750 mμ, to about 10,000 mμ. Up to 3,000 mμ the wavelengths are short enough to penetrate the surface upon which they fall and this region is known as penetrating infrared. It is these rays which are responsible for cooking steaks under infrared grills. A wavelength longer than 3,000 mμ will not in general penetrate the surface and heating is therefore on the surface only. The general warming of a body upon which non-penetrating infrared falls being by dissipation of heat from the surface to the interior.

Many industrial processes involve men working by furnaces which may be emitting heat by radiation from the visible spectrum through the penetrating infrared to the non-penetrating infrared and the control of this radiant energy so that it will not put an undue strain on the man is a special subject in itself which will be dealt with in Chapter 12.

SOUND

Sound waves are the vibrations of an elastic medium which provide the stimuli for hearing. They usually originate from the vibration of some object such as a tuning fork which sets up a succession of waves of compression and expansion in the medium which will pass freely through it. In air these waves travel at about 1,100 feet per second (760 m.p.h.). The speed of transmission through water is about four times greater than this. Sound waves will not pass through a vacuum. The alternations of pressure set up by a sound source do not cause the molecules of the medium to move but the pressure is transmitted in a wave outward from the source, the energy being passed on from molecule to molecule in a sphere of steadily increasing radius. It will be readily appreciated that a given amount of energy at the source will be transmitted to a steadily decreasing number of molecules so that the time will come when the energy is dissipated altogether and the sound will have died out.

A single sound or pure tone can be represented as a sine wave and complex tones are made up of a mixture of superimposed waves, but however complex the sound may be it is always possible to analyse it into its component sine waves. This analysis can be done mathematically and is known as a Fourier's analysis.

As with all wave forms, sound can be defined in terms of *frequency* and *amplitude*, the frequency determining the tone and the amplitude determining the intensity. The corresponding subjective sensations are pitch and loudness.

Frequency

The tone of a sound is expressed in cycles per second (c/s). It would be quite possible to define it also in terms of wavelength but this is not usually done with sound. Since sound waves can be set up by almost any vibrating force they can cover a very wide range from perhaps only 2 or 3 c/s to many thousands, but the frequencies which are audible to the human ear range from about 20 to 20,000 c/s. Sound above and below these frequencies may be audible to some animals. The basic tones of a normal musical scale cover the lower part of this frequency range from about 30 to 4,100 c/s. Middle C is usually taken as 256 c/s and each doubling of frequency produces an increase of one octave. (The fourth C above Middle C which is to be found only on some pianos represents a frequency of just over 4,096 c/s, a frequency which as we shall see later is of great importance when dealing with the effects of sound upon the ear.) For the purpose of describing industrial sound analysis the octave bands which are used do not coincide with those used in music, the lowest octave starting at 37·5 to 75 c/s. These are as follows:

Octaves used in music and in sound measurement.

Musical Octaves

    32   64   128   256   512   1024   2048   4096

'Noise' Octaves

    37·5   75   150   300   600   1200   2400   4800   9600   (19200)

Pure sine waves are not found in everyday life. They can, however, be produced in the laboratory. The nearest to a pure tone is the sound produced by a musical instrument which consists of a fundamental tone and various over-tones or harmonics. These overtones are simple multiples of the fundamental tones and it is the arrangement of these overtones which produces the peculiar quality which enables one to distinguish one musical instrument from another. Apart from those deriving from musical instruments, most sounds are made up of irregular frequencies which are not directly related and this makes their analysis more difficult. But it must be emphasised that they can usually be broken down ultimately into sine waves of different frequency and amplitude. The mixtures of sounds which are known as noise can be expressed as a sound spectrum in which the amplitude of the component frequencies is plotted against the frequency. This is similar to the spectrum analysis of coloured light.

Intensity

As has already been mentioned, the intensity of a sound depends on the amplitude of the constituent sine waves. The greater the amplitude the greater the pressure transmitted by the sound waves. This is known as the *sound pressure*, and is a measure of deviation from normal atmospheric pressure expressed in dynes per square centimetre. Additionally sound waves will transmit energy which is an integration of all the various pressures from peak to peak, transmitted by the sound wave. This is known as *sound energy* and is usually expressed in terms of watts per square centimetre. There is a constant relationship between sound pressure

Fig. 5. (a) A sound level meter (by courtesy of Dawe Instruments Ltd).

Fig. 5. (b) An octave band analyser (by courtesy of Dawe Instruments Ltd).

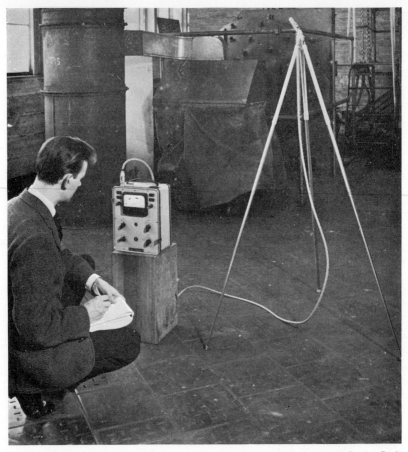

Fig. 5. (c) A sound level meter (by courtesy of Standard Telephones & Cables Ltd)

and sound energy, the energy being proportional to the square of the pressure. Sound pressure is normally used in research on hearing, in auditory research and for sound measurement in industry, while sound energy is used for research in acoustics.

Measurement of Sound Intensity

In practice we are not interested in the absolute sound pressure so much as in the ratio of one sound pressure to another. Since, as we shall see, human perception seems to be logarithmic, the range of sound pressures which can be 'heard' by the human ear is very great, and it is convenient to express these differences on a logarithmic scale. The unit normally used in instrumental measurement of sound pressure is the *bel*, which is the logarithm of a ratio of ten. This, however, is too large a unit for normal use and it is usual to divide the bel into ten parts, each part being a *decibel* (dB). A pressure of 0·0002 dyne per square centimetre has been adopted as the basic pressure for the English speaking countries, from which the decibel scale starts. This is roughly equivalent to the threshold of hearing at 1,000 c/s. Each increase, therefore, of 10 dB is equivalent to a ten-fold increase in sound energy and, over the range experienced in everyday life, is supposed to give a subjective impression of approximately doubling the loudness of the sound. Thus at 40 dB (which is about the sound level in a room in a private residence) the sound energy will have increased 10,000-fold, while the pressure will be 0·02 dynes/cm²; at 70 dB (which is about the level to be found in a typing pool) the sound pressure will be 0·63 dynes/cm² and the sound energy will have further increased 1,000-fold.

Measurements of the intensity of sound are generally made in industry by means of a Sound Level Meter (Fig. 5). For many purposes, however, this does not tell the whole story and it is now usual to use an Octave Wave-band Analyser in conjunction with the meter, in order to get the readings of the sound intensity in decibels for the various octaves shown above.

Subjective Sensations of Loudness and Pitch

The measures which have been described above are physical ones which can be recorded by suitable instruments. However, due to the variation in sensitivity of the ear at various frequencies, auditory sensations of equal loudness, for instance, would not agree with measurements made by an instrument. As a result, subjective scales must be based on the judgements of a large number of people.

Loudness Levels of equal loudness at different frequencies have been developed by Fletcher and Munson (1933) who obtained judgements by a number of subjects of sounds which were regarded as being of equal intensity to a reference tone of 1,000 c/s. Thus equal loudness contours were obtained with reference to loudness levels in decibels at 1,000 c/s. These equal loudness contours have been described in units of loudness level known as *phons*, so that the loudness of any tone is given a numerical value equal to the decibel value of a 1,000 c/s tone which is judged to be of equal loudness. From these contours (Fig. 6) it would appear that the ear is most sensitive at about 3,000 to 4,000 c/s and that the sensitivity

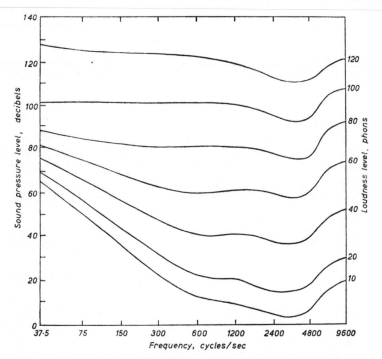

Fig. 6. Contours of equal loudness of pure tones (after Fletcher and Munson, 1933).

decreases rapidly below 700 c/s. Thus a tone of 3,000 c/s at 36 dB is judged to be as loud as a 100 c/s tone of 62 dB, and both are said to have a loudness of 40 phons.

In theory, judgements of relative loudness should be proportionate to the S.P.L. scale but, in practice, acoustical engineers found that this was not the case. So they developed a scale of loudness measured in *sones*, one sone being defined as the loudness of a 1,000 c/s tone of 40 dB (40 phons) a sound which is judged to be x times as loud as this reference is said to have a loudness of x sones. Loudness in sones and loudness level in phons are related by definition at 1 sone and 40 phons and a ten-fold increase in phons is related to a doubling of the value in sones.

Pitch Pitch is the subjective sensation which is dependent on the frequency of the sound. A scale for measurement of pitch has been devised based on a unit known as the *mel*, in which a 1,000 c/s tone of 40 dB is said to be 1,000 mels. Any tone which is subjectively judged to be twice as high in pitch as the reference tone is said to be 2,000 mels; and so on.

Beats Two tones of different frequencies but of reasonably equal intensity heard together, will produce beats if the difference between the frequency is slight. For instance, if there is a difference of 2 c/s between the tones, every half second the peaks of the frequencies of the two curves will coincide. This will cause a periodic increase and decrease in the loudness of the tones which will itself have a frequency of half a cycle per second. Thus the frequency of the beat is an indication of the extent of the difference in the frequencies of the two tones, the slower the beat the nearer the two tones will be together. If the tones differ in intensity by more than about 20 dB, the beats will cease to pulse and will become wavering. When they differ by more than about 40 dB beats will gradually cease due to the masking of one tone by another. Beats are used as the basis of piano tuning and will be well remembered by many as a characteristic of the note of some twin-engined aircraft during World War II.

The Doppler Effect If a sound source is approaching an observer (as, for instance, the whistle on a railway engine travelling at 60 m.p.h.) the frequency with which sound waves reach the ear will be greater than if the same sound source were stationary. As the sound source passes the observer and starts to move away from him the frequency of the sound will drop, which will cause a change in pitch as heard by the observer. This change in frequency, which is real and can be measured, is known as the Doppler Effect.

CHAPTER TWO

The Human Body I. Bones, Joints and Muscles

The bony skeleton of the body may be considered as the framework upon which the remainder of the body is built. It consists essentially of two major systems of levers – the arms and legs – joined together by an articulated column, the spine. In order that the bones which form the lever systems should be able to do useful work they are hinged together at the joints, the surfaces of which are covered by smooth cartilage to minimize friction. The forces which operate the levers are applied through the muscles, which contract to produce movement. The space between the muscles contains connective tissue in which blood vessels and the nerves are also to be found. It requires only an elementary knowledge of mechanics to realize that when any degree of force is required this should be obtained through the lever actions of the limbs. Just as no engineer would attempt to apply force through the bending of a piece of flexible metal hose, so the spine is equally unsuitable as a major means of obtaining force because of its lack of effective lever action.

The internal organs which lie in the chest and abdomen are only of interest in the present context in that they are the sources of supply for the fuel which is consumed when muscle does work and for this reason some attention must be paid to the circulatory system which is the means by which the fuel is distributed throughout the body. The head is of interest to us because it contains the two most important receptor organs: the eye and the ear, and the less important faculties, at any rate so far as most occupations are concerned, of taste and smell. The head also provides the bony container for the brain, the most important part of the central nervous system or central mechanism. In the discussion which follows we will deal only with those parts which will be useful in work design. For fuller information students should refer to an anatomical textbook such as that by Sinclair (1957).

THE BONES

There are in all some 206 bones in the skeleton, the majority of which are connected to their neighbours by joints which will permit movement. The bones most directly concerned in doing work are the long bones of the legs and arms and the miniature long bones of the fingers and toes. These bones are particularly characterized by having a shaft with two enlarged ends which form suitable surfaces for the joints. The surface layer of a bone is compact and dense surrounding a central portion of more open texture known as spongy bone. This spongy bone is made up of thin irregular plates which are so oriented that they strengthen the limb against the forces which they are usually called upon to bear. The development of these lines of strength in relation to the applied forces cannot fail to arouse the admiration of the engineer. The surface of the bone is in general smooth except for roughened areas which provide attachment for the muscle. These attachments may either take the form of a depression in the surface of the bone or be at the extremities of a projection which will permit the muscle to apply useful mechanical advantage. The smoothness of the bone surface may give a false impression of its solidity because in fact it is permeable, the surface being penetrated by a very large number of small canals about 1/20th of a millimetre in diameter, through which small arteries and nerves pass into the interior.

THE JOINTS

Before discussing the joints in detail it is necessary to consider the materials other than bone, which play a part in their construction. The most important of these are the stretch resistant *collagen fibres*. Also present are *elastic fibres* but these appear in relatively small quantities. Collagen fibres form the major constituent of *ligaments* and *tendons* which can therefore withstand pulling without being stretched. There are a few elastic ligaments which are made up from the elastic fibres; these are found exclusively in parts of the spinal column and help keep it in an erect position at the correct curvature. Collagen fibres are also a constituent of *cartilage*, which is of two main types. The first contains very few fibres and forms a thin layer over those parts of the bone which form the contact surfaces of the joint. The second contains a great many collagen fibres and this is generally known as *fibro-cartilage*. There is also an *elastic cartilage* which occurs mainly in the nose and the external ear. Fibro-cartilage forms part of such important structures as the inter-vertebral discs and intra-articular cartilage which are found in many joints. Muscle also plays an important part in some joint structure but this will be dealt with in greater detail later.

Of the various types of joints between bones, two are of importance in the present context. The first is the *synovial joint* (Fig. 7) which is to be found in the joints of the limbs. The second is known as a *cartilaginous joint*, the chief example of which is the joints between adjacent vertebrae in the spine.

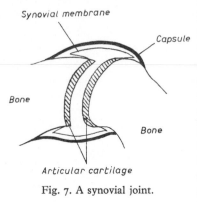

Fig. 7. A synovial joint.

In the synovial joints the two end surfaces of bone which would come into contact with each other are covered by *articular cartilage*, and the whole joint is surrounded by a *capsule* composed of *collagen fibres*, aligned in the different directions necessary to withstand the stresses to which the joint is likely to be subjected. The capsule is lined with membrane known as *synovial membrane*, which produces a fluid which acts as the lubricant for the joint and maintains a thin layer of this lubricant between the surfaces of the cartilage; it also provides nourishment for the cartilage and helps to remove any debris which may result from wear in the joint. Some synovial joints contain anular discs of fibro-cartilage, the exact purpose of which is not entirely understood, though they may assist in reducing friction in the more highly stressed joints, such as the knee, where two are to be found. Some joints have pockets of synovial membrane which may or may not be attached to the synovial cavity. These usually occur between the capsule and a muscle which is constantly in movement over it, and it is likely that they serve to reduce friction by preventing the surface of the muscle from rubbing directly on the capsule. Synovial membrane also surrounds some of the tendons, particularly those in the hand, wrist and ankle, to form a lubricated sheath or tube through which the tendon can move.

The bones of the joints are held together by the ligaments and the muscles. Since the ligaments are inelastic, in general they act as a means of limiting movement since it is only at the end of a movement that they become tight. In joints where movement is restricted, such as in hinged joints, ligaments are also to be found down each side of the joint which remain taut throughout the entire range of movement. They help to hold the bones of the joint tightly together and to resist any sideways movement which might damage the joint. The muscles operating a joint may be considered as being paired—one set flexing the joint and the other pair extending it. The forces exerted by these pairs of muscles are nicely balanced so as to obtain the required amount of movement, but being elastic both muscles of a pair are exerting force on the joint at the same time to give it stability. In addition, in some joints such as the shoulder, there are special muscles which are tight throughout the whole range of movement, whose main purpose is to control the extent of movement and to prevent overaction of the muscles operating the joint.

The direction and amount of movement to be found in a joint depend on the shape of the joint surfaces, the distribution of the muscles and ligaments and on muscle bulk. The joints of the extremities of the fingers and of the elbow and knee are simple hinge joints which permit movement in one plane. The limitations of these joints are important when determining the direction of the application of even a small amount of force. Joint surfaces which are more or less flat will permit movement in two planes at right angles to each other. This type of joint is to be found in the bones of the wrist and the ankle. The hip and the shoulder are ball and socket joints which permit a wide range of movement. The ball of the hip joint is almost completely enclosed in a deep socket; this gives great mechanical strength, but limits the range of movement. In the shoulder where a much wider range of movement is required, the socket is much shallower and as a result dislocation is quite common. The range of movement in synovial joints will be found in Chapter 4.

These synovial joints are beautifully engineered structures which work perfectly provided they are not subjected to strains for which they are not designed, and it should be the purpose of good engineering design to ensure that this does not occur. Damage may occur in a joint either by the sudden application of force in directions contrary to those which the joint was designed to withstand, or, possibly, through continued mis-use of the joint in a less acute form over a longer period. In the former the result may be a sprain or a dislocation. The synovial capsule may be torn which would cause a leakage of fluid into the joint which may fill and distend the joint cavity (e.g. 'water on the knee'). The capsule and its associated ligaments may also be stretched, particularly with a dislocation, and once this has occurred the ligaments are less able to withstand subsequent strains and recurrent dislocations are not unlikely. In the same way, repetitive unnatural strain over a long period may cause the ligaments to stretch slightly, which will ultimately result in lack of control similar to that resulting from a sprain or dislocation.

It has been suggested that a disability which is peculiar to the hand and wrist arises from the fact that the joints of the fingers are operated by muscles in the forearm acting on tendons which run in sheaths of synovial membrane. Where they pass over the bony protuberances of the wrist and knuckles, the membrane acts as a substitute for a pulley. This works well so long as the wrist is used in its normal position, but if force should be applied by the fingers with the wrist bent sharply, the lubrication afforded

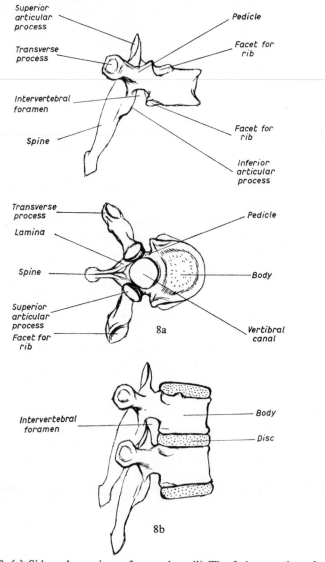

Fig. 8. (a) Side and top views of a vertebra; (b) The fitting together of vertebræ and intervertebral discs.

by the synovial membrane may be insufficient to prevent friction. If force is repeatedly applied with the wrist bent, friction may develop to a point where inflammation is set up to produce the industrial disability known as tenosynovitis. If this suggestion is correct, disability could be avoided by designing hand operations so that they are carried out as far as possible with the wrist straight.

The second type of joint which is of importance in certain contexts is the secondary cartilaginous joint which is found between adjacent vertebrae. As in the synovial joint the surfaces of the bone are covered with smooth cartilage to minimize friction, but between them is a disc of fibrocartilage. That part of each vertebra which is in contact with the intervertebral discs is known as the body and behind it lies the vertebral canal down which runs the spinal cord (Fig. 8). The inter-vertebral discs have a central elastic area surrounded by a fibrous shell and the whole apparatus of the spinal make-up is sufficiently elastic to act as a shock absorber. The discs account for approximately a quarter of the total length of the spine and, as the discs tend to degenerate gradually after middle age, stature will gradually decrease. Similarly the gradual compression of the discs throughout the day can cause the stature to be approximately one-half of an inch less in the evening, unless, of course, the day is spent lying down. The whole assembly is bound together by a broad strip of ligament back and front running from the skull to the coccyx. Individual vertebrae are joined by a series of elastic ligaments which assist in maintaining the normal curvature of the spine; they are probably the only ligaments which maintain a steady strain. The effect of the whole of this assembly is that it can bend forward readily, but not very far backward. Although the total amount of movement permitted by the mechanics of the spine is substantial the movement between individual parts of vertebrae is quite small.

In addition to bending, the spine will also permit rotation, the amount of movement varying between about 90° in the vertebrae of the neck to about 30° in the lumbar region. These rotational movements are important in controlling the extent of visual scanning, the combined movement of the neck, the lumbar region and the eyes enabling the full horizon of 360° to be scanned with the pelvis fixed in the sitting position.

The muscles which bend the body forward are mainly to be found in the walls of the abdomen, and in this position they are not able to exert a great deal of force. The body is straightened by a complex of muscles, the most important of which are the erector spinae, which are in themselves complex. These muscles act on bony protuberances which project backwards and are known as the spines of the vertebrae. If the body of a vertebra is considered as a fulcrum, each pair of vertebrae makes up a small lever. Nevertheless, the total effect is to produce an action which is less powerful than that of the larger levers which make up the limbs, and therefore, operators should, in general, not be called upon to exert force in postures which make it necessary for them to bend their backs. Knowledge of the mechanism of the spine is therefore of importance when designing equipment where lifting of any kind is involved, or where force is being applied by the arms. It is also of importance in the design of seating.

Should undue strain be put upon the mechanism of the back or should there be severe muscular fatigue, the result may be an aching back which can be caused by only moderate degrees of bad posture over a period. A more severe strain may cause damage to the muscles, to the elastic ligaments and, if the strain is sudden, the fibrous shell of the inter-vertebral discs may be torn to allow the softer, central portion of the discs, which is semi-fluid, to be squeezed out into the spinal canal, thus causing pressure on the spinal cord. This is commonly known as a 'slipped disc'.

THE MUSCLES

The muscles are made up of cells which have the ability to contract in one direction. Most of these cells are long and thin and are known as fibres. These fibres are collected together to make up the various muscles which are of three types, *striped or voluntary muscle* which supplies most of the voluntary movement, *smooth muscle* which makes up the viscera, and the unique *cardiac muscle* of the heart. For the purpose of work design, only the voluntary muscle requires attention in detail.

The fibres of voluntary muscle are long, thin and cylindrical, and may be as long as 30 centimetres. The individual fibres are connected together in bundles and these bundles in turn make up a muscle. Muscle fibres are also of two types; one of these is red and contains a pigment called myoglobin which, like the haemoglobin of the blood, has the power to combine with and liberate oxygen. Thus these red fibres form a kind of oxygen reservoir on which the muscle can draw when doing work in situations where the oxygen supply through the blood is inadequate to keep pace with the rate of consumption.

Muscles are not attached directly to bones; the ends of the muscle fibres change into collagen fibres which combine in bundles to form a tendon, which may be so short as to be almost invisible, or quite long as in the tendons which operate the fingers of the hand. The tendons are in turn attached to roughened portions of the bone surface or to the projections mentioned earlier. All the fibres of a muscle may not be connected to the same tendon so that one muscle may be attached to several separate parts of the body. Up to four separate points of attachment may be found and the nature of some muscles is indicated by their names, such as biceps or quadriceps.

When it contracts, a muscle fibre becomes about half its original length, so that the amount of movement which a muscle can produce depends on the original length of its individual fibres. On the other hand, the force which can be exerted by a fibre is irrespective of its length, so that the strength of the muscle will depend on the number of fibres which it contains. Thus the make-up of any muscle must be a nice compromise between the requirements of extent of movement and strength. Muscles in which the bundles of fibres lie parallel to each other have the greatest range of movement, but they are not capable of exerting a very large measure of force. On the other hand, the muscles with a large number of short fibres can exert a great amount of force but over only a comparatively short distance. In practice many muscles are a compromise between these two extremes in order to meet the three requirements of extent of movement, strength and minimum bulk. For instance, where strength is required without much bulk, the tendons may run up into the body of the muscle and the fibres may connect into it obliquely. This arrangement enables a larger number of fibres to act upon a tendon without increasing the bulk unduly.

The bulk of a muscle is of some importance. Obviously when the fibres contract they become at the same time greater transversely. If the resulting bulk is too great it may interfere seriously with the movement of a limb. Thus if the hand is brought up to the shoulder, the movement of the fore-arm is arrested, not by any of the mechanics of the elbow joint but by the bulk of the muscle in the forearm and the upper arm. If, after the movement has been arrested voluntarily, the wrist is pushed by the other hand towards the shoulder it will be found that further movement is possible. Thus the extent of this particular movement depends upon the muscular development of the arm and will vary from individual to individual. In the same way if the fingers were operated by muscles connected directly from bone to bone the amount of movement which would be possible would be comparatively slight because of the bulk of the muscles. The fingers are therefore connected by long tendons to muscles in the upper forearm near the elbow, these muscles being made up of long fibres permitting a wide range of movement. In this way, and in this way only, can the hands be made to operate as the useful implements which they are.

Individual muscle fibres operate on the all or none principle, thus when a graduated movement is required, more fibres are brought into operation successively. The greater the force required the greater the number of fibres which are brought into play. The number and timing of successive contractions are, in the voluntary muscles, monitored by the central nervous system through nerve fibres which may exercise control over a group of muscle fibres where gross movement only is involved or a single fibre where very fine adjustments are necessary, as for instance in the muscles which control the eyeball. It would be very rare indeed for all muscle fibres in the muscle to be in action at the same time, and it is possible that where a contraction has to be maintained for some time the fibres work, as it were, in shifts, some working while others are resting.

A nerve impulse acting on a fibre or group of fibres triggers off a complicated series of chemical reactions which cause and result from the contraction. Fundamentally adenosine triphosphate breaks down under the influence of an enzyme to adenosine diphosphate. This releases energy which enables the muscle to do work. The triphosphate must be regenerated before another contraction can take place and the energy required for this is supplied by the breakdown of glycogen to lactic acid, a poisonous by-product which in turn is removed by oxidation to carbon dioxide and water. The role of oxygen is thus to remove the by-products of the energy-producing reaction and this removal may continue for some time after muscular activity has ceased. It will thus be seen that the energy for muscular activity comes from a reaction which does not depend primarily on the presence of oxygen. This enables work to be done when the immediate supply of oxygen may be insufficient and permits the body to make sudden extreme effort which would be quite impossible were the energy to be obtained directly from the oxidation of some substance in the muscle fibre.

The oxygen may come either from that stored in the red muscle fibre or from the blood, and so long as the supply is adequate to prevent the build-up of lactic acid the work is said to be aerobic. On the other hand if the rate of work is such that the reserve of oxygen has been exhausted and the oxygen being supplied by the blood is inadequate the work is said to be partly anaerobic and the muscle builds up an 'oxygen debt' which must be repaid when the activity has ceased. In this condition there is an accumulation of lactic acid in the muscle and in the blood stream and this may cause a sensation of pain or muscular fatigue. For example if you hold your arm above your head when mending an electric fitting on the ceiling, the contraction of the muscle prevents the blood from flowing into the muscle. The lactic acid resulting from the work must be oxidized by oxygen from the reserve; when this is exhausted the lactic acid starts to build up and before very long the muscle becomes so painful that you have to stop whatever you are doing to allow the muscle to relax and so permit the blood to flow through it. Thus a muscle which is heavily contracted over a long period, even though the amount of dynamic work being done may be comparatively small, will be deprived of its oxygen supply by the effort of its own contraction. Thus work design which calls for the application of excessive force over a long period, whether operating a foot pedal, pulling a lever, working above the head or carrying a suitcase should be avoided. As far as possible all muscular activity should be intermittent so as to allow the blood to flow through the muscle in order to reduce the build-up of an oxygen debt or to facilitate the paying back of a debt which has been incurred.

It is recognized that postural muscles have to do static work, and it has been suggested that these muscles contain a larger proportion of red fibres than do those muscles which are concerned mainly with dynamic work, thus giving them a larger reserve of oxygen and probably also of fuel. Even with this advantage they also can run into an oxygen debt if a rigid posture is maintained for too long, such as when standing stiffly to attention. The same effect may be found in the typist who sits stiffly upright on the front edge of her chair while working and who may complain of backache at the end of the day. Thus it is important that the muscles which maintain posture should have an opportunity of relaxing and contracting, and work situations which compel an operator to remain rigidly in one position or to bend over a machine for long periods are thoroughly unsatisfactory and should be avoided.

When muscular exertion has been extended over a long period, even if the muscle does not build up an oxygen debt there may be an accumulation of intra-muscular fluid deriving from the products of combustion which the blood has been unable to carry away. This fluid lies between the muscle fibres and causes the muscle to swell. It may press both upon the fibres and upon the nerve endings thus producing the stiffness and soreness which is experienced after muscular activity. It may also press on the intra-muscular blood vessels thereby reducing the blood flow which in turn decreases the rate of removal of the fluid making the condition progressively worse. In most instances this intra-muscular fluid will be dissipated by rest after the activity has ended. If, however, insufficient rest is allowed, or if for some reason the dissipation is rather slow, then a fresh activity will cause an even greater increase in intra-muscular fluid in some instances. Thus a muscle which is continuously used, may become distended by intra-muscular fluid and there may be deposition of fibrous material which could ultimately interfere with the normal contraction of the muscle and cause permanent damage. For example the tibialis anterior muscle, which runs down the front of the lower leg, is used to hold up the toe either when the foot is resting on a pedal with insufficient return spring pressure to counteract the weight of the foot, or when it has to be raised above its normal position. Under these conditions this muscle may be used almost continuously throughout a working day, and intra-muscular fluid may accumulate. Unlike many muscles, however, there is very little space for it to swell, since it is held against the tibia by a layer of fibrous tissue, thus the pressure may become so high that some damage may occur to the muscle fibres, and the dissipation of the intra-muscular fluid during periods of rest may be impaired. Permanent lameness could result.

Heavy or repeated muscular exertion may cause some muscles to become hypersensitive, when they are more liable to contract than is a muscle which has not been exercised. In this state it may not even be possible to relax the muscle completely: it may have some continued residual activity when it should be at rest, and this may show itself as a tremor. This hypersensitivity, possibly combined with the pressure of the intra-muscular fluid on the fibres, may be one of the causes of the more common occupational muscular aches which take the form of localized contractions or muscular spasms which may cause local tender areas. In addition, whole groups of muscles may go into contraction spontaneously to give what is commonly called cramp, which can occur when some muscles are used continuously to exert a static force or on repeated movements of comparatively short range. Common examples of cramps of this kind are experienced by writers and telegraphists. Pain from this cause may occur only when the muscles involved are exercised and an individual may use other muscles in an effort to carry out a task without pain. In doing so he may throw an undue burden on these muscles and this in turn may lead to a further development of pain or cramp.

From the foregoing it will be realized that incapacity whether mild or severe can arise from two fundamental causes. First, the sudden exercise of force, and secondly, the continuous exercise of some muscles over a very long period. In the first instance, the muscles may be caught as it were unawares through an individual slipping or twisting in order to regain balance, or he may suddenly lift, perhaps in a bad posture, something which is too heavy. An individual who slips or falls will usually be said to have had an accident, and while this type of accident can often be foreseen and perhaps prevented, strains are always likely to occur from time to time and sometimes in most extraordinary situations. In contrast lifting awkwardly can, and should be, prevented both by the adequate design of the task and by giving an individual proper training in lifting.

The second type of disability, which develops through the over-exercise of some muscles, will almost always follow from the design of the task. A designer who has a knowledge of functional anatomy should be in a position to avoid compelling an operator to carry out tasks which continuously exercise only a few muscles or muscle groups.

THE BODY AS A SYSTEM OF LEVERS

In the introduction to this chapter we mentioned that in doing physical work the body functions as a series of levers. The component parts of these levers, the bones, joints and muscles have now been reviewed and we must look at all three as a whole.

As in mechanics, so in the body, three lever systems are to be found and these are illustrated in Fig. 9. The efficiency of these lever systems will depend on the position of the attachments of the tendons of the operative muscle to the bones which, in turn, will determine the mechanical advantage. In most joints this is not very great owing to the proximity of the points of attachment to the fulcrum of the lever systems. As a joint is moved the mechanical advantage may change through variation in the direction of pull of the muscles involved or in some instances due to a change in the mechanical advantage as a result of a rotation of the bone so that a tendon is wrapped round it. In every joint movement there is always one position in which each muscle is working at its greatest mechanical advantage, but this does not necessarily mean that all the muscles will be at their optimum position at the same time, so that the maximum force which can be obtained by any particular movement will, in general, be at the position at which the maximum number of muscles are working at their optimum. As an example of this

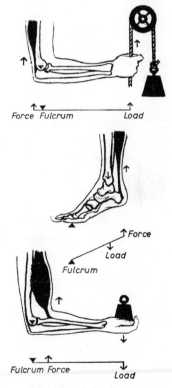

Fig. 9. Examples of the three types of lever.

effect, with specific reference to the forearm, the work of Provins and Salter (1955) is discussed in Chapter 4.

The mechanical advantage in a lever system is not the only factor which influences the force which can be exerted. The tension in the muscles is also affected by the position of the joint. We have seen that most muscles are complex structures made up of fibres of different lengths and as a result there will be limb positions in which more of the fibres are contracting than in others. There are experimental results that suggest that as a muscle becomes shorter its contractile power decreases. This suggests that less force can be applied at the end of a contraction when a muscle becomes short and fat, than at the commencement when the muscle is longer (Arkin, 1941).

The amount of force which can be exerted instantaneously, by the contraction of most of the muscle fibres in a muscle will far exceed the force which can be maintained or applied at regular intervals. It seems likely that in most movements the muscular activity is carried out by only a part of the fibres in the muscles involved, and that when any movement is repeated the different muscle fibres or even muscle groups will act one in relief of another, so, while it may be useful to know the maximum isometric force which could be applied, these forces must only be called for in extreme emergency and as a general rule the forces which should be needed for regular application should be very much less than maximum.

The efficiency with which work can be carried out will therefore depend on the efficient use of the lever systems of the body and it is important, even when relatively little force is applied, that the movements designed into a task should follow the natural movements of the limbs which will have to do the work. Due to the differential action of the muscles which has already been described, there are obviously optimum positions for carrying out different movements and these will become evident as movement is further discussed. It therefore follows that for optimum work design a knowledge of functional anatomy is of importance if inefficient or even crippling movements are to be avoided. Far too often individuals are called upon to exert even sub-maximal forces in positions which bring the effort required up towards the maximal level. For instance, a brake pedal of a car may require 90 or 100 lb for near maximum stopping (and this is not uncommon even in modern motor cars without power-assisted braking). Forces of this magnitude can be applied without difficulty if the pedal is so placed that the knee angle is about 160°, but if the seat is so high and the leg room so restricted that the knee angle is only 90°, the force required is at the limit which can be exerted by the leg.

CHAPTER THREE

The Human Body II. Metabolism and Heat Regulation

METABOLISM

We have already seen, when discussing the activity of muscle, that the energy which it uses comes from a complex chemical reaction. While muscle is one of the largest users of energy, there are innumerable other chemical reactions going on all the time to maintain the body alive, efficient and active. Food is the source from which the body draws the materials for these reactions, together with the oxygen which is breathed. The changes to which the foodstuff and the oxygen are subjected are summed up under the term *metabolism*, and the principal end products into which they are converted are water, carbon dioxide and heat, and it is in terms of heat in particular that bodily activity is described.

The three main kinds of foodstuffs are carbohydrates, fats and proteins. All three are compounds of carbon, hydrogen and oxygen but the proteins differ from the other two in that they also contain nitrogen. The carbohydrates and the fats are the main sources of energy for various bodily activities, while the proteins are used by the body to maintain the tissues, in which process the nitrogen plays an important part. When food is eaten it passes into the digestive system in which it is processed by a series of enzymes which act first in an acid medium in the stomach and then in an alkali medium in the intestines. The proteins are broken down into amino-acids, the required nitrogen is removed and the residual hydrocarbons are used as a source of energy with the carbohydrates and the fats. The carbo-hydrates are broken down into sugars, glucose principally, with some fructose and galactose. These sugars are then converted into glycogen, which provides the energy for muscular activity.

This is stored both in the muscles and in the liver. Not all the glucose is converted in this way. Some may be used immediately to supply energy for other reactions in the body and some may be converted into fat and stored in the tissue. Glucose is also stored as blood sugar in the blood-stream and in this form is a source of energy for the brain. The coma produced by overdoses of insulin (which rapidly reduces the blood sugar

concentration) is caused by the inability of the brain to continue functioning in the absence of adequate supplies of blood sugar. In general the fats
are not broken down as are the carbohydrates but are absorbed and pass
as tiny droplets into the venous blood. The blood carries these droplets
to areas of the body where they can be stored against future requirements.
When fat is drawn upon for use it passes to the liver where it is broken
down in the same way as are the carbohydrates (see Fig. 10).

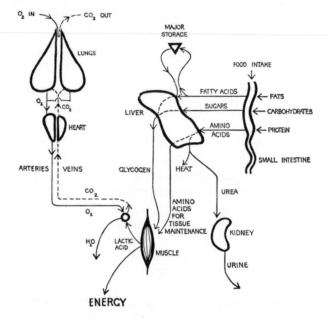

Fig. 10. Schematic diagram of the way food and oxygen contribute to the
production of energy in the muscle.

Thus far the food has provided a reserve store of fuel which can be
drawn upon for activity as needed. We have seen already how the muscle
uses this fuel to obtain energy; in the process lactic acid is formed, which
is removed by oxidation to carbon-dioxide and water with the liberation
of heat, which must eventually be dispersed from the body. The energy
expended in the complete absence of voluntary muscular activity is known
as the *basal metabolism* which represents the minimum energy expenditure
to keep the organism alive.

From the foregoing it will be clear that there are three functions upon
which estimates of the energy being expended by an individual may be
based. First, the food intake, secondly, the consumption of oxygen and

carbon dioxide production and thirdly, the amount of heat produced. All three have formed the basis of methods of measurement, which are described in Chapter 16.

HEAT REGULATION

It is common knowledge that the body remains at an approximately constant temperature. For this to happen, there must be a mechanism for the removal of metabolic heat so that, over a period, the heat generated and the heat lost are balanced. Since the body has virtually no control over the amount of heat generated, which depends on the activity being undertaken at a given time, the balance must be achieved by regulation of the heat lost.

During rest the peripheral muscles may account for about 20% of the heat generated, but during moderate exercise this may increase to as much as 75%, and the generation of heat will continue after exercise is terminated, especially when the work has been anaerobic.

To regulate heat loss the peripheral areas are allowed to cool or warm as necessary by adjustment of the thermal insulation of the tissues, the blood flow and the constriction or dilatation of the surface blood vessels. This results in an adjustable temperature gradient from the centre of the body towards the limbs and within the tissue of the body and the limbs themselves.

The amount of heat which can be lost from different parts of the body depends on the relation between the surface area and the volume, the degree of insulation afforded by subcutaneous fat and the distribution of the sweat glands. The surface area will influence the amount of heat which can be lost by radiation and convection, and the surface area and distribution of the sweat glands will influence the heat which can be lost by evaporation. At moderate temperatures and with moderate degrees of exercise, convection and radiation will account for about 75% of heat loss, and evaporation for approximately 25%, the main evaporation taking place through the lungs. At ambient temperatures above about 27°C or with heavy exercise, sweating will become the main means of disposing of surface heat and this effect is greatest when the body surface is entirely wetted. The evaporation of sweat is therefore the principal means by which the body will dispose of its excess heat, and the rate of sweating will depend on the air temperature and, even more important, on the relative humidity.

The heat generated in the body is carried to the surface by the bloodstream, and the first effect of increased heat production is an increase in the rate of blood flow and in the pulse rate. The surface veins of the skin become dilated, increasing the blood flow, and thereby raising the surface temperature and increasing the gradient between the surface and the environment. If this is insufficient to dispose of the excess heat, sweating will commence. Once the body surface is completely wetted, the amount of heat which can be evaporated will be limited by the latent heat of evaporation of water, and this for the average man will amount to about 8 Kcal/min (40 btu/min) for the nude man. Under these conditions about 1·5 litres per hour of sweat are evaporated: more sweat than this can be produced, and it has been suggested that the maximum tolerable limit for fit, acclimatized young men is about 4·5 litres in four hours, but this rate cannot be kept up indefinitely, since the sweat mechanism may become fatigued. When a man is clothed the rate of evaporation will be reduced. Should a man be working in radiant heat he would have to get rid of the heat absorbed by radiation as well as that generated within the body, and he would not therefore have to be working at a rate as high as 8 Kcal/min for his evaporation to reach a maximum.

If the rate of heat gain becomes greater than the loss, the body temperature will rise, and it is suggested that the allowable rise in rectal temperature should be 1 °C (or an increase in heart rate to 125 beats/min) which will permit it to rise to 39·0 °C (102·2 °F). If exercise and/or heat exposure are terminated at this point, the rate of recovery will depend on the ambient temperature and the relative humidity in which the recovery takes place; should the recovery not be complete when work is restarted, the allowable rise and the period of work will be correspondingly less. If the body temperature is allowed to rise too much, the metabolism will increase progressively and may set up a cycle of temperature increase during which thermostatis fails and heat stroke results. Should the temperature rise to 42 °C (108 °F) in the brain, the nervous centres will cease to function and death will follow. Conditions in the United Kingdom under which this may occur are probably not very frequent, though in industries such as iron and steel where men may be subjected to a high degree of radiant heat, or in operations such as mine rescue, or under conditions of very high humidity such as may be found in dye works or laundries, heavy work for very long periods may cause an individual to approach the limit of tolerance. Methods of determining these tolerance limits will be discussed later.

Loss of heat by sweating depends on the body surface area and this is usually taken for the 'average man' to be 1·8 square metres; of this approximately 7% is accounted for by the head, 19% by the arms and hands, 29% by the legs and feet, and 35% by the trunk. Thus a man clothed in long trousers, with a shirt with sleeves buttoned to the wrist, will expose only about 12% of his surface area; if he were to wear a singlet and trousers, the area exposed would be increased to about 30%. Should he wear brief shorts instead of trousers then the area exposed would be doubled to 60%. Although it is not customary to do so in this country, the wearing of shorts in a hot environment might thus be very advantageous. In this connection, however, it must be remembered that under conditions of high radiation the clothing itself (especially if it is made of wool) acts as a barrier to the radiant heat. Surface area is not the only determinant of heat loss, since some parts of the body sweat more heavily than others (the head and the back of the hands most readily, and the palm and sole of the foot least), but even so the limbs remain the main means of disposing of excess heat and their exposure to the atmosphere will greatly facilitate heat loss.

Cold

If the loss of heat from the body is greater than heat production the body temperature will fall. In order to prevent this the surface blood vessels will become constricted, keeping the blood away from the surface of the body, and sweating will cease. The movement of air along the surface of the skin is reduced by the erection of the hairs to give 'goose-flesh'; shivering will commence so as to produce muscular activity which in turn will generate additional heat. If the cold is sufficiently severe the surface will become numb and the sensation of cold will gradually be reduced. If the body temperature is reduced to 30°C (86°F), shivering will decrease and will cease by the time the body has reached 27°C (80°F), and the blood will steadily become more viscous so that the heart will have increasing difficulty in pumping it around the body. (This can put a strain on the heart so that some older people are less able to withstand extreme cold.) Below 27°C the individual becomes sleepy and at 25°C death may result.

Climatic conditions in the United Kingdom make it unlikely that individuals in industry will be subjected to temperatures extreme enough to cause the symptoms described above, except in the cold storage industry. On occasions those working in outdoor occupations such as farming or transport may be subjected to severe cold, and failure to survive these conditions will be largely due to inadequate foresight and clothing.

BODY TEMPERATURE

The temperatures which have been discussed above are those of the deep core of the body. Skin temperature, on the other hand, can vary very widely – by as much as 20°C – and may fall, even under normal conditions, to below 30°C (86°F). Under conditions of hard physical exercise the skin temperature may rise substantially above the deep body temperature. Skin temperature measurements are, therefore, not a reliable indication of events in the body, and for this purpose rectal temperature is usually used, although this may also be rather more variable than the deep body temperature itself.

The deep body temperature varies cyclically throughout the 24 hours over about 1·5°C (3°F), being for most people at its lowest in the early hours of the morning and at its highest in the early hours of the afternoon (Fig. 11). During periods of low body temperature, many bodily activities are decreased, mental activity is diminished, digestion slows down, the

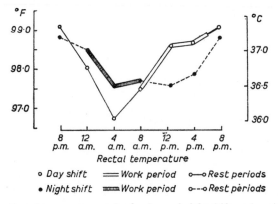

Fig. 11. Rectal temperature cycles for day and night shift workers (adapted by B. Metz from Van Loon, 1963).

secretion of urine is decreased; all of which is conducive to a good night's sleep. On the other hand, when bodily temperature is higher the body is more active, a condition which is favourable for work.

It appears that in many individuals this temperature variation is produced by their activity patterns. Thus if a man is normally working during the night and sleeping during the day, his diurnal temperature rhythm may be exactly 12 hours out of phase with that of a man who is normally working by day and sleeping by night. Should there be a change in activity pattern, as when transferring from day-shift to night-shift, in most people the temperature cycle will change to be in accordance with the new activity pattern. However, this change may take some time, perhaps up to four days, so that in the common shift pattern of a week on and a week off nights, an individual may, for the first three or four days of each week, be working when his bodily mechanism demands rest and be resting, or trying to rest when his body is most active. On the fifth, and probably last day of his shift, the temperature cycle may have adjusted to new conditions and then he will go back on to the opposite pattern again. The extent to which the temperature cycle influences the health and efficiency of men on shift work is not by any means clear but the evidence is reviewed in Chapter 19.

CHAPTER FOUR

The Human Body III. Body Size. Limits of Movements and Functioning of Limbs

The design of all equipment must be considered in relation to both the size of the individuals who are going to use it and the movements which they can make without difficulty or strain. Individuals vary greatly in their size from dwarfs to giants (though luckily these extremes occur very rarely indeed), on the other hand the limits imposed on movement by the joints and tendons show much less variability. For our purpose it is necessary to have some information about the body measurements of the population which is likely to use the equipment, and the systematic measurement of body dimensions, using specialized instruments, is known as anthropometry (the measurement of man). Broadly speaking anthropometric data may be used for two purposes. First, to determine the size and shape of the equipment which a man is to use, and second, to determine the space in which a man is to work. The dimensions which are required for these two purposes will not always be the same. Data on the limits of body movement and on the forces which can be exerted by a man will come from studies in functional anatomy or, as it is sometimes called, biomechanics. For convenience we will deal with the two separately, although in the practical situation the body dimensions and the functioning of the limbs must be considered at the same time.

ANTHROPOMETRY

Most of the anthropometric data which have been published have come from large or small scale surveys carried out for a special purpose, in many instances on a selected population. The difficulties of organizing surveys with adequate numbers are very great and it is not surprising therefore that the majority have been undertaken for use in the design of military equipment. Some surveys on civilians have also been carried out, but they have been mainly for the purpose of designing clothing or footwear, and many dimensions which would be useful to machine designers have not been taken. A further difficulty is that the dimensions measured have usually been those of the static individual whereas the designer will want dimensions which relate to an active individual. For instance, it is not much use knowing the reach of the hand to the tip of the fingers when what is really required is the best distance from the body at which a control can be grasped and operated efficiently. Some work is being done on the anthropometry of the active man and this may perhaps be called functional anthropometry in the same way as we have distinguished functional anatomy from the descriptive anatomy of the cadaver.

In recent years the use of photography has been developed as an alternative to the taking of measurements directly from the subject. To be successful, the photographic procedure should be carefully planned in order to produce pictures on a standard scale (Tanner and Weiner, 1949). An important advantage of this method is that the records are permanent, and measurements may be taken at any time. This means that if the need for certain dimensions is not realized at the time of the original survey, they may be obtained later from the photographs. The method also takes up much less time for the subjects than the normal method of anthropometry and thus enables very large numbers of individuals to be dealt with speedily. In addition it may be possible to get data which cannot readily be obtained from measurement of the living body. A large library of photographs of this kind has been collected in the Department of Growth and Development, Institute of Child Health, The Hospital for Sick Children, Great Ormond Street, London w.c.1. Cinematography may also be used in anthropometry to determine some functional dimension. Darcus (1948) for instance, used it to study the complex movement of the head during visual tracking.

Photography has also been used by Dempster et al. (1959) in functional studies of the envelope described by the hand when gripping a handle maintained in a constant angular orientation in relation to reference co-ordinates. The region thus defined they call a 'kinetosphere'. When kinetospheres representing a related series of hand orientations are superimposed they produce a 'strophosphere' which defines the area in which certain types of movement, for instance rotation, can take place. The method used was to photograph a flashing neon lamp on the handgrip, so that the path moved by the hand in relation to the body was traced on the film.

Since this technique is somewhat time consuming, only 23 subjects were used in the investigation reported. The results therefore give only mean dimensions and not ranges, and the means are useful only to the extent that the subjects used constituted a sample representative of a general

Table 1. Mean Stature from Various Anthropometric Surveys

Source	Subjects	Date	N.	Country	Mean Stature ins.	Percentiles 5th	95th
			Males				
21	Company Directors	1959	2,756	U.K.	70.0	65·0	73·3
1	Air Force	1943	2,961	U.S.	69·2	65·4	73·1
2	Air Force	1950	4,062	U.S.	69·1	65·2	73·1
3	Civilians*	1945	1,959	U.S.	69·0	64·5	73·8
4	College Students	1928-30	23,122	U.S.	68·68	—	—
5	Air Force	1952	3,331	U.S.	68·54	64·2	72·7
6	Army	1946	24,508	U.S.	68·47	64·3	72·64
7	Civilian Drivers	1953	306	U.S.	68·35	64·64	72·48
7	Army Drivers	1953	2,380	U.S.	68·23	64·21	72·50
8	Civilians – all ages	1951	9,957	Ireland	68·19	—	—
9	Army (white)	World War II	60,695	U.S.	68·14	64·6	72·3
10	Army (white)	World War II	385,937	U.S.	68·01	—	—
9	Army (coloured)	World War II	7,300	U.S.	67·89	64·3	72·0
11	Naval	1953	399	England	67·9	—	—
10	Army (coloured)	World War II	78,638	U.S.	67·7	—	—
12	Civilian Administrators	1941	339	Scotland	67·8	—	—
13	Air Navigators – age 18-30	1955	7,084	France	67·45	—	—
12	Civilians, semi-skilled	1941	572	Scotland	66·4	—	—
14	Civilians – all ages	1943	27,515	U.K.	66·02	—	—
			Females				
3	Civilians*	1945	1,908	U.S.	64·9	60·8	69·1
15	Air Force	1952	851	U.S.	64·0	60·3	68·2
6	Army	1946	8,549	U.S.	63·92	59·9	68·0
16	Civilians – all ages	1947	5,000	Holland	63·8	—	—
17	College Students	1928-30	17,127	U.S.	63·57	—	—
20	College Students	1927	460	U.K.	63·4	—	—
18	Civilians	1941	10,042	U.S.	63·16	—	—
19	Civilians – all ages	1951	4,995	U.K.	63·02	—	—
14	Civilians – all ages	1943	33,562	U.K.	61·92	—	—
20	Factory Women (20½-55½)	1927	1,645	U.K.	61·45	—	—

*Most subjects wearing shoes.

Sources of the Data

1 Ashe *et al.* (1943)
2 Hertzberg *et al.* (1954)
3 Hooton (1945)
4 Diehl (1933a)
5 Daniels *et al.* (1953b)
6 Quartermaster Corps (1946)
7 McFarland *et al.* (1953)
8 Hooton and Dupertuis (1951)
9 Army Services Data (World War II)
10 Karpinos (1958)
11 Geoghegan (1953)
12 Clements and Pickett (1952)
13 Ducros (1955)
14 Kemsley (1950)
15 Daniels *et al.* (1953a)
16 Sittig and Freudenthal (1951)
17 Diehl (1933b)
18 O'Brien and Shelton (1941)
19 Kemsley (1957)
20 Cathcart *et al.* (1927)
21 May and Wright (1961)

population. The development of this photographic approach to functional anthropometry, while not entirely new (the present author used a similar but simpler approach during the war to define space requirements in armoured fighting vehicles), gives an indirect method of arriving at three-dimensional limits related to various movements and positions of the limbs.

Another method for obtaining similar functional data has been developed by Dempsey (1953). His apparatus consisted of a wooden frame with ten horizontal and five vertical rods in line with the seat reference point which could be slid in and out by the subject. On the end of each rod, a knob switch or similar control could be mounted. The subject was seated and was slowly rotated about the seat reference point through a number of positions, in each of which he adjusted the length of the rods for comfort-

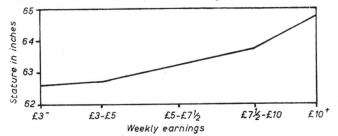

Fig. 12. A relationship between women's stature and their earnings (after Kemsley, 1957).

able manipulation of the control. In this way an envelope for a particular control movement could be built up.

In addition to obtaining the data directly from the subject or from photographs it is also possible to obtain certain dimensions by statistical means from other known dimensions, since a number of body measurements are quite closely correlated. Thus, when these relationships are known, a number of dimensions may be obtained from key measurements such as the stature, weight and girth which are much more easily obtained from large populations than detailed measurements. In this way functional measurements which have been obtained on a comparatively small selection of individuals can be extended to cover much larger sections of the community.

The differences in the results of some anthropometric surveys are shown in Table 1 for an important basic dimension, the stature. It will be seen that there is no great degree of unanimity even among the American figures, which predominate. People in the United Kingdom are somewhat shorter than white Americans, and even within the countries themselves there may be regional differences. Furthermore, there appear to be differences in different occupational groups. Thus the problem of deciding what figures to recommend for the use of designers is made difficult by the existence of national, regional, occupational and age differences which have been demonstrated in the different surveys. For instance, in Clements and Pickett's (1952) investigation of the stature of Scotsmen it was found that the administrative classes were taller than semi-skilled or labourers, the mean difference amounting to 1·4 in., while a sample of Company Directors whose average age was 51 years was found by May and Wright (1961) to be 4 in. taller on the average than a sample of men from social classes III, IV, and V (Kemsley, 1950). Occupational differences are also to be found in women as is shown in Fig. 12. Age also appears to have an influence. It would seem that stature is likely to reach its maximum at about the age of 20 and to decrease at the rate of about 0·03 in. per ten years in men (Clements, 1954). A study of a small number of women also revealed a difference of about 1·3 in. between women of an average age of 23 and women of an average age of 55 (Kemsley, 1957). Whether this change with age is due to an actual reduction in stature or is due to a general tendency for the height of the population to increase is not entirely clear.

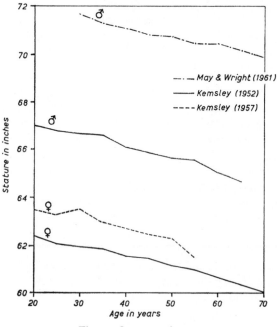

Fig. 13. Stature and age.

The extent of this increase is indicated by a report of 1875 which suggests that at that time the stature of American white males examined for military service was in the region of $67\frac{1}{2}$ in., which is less than the figures for American males given in Table 1. Thus, although there can be very little doubt that people do get shorter as they get older, it is not possible accurately to quantify the shrinkage on the basis of comparisons of individuals of different ages. This can only be done on the basis of longitudinal studies. An indication of the extent of the changes of height with age is given by Fig. 13.

Some of the problems we have discussed may to some extent be mitigated by the high probability that whatever the mean stature of a sample, any given body dimension of length will be very nearly a constant proportion of the stature. This means that once stature has been determined from a sample, dimensions not available in the sample can be obtained by proportion. This method is used by Barkla (1961) to obtain dimensions for the design of seats. For instance, he derives from various surveys the following proportions of sitting height to standing height for males: 0·526; 0·529; 0·530; 0·530; 0·522; 0·520; 0·522; 0·528; 0·529; 0·534; 0·520. If we take the extremes of these values for a man of 68 in., the difference between the two derived sitting heights is rather less than 1 in. (which may or may not be of importance according to the context). For practical purposes a reasonably reliable estimate of the average proportion would be of 0·526, giving a sitting height of 35·77 in., from which the extremes deviate by 0·57 in. and 0·41 in. respectively for a 68 in. man. To make it possible to apply the data given in Table 2 to individuals whose stature differs from the mean, the best estimate of the proportion of the various body dimensions is included where appropriate.

The method of proportions cannot so easily be applied to dimensions which represent the bulk of the body. Thus a man 5 ft 9 in. tall may have a waist circumference of about 30 in., compared with 41 in. for a stout man of the same height, while the latter's hip breadth is likely to be $2\frac{1}{2}$ in. greater. On the other hand the arm length and the sitting height of these two individuals is likely to be much the same. Age will also be an important factor in measurements which relate to bulk. Women over 55 years of age will, on the average, be 6 in. larger round the waist and $3\frac{1}{2}$ in. round the hips than women aged 20 to 24 years. They are also likely to be about 67 lb heavier. Women who have borne children will, when allowance has been made for age, tend to be shorter and plumper than spinsters (Kemsley, 1957).

From what has been said it will be seen that it is impossible to arrive at dimensions for an 'average' man, and therefore we should consider just how important it is that a table of dimensions which may be used by designers should be an exact representation of the population for which the equipment is being designed.

Ideally, equipment should be made adjustable to suit a predetermined proportion of the population, and for this purpose the percentiles are often more important than the mean. But while adjustments of this kind can be provided in such things as the seat and steering wheel of a motor car it is usually impossible to provide adjustments in controls of a heavy machine tool. When equipment has to be non-adjustable, the designer should choose dimensions suitable for a pre-determined type of user. Whether this be for the smaller user or the larger user will depend on circumstances. For instance if a reach is involved, a dimension at which a small person can reach should be used. Whereas if head-room or leg-room is being considered, then the larger individual must be allowed for. The use of critical dimensions would seem to be of importance only when space is very restricted as in the cockpits of single seater aircraft or racing cars, though in some instances artificial restrictions are introduced by designers who seem to be unaware of the physical need of the human being. The modern motor car is a typical example of this, the rear seat passengers often being given insufficient room for their legs in order, apparently, that the luggage may be carried in greater comfort.

We might, then, reasonably conclude that in most industrial situations anthropometric dimensions are not usually critical provided that certain limits are not exceeded. Thus the mean height of a population may be relatively unimportant if ample head room is given in any event; but the range or 95th percentile is important in indicating the lower limit for the head room. There will be times, however, when more exact dimensioning may be required. For this purpose measurements may be made on a sample of the relevant population, data may be extracted from published work, or data such as those given in Table 2, may be used. If tolerances are small or the inter-relationships between dimensions are not self-evident, fitting trials may have to be undertaken. In this case some caution will be required because people who are inexperienced at making subjective judgements of comfort, usability and so on, instead of judging a new situation on its merits are apt to produce results which are related to a situation with which they are familiar. For this reason more objective assessments should be made where possible. Seat heights, for example, can be arrived at with the aid of electromyographic measurements or measurement of the pressure between the thigh and the seat (Lundervold, 1958). Observation of the behaviour of the sitters is also instructive: sitting on the front edge of the seat and not using the back rest, or sitting slumped with the buttocks

Table 2.   Anthropometric Measurements for a Civilian Working Population

	Males Proportion of Stature*	Mean	Percentiles 5th	95th	Females Proportion of Stature*	Mean	Percentiles 5th	95th
Standing in Shoes*								
Stature	—	69	65	73	—	65	60½	69
Eye level	0·936	64½	61	68½	0·938	61	56½	64½
Elbow height	0·608	42	39	45	0·603	39½	36½	43
Shoulder height (maximum height for controls)	0·811	56	52½	59	0·805	52½	49	56½
Symphysis height (minimum height for controls)	0·507	35	32	37½	0·490	32	30	35
Gluteal furrow	0·468	32	29	34½	0·437	29	27	32
Lower leg length (crotch height)	0·485	33½	30½	36	0·462	30	28	33
Sitting								
Height above seat	0·522	36	33½	38	0·532	33½	29½	34½
Eye level above seat	0·477	31½	29	33½	0·468	29½	25½	30½
Elbow height	—	9½	7½	10½	—	9	8½	10
Buttock to back of knee	0·280	19	17	20½	0·286	18	16	19½
Buttock to patella (knee cap)	0·346	23½	21½	25	0·350	22	21½	24
Bitrochanteric (buttock) width	—	14	13	15½	—	14½	12½	16
Minimum height of controls above seat	—	1	—	—	—	1	—	—
Shoulder height (max. height of controls above seat)	0·338	23	21½	25	0·333	21	19	22½
Popliteal (lower leg under knee) height (no shoes)	0·242	16½	15½	17½	0·246	15½	14½	16½
Patella height (no shoes)	0·316	21½	20	23½	0·310	19½	17	21½
Breadth across both knees	—	8	7	9	—	7½	6½	9
Standing or Sitting								
Reach for operation of controls from c.l. of body at shoulder height	—	(27)	24	(29½)	—	(24)	21	(26½)
Maximum span at working level	—	(60)	55	(64½)	—	(55)	50	(59½)
Normal span at working level	—	(48)	43	(52½)	—	(44)	39	(48½)
Maximum forward reach from front edge of bench	—	(20)	18½	(21½)	—	(17)	15	(18½)
Normal forward reach from front edge of bench	—	(13)	12	(14)	—	(11)	10½	(12½)
Shoulder width	—	18	16½	19½	—	16	15	18
Shoulder to elbow	0·213	14½	13½	16	0·210	13	12	14½
Elbow to finger tips	0·272	18½	17	20	0·270	17	16	18½
Foot length	—	10½	10	11½	—	9½	8½	10
Elbow width	—	17½	15	19½	—	16	14	18½
Dimensions relating to equipment								
Clearance between seat and work surface	—	6½	5	7½	—	6½	5	8
Minimum leg room	—	—	—	33	—	—	—	30
Back of seat to front edge of bench	—	—	—	14	—	—	—	14
Clearance between floor and work surface	0·338	23	21½	25	0·350	22	19½	24

*Standing dimensions and proportions are based on stature to which has been added 1 inch for shoes in males and 2 inches in females. When using proportions for other statures heel height must be added before calculating standing dimensions.

The dimensions given in this table are the author's 'best estimate' from available anthropometric surveys. Dimensions are given to the nearest half inch; to do otherwise would imply an accuracy which the data do not possess; they are not necessarily accurate to the nearest half inch but they are probably reasonably consistent within themselves.

forward, are signs that a seat is too high. Extensive fitting trials for motor vehicles have been carried out at the Regie National des Usines Renault by Wisner (1961). The equipment which he used is shown in Fig. 14.

In the initial stages of design it is often useful to be able to set out the geometry of a projected work place. For this purpose some of the dimensions given in Table 2 need to be modified. For instance, in determining

Fig. 14. A test rig for anthropometric fitting trials (from Wisner, 1961 – by courtesy of the Controller of Her Majesty's Stationery Office).

the loci of the centre of a handwheel when a range of shoulder and elbow angles are to be accommodated, the geometry must be based on the approximate hinge points of the limbs and not on the total length of the limbs (Murrell, 1963 a; Wisner and Rebiffè, 1963).

For a side elevation the hinge points for the shoulder and hip may be placed on the centre plane of the body at 5 in. from a seat back or from the buttocks, the shoulder hinge at 2 in. below shoulder height and the hip hinge 3½ in. above seat when sitting or at 47% of the stature when standing. In the frontal plane the shoulder hinge may be located 1½ in. from the shoulder width and the hip hinge 2½ in. in from the hip width. The shoulder hinge – elbow hinge may be obtained by subtracting 1½ in. from shoulder – elbow dimension, and elbow hinge – finger tips by subtracting 1 in. If the hand is closed an additional 4 in. is subtracted. For the hip hinge – knee hinge subtract 6½ in. from the patella – buttock dimension (in some instances it may be useful to use seat reference point (S.R.P.) to knee hinge in which case only 1½ in. is taken off) and for knee hinge – sole of foot 1½ in. is subtracted from patella height.

An alternative method is to use manikins of which a number have been proposed (e.g. Lovell, 1954 or Wisner, 1961); these should be of several sizes in order to cover the range of potential users. But whichever method is used in the initial layout it should be taken only as a general guide as to whether a design is likely to be functionally correct and should be supplemented by comprehensive fitting trials with individuals representative of the range of population for which the equipment is intended.

In addition to the difference between individual men to which we have referred, there are also quite big differences between men and women, and it is important when designing a machine which is to be used exclusively by women that women's dimensions are used. When the sex of the user is not known beforehand, the situation becomes more difficult since the range which will have to be accommodated will have to be rather greater. For instance, to cover the 90% range of men and women, the elbow height will have to extend from 45 in. down to 36 in., instead of to 39 in. for men only. In some machine tools this difficulty can be overcome by standing the shorter person on a raised platform, but this will not overcome difficulties of reach. Therefore, if the machine tools are likely to be operated by women as well as by men the dimensions for women's reach must be used rather than those for men.

In some situations caution will be necessary when combining male and female data. In settling the dimensions of motor vehicles for instance it might seem appropriate to bring together the whole range of male and female dimensions to give the ranges which should be allowed – but as Wisner (1961) points out it cannot necessarily be assumed that the female population as a whole will be representative of a population of women drivers, nor, in setting up percentiles, can it be assumed that in a driving population there is the same proportion of females to males as there is in the population as a whole. Table 3 shows the proportion of women to men drivers in France in the ten years from 1950 to 1959. Clearly a decision would have to be made on whether ranges of adjustment should be based on percentiles related to the dimensions of all men and women or to the dimensions of *drivers* only. If the latter view is taken, fewer of the smaller women will be catered for because of the smaller proportion of women drivers in relation to men. In order to make the necessary calculations and to obtain the required dimensions, percentiles other than those given may be required. These are easily obtained by plotting known percentiles (or

Table 3.   Proportion of Women Among the New Licensed Drivers in France
from 1950 to 1959
(after Wisner, 1961)

Year	Total	Men	Women	Percentage Women
1950	454,269	412,741	41,528	9
1951	470,931	416,584	54,347	11
1952	517,678	442,984	74,694	14
1953	529,165	441,183	87,982	16
1954	544,572	441,343	103,229	19
1955	579,597	453,363	126,234	22
1956	599,412	434,239	165,173	28
1957	610,883	454,527	156,356	25
1958	665,521	487,883	177,638	23
1959	663,291	474,316	188,975	29

standard deviations when these are available) on normal probability paper against the required parameter. In the same way the proportion of any population for whom a given dimension is suitable (or unsuitable) can be obtained. For instance, if the front seat of a car is 19 in. long only about 12% of adult women can sit reasonably upright without pressure at the back of the knee, the remainder – 14 million of them – are likely to be uncomfortable if not incapable of complete control of the vehicle. This information is derived from the percentiles for buttock to back of knee given in Table 2 assuming an adult female population of 16 millions.

As an example of the extent to which anthropometric data are used (or not used!) in the design of road vehicles, one of a series of studies which have been carried out since 1949 by the Harvard School of Public Health, Department of Industrial Hygiene, will be quoted. This group concentrated first on evaluation of trucks and buses but later studied private passenger cars (McFarland and Domey, 1960). In the course of their studies this group have set a series of dimension standards and it was against these standards that the models produced in 1957 were evaluated. Only male dimensions were used (they appear not to have had female dimensions available). Seventeen vehicles in all were examined and it is evident from the data which are given in the report that in a number of respects the vehicles did not conform to the recommended dimensions. In some instances deviations would be in a direction which would not be of importance but in other instances a deviation could cause inefficiency, danger or discomfort. For example, it is recommended that the seat depth should be between 18 and 19 in. for a man. The front seats of four of the vehicles were at the upper limit and the rear seats of two of the vehicles were above the upper limit. Since the buttock to back of calf dimension for women is about 2 in. less than that for men, the front seats of only two of the vehicles and the back seat of only six of the vehicles would fail to cut into the back of the knees of women drivers or passengers. Against this, a dimension of 17 in. which might be considered suitable for women was said by the authors to be too shallow for men. In considering roof clearance in relation to seated height, all the vehicles had sufficient roof clearance to cover men up to the 95th percentile, but 12 of the vehicles would have been too low for the tallest men in the sample of 313 civilian truck drivers which had been measured by the authors. In any event this would have meant that for a number of drivers their heads would have been pressing against the roof and no clearance would be allowed for headgear or for vertical movement due to bumps. The authors recommend a clearance of 40 in. and only one of the vehicles examined had this clearance, the remainder being too low. In this instance, because women tend to be shorter than men, the clearance would probably be adequate for them unless they were wearing exceptionally tall hats. Of even more importance is the clearance between the top surface of the brake pedal and the lower edge of the steering wheel. The minimum distance recommended is 26 in. and only seven of the vehicles had this clearance, which means that on the others it is possible that the driver would hit his knee when moving his foot from the accelerator to the brake. No less than four of the vehicles had a distance 2 in. less than that recommended. Finally the steering wheels on most of the vehicles were adjustable; it was recommended that when the wheel was adjusted to the middle of its travel, there should be a minimum distance of 15 in. between its lower rim and the seat back, to allow for accessibility, for the abdomen of the driver and to avoid damage in the event of impact. Only one vehicle complied with this requirement. In one instance the range of movement allowed a clearance of 13 to $7\frac{1}{2}$ in. while another had $13\frac{1}{4}$ to 9 in., so that a driver who was putting on weight would have great difficulty in getting into these vehicles and would probably be rubbing his abdomen against the near edge of the steering wheel. This report shows very clearly that insufficient attention was given to the available anthropometric data in the design of American motor cars in the year 1957.

Body Weight

A knowledge of the weights of different parts of the body is of importance in designing controls on which a limb must rest, for instance, a foot pedal or a 'dead man's handle'. The return spring pressure on this type of control should be sufficient to support the weight of the limb being used when it is at rest so that no muscular effort is required to keep the limb from activating the control in the rest position. Obviously the proportion of the weight in different limbs will vary with the build of the individual and between sexes, and from the figures available it seems that the biggest variation will be found in the trunk and the thigh. In most instances determinations of the proportions of body weight have been undertaken on the living body (Lay and Fisher, 1940) but the results so obtained appear to agree fairly well with those obtained by Fisher (1906) who dissected a frozen cadaver.

Clearly only part of the weight of some of the limbs will need to be supported by a control. For instance, when the foot of a seated operator is resting on a pedal the weight to be supported will be that of the foot, the lower leg and part of the weight of the thigh. Although the limbs cannot be treated strictly as if they were levers with friction-free joints, it will be sufficiently accurate for our purpose if the centre of gravity of the thigh, lower leg, upper arm and lower arm and hand are taken as being at a distance of 45% of the length of the limb from its upper extremity and that of the foot at approximately one third of the distance along its length from the heel. The proportions of the body weight found in each portion of the body together with the weight in pounds for a standard man of 65 kilos (144 lb) is given in Table 4.

Table 4. The Weight of Body Members

Part of the Body	Percentage of Whole Body		Weight for a Standard Man (lb)
Head	7·0		10
Trunk	44		63
Thighs	22	for two	32
Lower legs	10	,, ,,	14½
Feet	4	,, ,,	6
Upper arms	7	,, ,,	10
Lower arms	4·5	,, ,,	6½
Hands	1·5	,, ,,	2
	100		144

BODY MOVEMENT AND STRENGTH

In this section we shall deal with the forces which can be applied by the body parts in different positions and with the limitations on the movement of the limbs which may be imposed by structure of the joints or by bulk of the muscle. These are dealt with together because they are interrelated. Their importance lies in the limitations which they impose on the design of controls, since the position and the range of movement of a control, and the force required to operate it, will depend on these functions of the limbs and joints.

The decision whether a particular control movement is satisfactory or not can be difficult since many control movements are not made by one joint alone but are the sum of the movements of several joints. This is particularly true of movements made by the arm and hand. Unfortunately, most of the information available from research is from the movement and strength of a single joint. This is not very surprising since the number of composite movements which can be made with the arm must be very high indeed. In spite of this it is possible to draw a number of general conclusions which should be valuable to the designer.

General Principles

The accuracy with which a muscular movement can be carried out will depend partly on the length and volume of the muscles involved and partly on the number of muscle fibres controlled by each motor nerve ending. As a result leg movements are somewhat less accurate than movements of the arms and hands. From this it follows that where great accuracy of control is required the hand and arm should be used rather than the foot. The accuracy with which controlled movements can be made can depend on the position of the limbs. There is very little evidence on this point relating to the legs, but with the hand and arm, the further the hand is away from the body the less accurately can movements be made. The solid area in which an accurate control movement can be made by the hand is quite small, the greatest accuracy being obtained when the hand is in close to the body at approximately elbow level.

In the same way as accurate control of a limb is dependent on one group of muscles acting as antagonists to another, so also control movements which are the result of one limb acting in antagonism to another are more accurate than those made by one limb alone.

In the laboratory most investigations of the forces which can be applied by the limbs or the body have been made using isometric force, that is when there is virtually no movement of the part being tested, measurements in this situation being typically made by means of strain gauges. In practice isometric force comes into play in most control movements, even though free movement precedes or follows the application of isometric force. For instance, when the brake pedal of a car is used there may be free movement in the pedal before the pressure is applied. When the brakes are actually operating the amount of movement in the pedal is negligible. In the same way if a weight is being lifted, isometric force is applied to the weight until the force exceeds the weight of the object, at which point the object will start to rise. If the weight of the object is greater than the force which can be applied to it the force remains isometric and the object remains on the ground. Therefore studies of isometric forces will give useful information about the torques or the pressures which can be applied to controls in different positions. But Darcus (1954) has suggested that although these may show the optimum position at which the greatest force can be exerted, it does not necessarily follow that these are the best positions when the forces are sub-maximal. Caldwell (1961) has, however, shown that the limb positions which are most favourable for the production of maximal forces tend to be also those most favourable for the maintenance of sub-maximal forces. Darcus further suggests that where maximal forces have to be exerted, an individual will unconsciously adopt a position which will bring his limbs to the position in which the greatest force can be applied.

Most limbs have a neutral position in which, when they are unsupported, muscle activity is at a minimum. The maintenance of this position is therefore the least fatiguing and probably the most comfortable. When the limb is supported, comfortable positions may not coincide with the neutral position of the unsupported limb: for example, when the leg is unsupported the neutral position of the knee joint is with the leg in a straight line with the thigh, but should the subject be sitting or lying, the most comfortable and probably the most relaxed position is with the knee bent at an angle of approximately 110°.

With these points in mind we will discuss the range of movement and the force which can be applied by the different parts of the body separately. Physiologists use a number of terms to describe the movements of the limbs and it would seem to be best to use these rather than to try and translate them into plain language. They may be explained by considering the movement about the shoulder joint. Consider a man standing upright, with his arms hanging at his side, with the palms inwards. The raising of the arm sideways to the horizontal position is known as *abduction*, and the opposite movement is known as *adduction*. These two terms are generally used to describe the movement of a limb away from or towards the centre line of the body. The movement of the arm above the horizontal is known as *elevation*. The raising of the arm forward to the horizontal is *flexion* and the reverse movement back again is known as *extension*. If this movement is continued backwards it is called *hyper-extension*. If the forward movement is continued above the horizontal it would be described as *forward elevation*. In all these movements the shoulder is acting as a

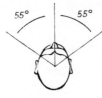

Rotation

(a) Flexion
(b) Hyperextension

Lateral bending

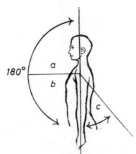

(a)Forward elevation (b)Flexion
(c) Hyperextension

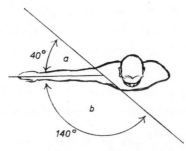

(a) Hyperextension   (b) Adduction

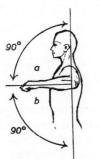

Rotation in abduction
(a) External   (b) Internal

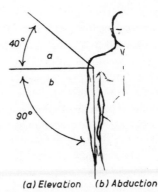

(a) Elevation   (b) Abduction

Fig. 15. (a) Movements of the head; (b) Movement about the shoulder.

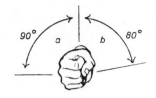

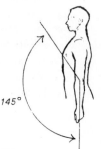

(a) Supination (b) Pronation

145°

Flexion

(a) Adduction
(b) Abduction

(a) Dorsiflexion (b) Palmarflexion

Fig. 15 (*c*) Movement about the elbow; (*d*) Movement about the wrist.

hinge in the direction in which the arm is being moved. If the elbow is bent so that the forearm is horizontal and the upper arm vertical, the hand can be moved to a position against the stomach or outwards to the side. This movement is achieved by a rotation of the shoulder joint and is known as *internal* and *external rotation* from the neutral position. If the arm is then abducted to the horizontal and the elbow is flexed to 90°, the hand being forward, a movement downward from the horizontal is *internal*, and upward is *external rotation in abduction*. If the arm is now adducted to the side and the elbow flexed so that the lower arm is horizontal, the neutral position of the hand is with the palm facing inward. From this position the hand can be rotated so that the palm is either upwards or downward. This movement is not a rotation in the sense that the movement of the joint of the shoulder has been described, since it is obtained by crossing the two bones of the lower arm. It is not, therefore, described as rotation, but the movement of the palm upward is described as *supination* and downward is *pronation*. These terms are illustrated in Fig. 15.

The Head

The movements of the head are shown in Fig. 15(a). At first sight it may seem somewhat surprising that the movement to right or left is only 55°. The movement of the eyes is greater (90° to 100°) and this will give the subjective impression that the head is turned further than it really is, so that the eyes and head together give a coverage for foveal vision of more than 180°. Although the head will bend backward 50°, this is, on the whole, a rather uncomfortable position, at the extreme it becomes difficult to swallow. A more comfortable position is to move the head forward and it is generally recommended that displays which have to be watched for long periods should not be placed above the eye level of an individual sitting upright (Fraser, 1950).

The Trunk

Most movements of the trunk are made in conjunction with other parts of the body and tasks which prevent an erect posture should be avoided. The rotation of the spin is approximately 40° to right or left and this movement should not be used with flexion to give a twisting lift, a common cause of back injury (Glover, 1960).

The Upper Limbs

The ball and socket joint of the shoulder has the greatest range of movement of any joint in the body and when to this is added the flexing and rotational movement of the elbow, there is only a comparatively small area behind the body which cannot be reached by the hand. This does not mean, however, that controls may be put in almost any position surrounding the body. The areas in which accurate control movements can be made are quite small and the areas in which force can be applied efficiently are not very much greater. The average range of movement of the shoulder is given in Fig. 15(b), of the elbow in Fig. 15(c), and of the wrist in Fig. 15(d). The neutral position for the arm when a person is standing is with the arm hanging straight down at the side with the palm inward. When a person is sitting it would seem that, provided the forearm is supported, the most comfortable position is with the elbow flexed at 90° and the palm facing in. Although the neutral position of the hand is usually taken as being when the fingers are straight, it seems likely that the most relaxed position is with the fingers flexed.

In most instances when forces are applied by the upper limbs, they are not acting alone, but in conjunction with other parts of the body. For instance, the signalman in an old type of signal cabin will pull or push levers using the whole of his body weight, or the oarsman using a sliding seat will obtain the major part of his drive through straightening the legs and swinging the trunk backwards. It is only at the end of the stroke that the arms come into play. There are so many combinations of arm action with body weight that if we are to obtain any indication of the forces which can be developed by arm action alone we must consider only measurements

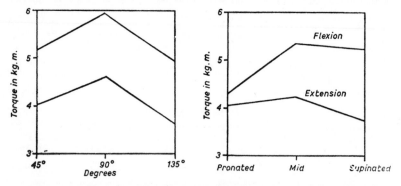

Fig. 16. (a) The effect of elbow angle on maximum torque (after Provins and Salter, 1955); (b) The effect of forearm position on maximum torque (after Provins and Salter, 1955).

which are obtained when the operator is sitting with an adequate back rest and, for pulling movements, with an adequate foot rest also. Therefore the forces which are given in the discussion which follows apply to an operator in this position.

Provins (1955a) has studied the isometric forces which can be applied to a hand wheel in different positions, using the elbow joint alone or the shoulder alone. He shows that the force which can be applied by the shoulder is about one-third greater than that which can be applied by the elbow, and that this seems to be independent of the degree of flexion of the elbow joint between 0° and 90°. He comes to the conclusion that 'the maximum torque which is exertable on a handwheel, or the maximum length of time which a lesser force can be maintained is greatest when, with the elbow bent at a right angle, the plane of the handwheel is tangential to an arc about the shoulder joint' and 'the absolute position of the hand-wheel (within rather broad limits) or the direction of attempted rotation

appears to be of little importance'. He gives the maximum torque for his
subjects as 10·4 Kg/m for the shoulder and 7·63 Kg/m for the elbow.
Provins and Salter (1955) have investigated the torque which can be
exerted at the elbow joint by flexion or extension against a strain gauge.
They found that the torque which could be applied by the non-preferred
hand was about 80% of that which could be applied by the preferred hand.
The maximum torque was exerted when the elbow was at a right angle
with an almost equal decrease of torque at angles of 135° or 45°. In all
three positions the strength of flexion was almost half as great again as that

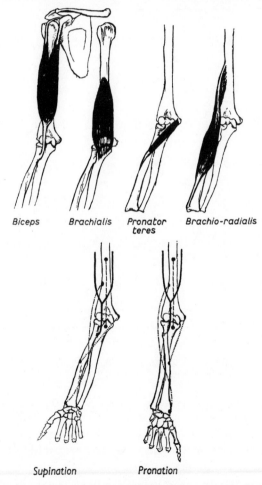

Biceps   Brachialis   Pronator   Brachio-radialis
                              teres

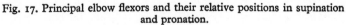

Supination   Pronation

Fig. 17. Principal elbow flexors and their relative positions in supination
and pronation.

of extension. The effect of rotation of the forearm at the elbow joint was also shown to be important. In both flexion and extension the neutral position was found to be superior to both supination and pronation. The former was superior to the latter in flexion and inferior in extension. The torques which were obtained in these experiments are shown in Fig. 16.

Provins and Salter also give in their paper an interesting theoretical discussion of these results based on the mechanical advantage of the various muscles which are operating on the elbow joint in the three positions of

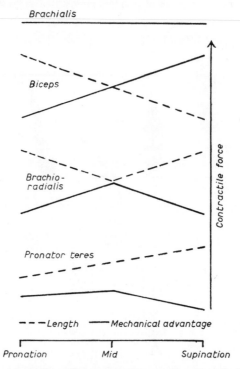

Fig. 18. Suggested effect of forearm position on the contractile force of the elbow flexors (after Provins and Salter, 1955).

rotation. The four muscles principally involved are the brachialis, biceps, brachioradialis and pronator teres. The layout of these muscles is shown diagrammatically in Fig. 17. Three of these four muscles are affected by the rotation of the forearm, changing both their mechanical advantage and their length. The fourth, brachialis, is unaffected. When the forearm is rotated from the fully supinated to the fully pronated position, the tendon of the biceps will wrap itself round the radius, reducing the lever effect and the mechanical advantage. At the same time the biceps will become progressively longer. During the same movement pronator teres will become shorter, while the length of the brachioradialis is probably greatest when the arm is fully pronated or supinated. Both these muscles have their greatest mechanical advantage in the neutral position (Fig. 18). Since a shorter fat muscle can exert more force than a longer thin muscle, rotation from pronation to supination will increase the contractile force of the biceps. The length of the brachioradialis is greatest and the mechanical advantage least in the extreme positions, so that the greatest force which this muscle can exert is at the neutral point. The pronator teres is at its shortest but its mechanical advantage is rather poor in the pronated position. In the mid-position its mechanical advantage is better but the muscle is somewhat longer, while in the supinated position the muscle is at its longest and its mechanical advantage is again poor. Thus we see that all three muscles will exert the least contractile force in the pronated position while being at their combined best in the mid-position. In the supinated position the increase of contractile force of the biceps will to some extent offset the decrease in the other muscles, so that the force produced in the supinated position is somewhat greater than that in the pronated position, although both are less than in the neutral position. This would explain the results which have been found in the experiment.

Provins (1955b) has further studied the forces exerted by the elbow and shoulder and his results suggest that the torque exerted by both arms together can be taken to be the sum of the two sides acting separately. The results also show that when the body weight is not used, greater force can be exerted by the elbow of a sitting than a standing subject, since in this position there is fixation of the trunk. Two points should be noted in connection with this work of Provins. First, that it was done on a small number of subjects, of which two or three were women and second that the torques which he gives are those at the hand grip. The actual torque which could be applied to a shaft would depend on the radius of the crank or wheel being used. In the absence of any better data these torques may be used for guidance, with these limitations borne in mind. The position of the hand will also influence the weight which can be lifted by the action of the elbow with the forearm flexed at 90°. If the palm is supinated, approximately 55 lb (24 kilos) can be lifted, compared with 40 lb (18 kilos) if the hand is pronated. This is in accordance with Provins' suggestion that in the supinated position, the biceps, which does the work in this position, is at its shortest, and has the greatest mechanical advantage, whereas in the pronated position, the main work is probably done by the extensor carpi which has a poor mechanical advantage.

Pulling and pushing movements in a horizontal direction on vertical levers will involve both the elbow and the shoulder joints. Hugh-Jones (1947) has shown that for a seated operator with an adequate foot rest the greatest pull is exerted when the arm is fully extended. When the handgrip is 9 in. above seat level and about 33 in. from the back of the seat the mean isometric pull is about 80 lb and this pull decreases as the hand nears the body to 70 lb at 29 in. and 60 lb at 23 in. Maximum push by the preferred arm is obtained when the arm is not quite straight, the push being obtained by straightening the elbow from about 165°, thus at a distance of 29 in. from the seat back a push of approximately 125 lb can be expected, but if the lever is nearer at 25 in. or further away at 33 in. the push drops to about 105 lb. The effect of height of handgrip above the seat is shown in Fig. 19.

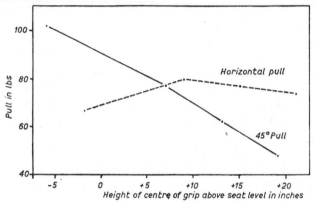

Fig. 19. Effect of control height on pulling force (after Hugh-Jones, 1947).

Results obtained by Hunsicker (1955) expressed in terms of elbow angle are shown in Fig. 20. While they confirm Hugh-Jones's finding that force decreases as the hand nears the body, they do not show that pull is greater than push, or that the greatest push is obtained by straightening the arm from about 165°. Against this McAvoy (1937) agrees with Hugh-Jones that the maximum push is developed with the arm bent and the maximum pull with the arm straight. It will thus be seen that the maximum pull or push can be obtained over a very short distance of travel, about 3 in., and only when the direction of movement of the controls is towards the shoulder of the operator. Much of the research on pulling and pushing movements with the arms has been summarized by Orlansky (1949) and recommendations based on these research results will be given in the control design section.

Study of the rotational movements of the forearm, as when rotating a handgrip or tightening a tap, shows that in pronation the maximum torque of about 8 ft/lb is achieved when the hand is fully supinated, force decreasing as the hand is turned from this position to the fully pronated position. The reverse is true of supination, the maximum force being

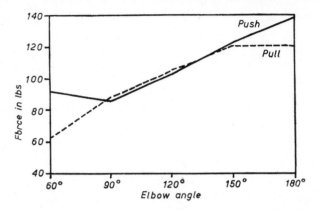

Fig. 20. Mean values for pushing and pulling for 55 subjects (after Hunsicker, 1955).

applied at the end of the movement, thus a greater force can be applied if the right hand is rotated anti-clockwise from the supinated position than clockwise from the pronated position (Salter and Darcus, 1952).

From these results it is possible to make some recommendations on the position of the upper limbs for the operation of controls. It would seem that for a wheel type of control where a high degree of force is called for, the shoulder and elbow should be used, implying that the wheel should be horizontal. This, however, does put the wrist in rather an unsatisfactory position and force might more easily be applied to a handgrip sticking vertically up from the rim of the wheel which would enable the hand to assume a neutral position. On the other hand, where the forces to be applied are rather less, it would appear that a wheel in the vertical position is to be preferred, placed in relation to the body so that the elbow is flexed at 90°. These findings seem to be followed in some motor vehicles; heavy vehicles such as buses often have almost vertical steering columns, whereas some private motor cars have the column very nearly horizontal. The main disadvantage of the wheel in the latter position is that for maximum efficiency the length of the column should be adjustable, it being clear from the above research that a vertical wheel held at arms length is unsatisfactory for normal motoring.

The results also suggest that the handgrip on a lever which is to be pulled or pushed should be arranged so that the hand is in the neutral position; that the position for maximum pull is at arms length or for maximum push when the elbow is bent at 165°. In practice this is rather difficult to achieve if there are more than 2 or 3 in. of travel on a lever, and suggests that the distance of travel of such controls should be quite short.

The Lower Limbs

When standing, the neutral position for the knee joint is with the legs in a straight line with the body, and in this position the knee angle is given as 180°; however, it is probable that when sitting or lying down the most comfortable position is with the knee flexed to between 70° to 130°.

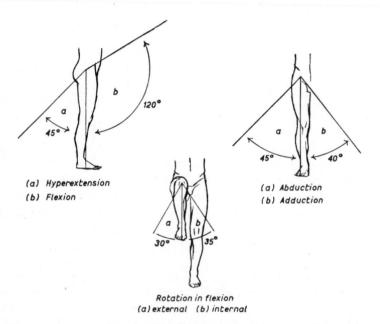

(a) Hyperextension
(b) Flexion

(a) Abduction
(b) Adduction

Rotation in flexion
(a) external (b) internal

Fig. 21. Movement about the hip.

Like the shoulder, the hip is a ball and socket joint, but owing to the need for greater strength, movement is restricted to a much greater degree than it is in the shoulder. The average flexion of the hip is about 120° (see Fig. 21), but this in most people can only be achieved when the knee is also flexed. Hyper-extension is about 45°. The leg can be abducted to about 45° and any position up to this limit appears to be comfortably maintained at rest when seated. Adduction up to 40° however, is not a position which can be maintained comfortably; in general, forces should not be applied when the leg is adducted. Rotation is greatest when the leg is flexed to about 90° at the hip, when there is approximately 60° external, and 30° internal rotation (rotation in flexion). If the hip is extended to the normal position the rotation falls to about 35° external and 20° internal (rotation in extension). The knee is a hinge joint with about 135° of flexion (Fig. 22), unlike the elbow it has no rotation. The tibia in the leg is larger than the fibula which lies against it, so that the bones cannot cross as can the radius and ulna in the arm. This is probably just as well because if it were possible to turn the sole of the foot upward as one can the palm of the hand, walking would probably be an unstable process. It will thus be seen that the lower limbs have been developed for stability and strength, while the upper limbs have been developed primarily for mobility.

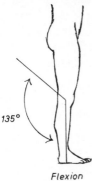

Fig. 22. Movement about the knee.

The neutral position for the ankle is with the outer edge of the foot at 90° to the axis of the leg. From this position the ankle can be flexed 20° or extended approximately 35° (see Fig. 23). This range of movement is very important when designing foot pedals. The ankle joint can also be moved approximately 30° to right or left. but this movement should not be called for in the design of controls.

Force is normally applied by the leg through pushing, though it is possible for the seated operator to apply a force by swinging the leg. The force which can be applied downwards by a standing operator will depend

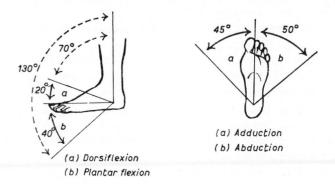

(a) Dorsiflexion
(b) Plantar flexion

(a) Adduction
(b) Abduction

Fig. 23. Movement about the ankle.

on his weight. If the force required is greater than the operator's weight, he will simply stand on the control without having any effect. A seated operator on the other hand, when supplied with an adequate backrest and pushing in an almost horizontal position, can when the leg is straightened from a knee angle of about 165° produce forces in the region of 700 lb for a single instantaneous thrust of short duration. This push falls very rapidly as the knee becomes more flexed dropping to about only 60 lb at 90°. The point of contact of a pedal with the foot is of importance. For maximum thrust the pedal should be pushed with the heel or instep. If

Table 5.   Summary of Principal Limb Movements

Limb	Joint Involved	Movement Direction	Extent
Head	Spine	Rotation	110°
Arm	Shoulder	Abduction and elevation	90° + 40°
Arm	Shoulder	Hyper-extension, flexion and forward elevation	45° + 90° + 90°
Lower arm	Elbow	Flexion	145°
Hand	Wrist	Dorsiflexion	65°
		Palmar flexion	75°
		Adduction (out)	45°
		Abduction (in)	40°
		Rotation	180°
Leg	Hip	Flexion	130°
		Hyper-extension	50°
		Abduction	45°
		Adduction	40°
Lower leg	Knee	Flexion	160°
Foot	Ankle	Dorsiflexion	20°
		Plantar flexion	40°

the toe only is used the maximum thrust drops to about one third, since the muscles of the foot are unable to resist the strain and the foot is bent backward by the force being applied by the leg.

For convenience the most important movements of the body are summarized in Table 5. The adaptation of these data to control design will be dealt with when individual controls are discussed. When considering what forces are acceptable in operating controls, it must be remembered that most of the experiments which are reported in the literature have been done on young men and women, and therefore the reported maximum loads are usually for individuals of under 25 years of age. It is, however, commonly recognized that physical strength does not remain constant throughout life and it has been shown experimentally that it is likely to reach its maximum at about the age of 25 and to decline gradually thereafter both in men and women. In general it can be said that the strength of a man of the age of 60 will be about 15% less than his strength at the age of 25. It must be remembered however that there are substantial individual differences in physical strength and that unless especially selected populations are being designed for, the loads which are called for in the design of equipment should allow for the weaker portion of the population which is likely to use the equipment.

Speed of Movement

When an individual responds to a signal for action, the response is made up of two parts, the response initiation time and the physical movement or motor activity. In many psychological experiments on these functions the subject is required to respond by pressing or releasing a key. In these circumstances the movement is so short and occupies so little time relative to the response initiation time, that it can be and usually is ignored. In other instances a larger movement is required and the timing of the response initiation time is often, but not invariably, taken from the appearance of the stimulus signal to the commencement of the movement which is indicated by the release of a micro-switch or the moving of a stylus, or in some similar manner. Unfortunately not all experimenters have separated the response initiation time and the movement time in this way, especially when the movement time is relatively short in relation to the response initiation time, but have put the two together under the more commonly used term 'reaction time', e.g. Goldfarb (1941). As a result the term reaction time is ambiguous and for this reason latency or response initiation time are preferred by some researchers and the latter will be used here.

The laboratory situation which we have been describing represents the kind of experiment in which the subject makes a discrete response to a discrete signal and this is our major source of knowledge on the various factors which can influence both components. In real life, however, discrete movements are not as common as continuous response to a stream of stimuli, but it is suggested by some workers that an apparently continuous response is in fact made up of a rapid succession of discrete responses. Studies of continuous response during tracking with and without preview have been made in the laboratory, Welford (1958) and Griew (1958b), in which the subjects were required to follow a random wavy line which was being traced on a moving strip of paper. The target line and the response line drawn by the subjects were available for subsequent study. An essential feature of this type of continuous response is the unpredictability of the stimulus. In other situations where the stimulus becomes predictable it is possible for the subject to learn the characteristics of the response required. The discussion which follows will therefore deal with the situation in which the stimuli, whether discrete or continuous, are unpredictable. The situation in which they are predictable is already covered by texts on motion study and industrial training.

Response Initiation Time (R.I.T.)
The factors which influence the response initiation time are the type of physical stimulus used, the intensity of the stimulus, the duration and size of the stimulus, the number of stimuli and the degree of preparedness of the subject to respond. The information content of the stimulus will be discussed very briefly in Chapter 6, in terms of information theory and will not therefore be referred to here.

The R.I.T. can depend upon the sense modality involved. In general the speed of response to sound or touch is shorter than to light (Teichner, 1954), but the absolute differences obtained will depend on the specific circumstances under which the activity is carried out. R.I.T. is influenced by training, so that statements on the relative speeds of response to, say, sound and light must take the state of training of the subjects into account. In one experiment (Forbes, 1945) the mean R.I.T. to an auditory signal was 192 msec and to a visual signal was 289 msec. The standard deviation of R.I.T. to the auditory signal was 36 msec compared with 43 msec for the visual signal. Results from other workers give for auditory and visual 'reaction times' 153 and 174 msec (Todd, 1912) respectively and 217 and 258 msec (Baxter, 1942) respectively for simple response. These 'reaction times' almost certainly include some movement time and other experiments on one sense modality alone suggest that when R.I.T. only is measured the response can be made to sound in 120 to 150 msec and to light in 150 to 180 msec (e.g. Seashore and Seashore, 1941). Baxter and Travis (1938) measured the R.I.T. in the vestibular sense modality and found that the response to perception of movement from rest was significantly faster than perception of change of direction of movement. Sometimes signals are received through more than one sense modality at the same time. If these signals are of the same strength, the R.I.T. is likely to correspond to that of the sense modality which, operating alone, would give the shortest R.I.T. If, however, the stimuli are not of the same intensity, the R.I.T. will correspond to that of the strongest stimulus.

The nature of the response can also influence the time to respond. Manual response to auditory stimuli was found by Venables and O'Connor (1959) to be faster (270 msec) than verbal response (370 msec). The corresponding times for visual stimuli were found to be 320 and 410 msec. For the effect of temperature on R.I.T. see Chapter 12.

For a stimulus to be perceived, it must differ from its surroundings and the speed of response will depend upon the stimulus intensity. There is some evidence also which suggests that the shape of the stimulus, in the sense of being definite and conveying immediate meaning rather than being hazy and of doubtful meaning, will influence the time to respond. These phenomena are well known to vehicle drivers: in fog for instance the presence of another vehicle on the road is more readily appreciated when it displays bright headlights rather than less bright and smaller parking lights. Not only is the stimulus more intense but the two head-lights being large and spaced at a recognizable distance apart will be instantly identified as belonging to a car. Parking lights could belong to two bicycles. This effect is probably at its greatest near the threshold of discrimination. So long as the stimulus does not approach the threshold it seems that the duration and size of the stimulus will have little influence on the response initiation time, on the other hand stimuli which are too great in magnitude may have an adverse effect, particularly if they are unexpected. Expectedness, in fact, plays an important role in determining response initiation time; in the experimental situation a warning is often given before the stimulus is displayed and the period which elapses between the warning and the stimulus is often called the fore-period. (An every day example of a fore-period is the amber of the traffic lights which warns the driver that the lights are going to change either to green or to red.) Telford (1931) experimenting with an auditory stimulus found that simple reaction time was at a minimum of 240 msec when the fore-period was 1 sec in duration; with a 2 sec fore-period the reaction time was 250 msec, and with a 4 sec fore-period it was 273 msec. With a short fore-period of only 0·5 sec the reaction time was longest at 333 msec.

Experience in operating equipment may often enable an operator to predict when events are going to occur and these predictions will enable him to respond more quickly to the cues when they arrive. In many instances a means of giving a warning to the operator that signals to which he must respond will occur can be with advantage designed into the equipment, but if the warning occurs too soon or too long before the event some of the advantage will be lost.

It not infrequently occurs that before a response is made a subject must coordinate a number of stimuli which may be in the same or in different sense modalities. So many combinations of stimuli are possible that experiments upon them are usually specific but they all point to the conclusion that the more complex the group of signals to which the operator must respond, the longer his R.I.T. is likely to be. In one experiment, Lemmon (1927) required his subjects to respond to one of two lights by lifting a finger from a key corresponding to the side on which the light appeared. Under these conditions the R.I.T. was 290 msec. When, however, he displayed more than one light on each side and the subject had to respond to the side showing the greatest number of lights, the R.I.T. increased to 650 msec when three lights were shown against four, and to 741 msec when four lights were shown against five.

Complexity of the response may also under certain circumstances have an effect on R.I.T. Griew(1959) reports an experiment in which groups of younger men (aet. 20 to 26 years) and older men (aet. 50 to 57 years) were required to move a stylus from a starting disc to a target designated by one of two, four or eight lights. The target was either a plain disc or a disc with a hole in it, the subject in this case being required to hit the disc and then enter the stylus into the hole – a somewhat more complex response. No difference between the two conditions was found with the younger subjects but with the older subjects the R.I.T. was significantly longer (at the 1% level) when the more complex response was demanded, the difference being about 70 msec.

There are a number of factors within the individual himself which may influence his speed of response and of these age and sex would seem to be the most important. The times which have been given in the preceding paragraphs have been obtained from men, usually young; under similar conditions the R.I.T. for women would normally be longer than that for men and R.I.T. of both is influenced by age. In one study by Bellis (1933), men and women were required to respond to auditory and visual signals by pressing a telegraph key. The results, shown in Fig. 24, indicate that the shortest 'reaction times' fall between the ages of 21 and 30 years and that women are slower than men especially at the extremes of age. Similar results in relation to age have been obtained by a number of other workers

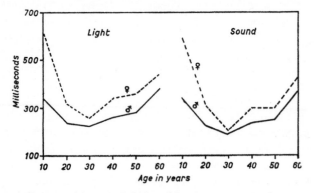

Fig. 24. Changes with age in R.I.T. to light and sound (after Bellis, 1933).

and their results have been summarized by Welford (1959b). In tests of R.I.T. involving response to auditory stimuli with different limbs Seashore and Seashore (1941) showed that men were faster than women. Differences have also been found in studies of braking on cars. Data from the Seashores' experiment are given in Table 6, which in addition to showing the sex difference also suggests that response is faster with the hands than with the feet though there appears to be no difference between the hand and the foot on either side. These results are confirmed by Baxter (1942) who found no difference between the right and the left

Table 6.   Mean R.I.T. in Relation to Limb and Sex
(from Seashore and Seashore, 1941)

Limb	Mean R.I.T. in msec	
	Males	Females
Right hand	147	171
Right foot	174	197
Left hand	144	168
Left foot	179	200

hand with either auditory or visual stimuli. The age of his subjects is not given but results from Miles (1931) suggest that in the young and middle age groups the differences between the hand and the foot, if they exist at all, are very small but differences may appear with people over the age of 50. From the foregoing it would seem that differences between the limbs used to respond are so small that they may well be of little practical importance, on the other hand sex differences and age differences may be of importance and will have to be taken into account in equipment design.

Characteristics of Movement

The time required for and the precision of a movement can be influenced by a number of factors, including the distance moved, the plane and angle of movement in relation to body, the manipulation which will be required at the end of the movement, and whether or not the movement can be monitored by vision.

Free movements are those in which the limb can move without constraint from one position to another. Such movements may be single positioning movements in which the limb makes a once and for all action or they may be serial movements in which a succession of free movements take place between one manipulation and another. We are not here concerned with the manipulations which take place at the end of a free movement, nor with movements which are subject to constraint by controls or other devices. The characteristics of a movement under these circumstances are determined by the manipulation or control movement required and will as a rule be specific to a particular situation. Where appropriate these movements will be considered when the operation of controls and other devices is discussed.

The relationship between time and distance moved has been extensively studied in the preparation of various pre-determined time systems and the results have been incorporated in the data for these systems. These data are published in the form of standard times and they show, as one would expect, that the time taken to move different distances is not proportional to the distance, since a larger part of the time over shorter distances is occupied in starting and stopping the movement. In one laboratory study

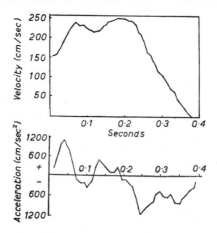

Fig. 25. Changes in velocity and acceleration during a response movement
(after Murrell and Entwisle, 1960).

(Brown and Slater-Hammel, 1949) a movement over 2·5 cm took 270 msec compared with 430 msec for 10 cm and 670 msec for 40 cm. That free movement does not, as is often suggested, consist of an acceleration phase followed by a steady phase followed by a deceleration phase was demonstrated by Murrell and Entwisle (1960) in a high speed cyclo-chronographic study. This technique enabled them to measure accelerations over short intervals of 10 msec throughout the movement and they found that there was a very rapid increase in acceleration followed by a decrease of acceleration and a second peak of acceleration within the first 100 msec of the movement. A typical velocity and acceleration curve from their results is given in Fig. 25. It has been tentatively suggested that this 'dip' in the acceleration may be associated with an overlapping of decision and movement as has been proposed by several authors (e.g. Griew, 1959). Confirmatory work on this point is however needed.

The effect of direction on movement has also been studied by Brown and Slater-Hammel (*loc. cit.*) who showed that an inward-outward movement made by the right hand was faster than a right-left movement which in turn was faster than a left-right movement. The differences between these movements are however only in the region of 50 msec over a distance of 10 cm, so they will be of practical importance only when speed is of major concern. When movements between predetermined locations are under visual control, they will take about 17% longer than when the movement is between mechanical stops (Barnes, 1936). Movements over short distances which require accurate positioning at their terminations, appear to be made with greater accuracy when the hand is moved out from the body, than when the hand is moved to a position immediately in front of the body. The optimum angle of movement being 60° from the centre line of the body (Briggs, 1955).

Smith and his associates, in a number of studies carried out at the University of Wisconsin using a device called the Universal Motion Analyser (Smith and Wehrkamp, 1951), have shown that the time required to make a movement can be influenced by the manipulation which has to be made at the end of the movement. In these studies the movement pattern remained the same but the devices which had to be manipulated at the end of each movement could be changed. There were 34 such devices on the panel. In one study the subjects were required to turn a switch through 40°, 80° or 120° (Harris and Smith, 1954). It is not surprising that it took longer to turn the switch through 120° than through 40°, but they also found that the travel time between each switch was greater for the 80° turn than the 40° turn and even greater for the 120° turn, in spite of the distance travelled remaining the same, these differences being significant at 0·01 level.

In another study (Wehrkamp and Smith, 1952) the subjects were, in one part of the experiment, required to turn a switch and in another to pull a pin out of a hole, the switches and the pins occupying identical positions on the panel. The travel time for turning the switch was 32% faster than that for pulling out the pins. It is implicit in predetermined time systems that each element in a task is independent of every other element and that the times for these elements are therefore additive. This work by Smith and his collaborators suggests that this is, in principle, incorrect and that the time taken for one element can be influenced by the element which precedes it and by that which follows it. Were the differences in travel time small, they might perhaps be ignored but when they are as high as those found in Smith's work, they can introduce errors when movement time forms a major part of a work cycle, and this could render synthetic times either unrealistic or unreliable.

So far we have been considering movements which can be monitored by vision. Sometimes, however, movements have to be made without visual control and Fitts (1947) has shown that such 'blind' positioning movements were made most accurately when they are in a direction straight ahead of the subject and below shoulder height. It appears also from other studies that when the distance of movement is short there is a tendency to overshoot but when the distances are longer there will be a tendency to undershoot. When vertical downward movements are made there will be a tendency to overshoot whether the distance of movement be short or long.

In these various studies of movement there does not appear to have been extensive practice before the trials were given so that it is difficult to say what the effect of training would be on the results. It is not unlikely that if extensive practice were given until the required movements became habitual, some of the differences might disappear. It seems reasonable to conclude that although these differences have been shown in the experimental situation, their importance should not be over-emphasised in equipment design. However, if it is possible to choose the pattern of movement to be followed, the direction and extent of the optimal movements may, with advantage, be incorporated when equipment is designed. Other principles which may need to be taken into account are very clearly stated in numerous textbooks on motion study and they will not be repeated here.

In the preceding discussion we have seen something of the methods which are used to obtain biometrical data. The use of these data in relation to specific design features such as controls, equipment arrangement and so on will be given where relevant, when these features are dealt with.

CHAPTER FIVE

The Human Body IV.
The Nervous System

The muscles alone cannot initiate activity except under certain unusual conditions of hypersensitivity which have already been described. The initiation and control of muscular activity is performed by the nervous system, and this in itself is made up of a number of sub-systems of which the most important are: the *sensory pathways* which convey information from the internal or external sense organs to the central mechanism for processing, the *motor pathways* which convey instructions from the central mechanism to the muscles, the *autonomic pathways* which control the involuntary action of the heart, the gut and some glands, and look after all the processes in the body which have to continue if the animal is to remain alive, and finally the *central mechanism* which consists of two parts, the *brain* and the *spinal cord*. The brain is the controlling element of the whole; it receives information through the sensory pathways, processes it and decides what courses of action are to be taken. These decisions then have to be broken down into a series of messages which are sent out over the motor pathways to the muscles which will be required to take part in the action which has been decided upon. These messages are dealt with in the spinal cord which sorts them out and despatches them down the correct channels.

THE NEURONE

The nervous system is based on the neurone which consists of a nerve cell together with two or more nerve fibres. A fibre which brings messages to the cell is known as a *dendrite* and that which takes messages from the cell is known as an *axon*. A neurone may have a large number of dendrites, but will always have only one axon (Fig. 26). The *sensory neurones* are the simplest form of neurone, each having one dendrite and one axon; they transmit information to the brain from the sense organs. Most of the other neurons have a number of dendrites which branch out from the cell rather as do the roots from a tree. They are classed as *multipolar neurones*. In general they fall into two main groups—those which lie entirely within the central mechanism and serve to link up other nerve cells; these are called *connector neurones*, and those which leave the central mechanism and supply the muscle fibres. These are known as *motor neurones*. Thus in the complete system, the information is received through the sensory neurones, conveyed within the central nervous system by the connector neurones and the resulting cues for action are conveyed to the muscle fibres by the motor neurones.

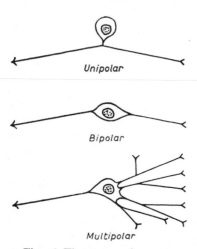

Fig. 26. Three types of neurone.

In order, it is believed, to insulate one nerve fibre from another and so prevent short circuiting of messages between them, the fibres are coated with a material called *myelin*, and when a fibre leaves the central nervous system the myelin is covered with a thin sheath of membrane which serves to contain and protect it. This sheath is nipped in about every half millimetre along the length of the fibre, producing, as it were, a string of sausages. The exact purpose of this is not known, but it is possible that these constrictions act as valves which would prevent the myelin from leaking out from the interior of the sheath should the nerve be damaged.

It is generally believed that each neurone is complete in itself and is never connected directly with any other neurone, there always being a gap between the axon of one neurone and the dendrite of another, which is known as a *synapse*. Exactly how the messages jump across this gap from an axon to a dendrite is not clearly understood. It may be that the impulse jumps as does an electric spark, or that some chemical transmission is involved. It is known, however, that the impulse is transmitted unaltered and that the passage across the gap takes approximately one thousandth of a second. The axons of a number of neurones may end in a single synapse, and since the dendrite leading from the synapse can receive only one message at a time, each synapse has a refractory period which prevents the cell from being overloaded. This refractory period limits the rate at which the dendrite can transmit signals, and those which are received at the synapse at a rate greater than this may be inhibited or blocked.

A system of inter-communication between nerve fibres which is peculiar to the sensory system is known as the *ganglion* in which the nerve cells are joined together in small clusters. In a ganglion the component cells may be in communication with each other so that messages may be interchanged much in the same way as they are in a telephone exchange. Messages can thus be transmitted up the most appropriate route which permits a much wider range of behaviour than would be possible were the sensory nerve fibres connected uniquely and individually to the central mechanism as is the case with some of the lower orders of life. From the ganglia the dendrites spread throughout the body to any part which is sensitive to stimulation.

The Receptor Neurones

Dendrites which form part of the sensory mechanism may branch close to their ends and the end of each twig becomes modified and is known as an *end-organ*. These end-organs are in effect transducers in that they change the energy which they receive into nerve impulses. They are of a number of different types. For instance, end-organs which are on or connected with the surface of the body may be, *inter alia*, beads, plates or coils. Fibres of the beaded type form a complex network throughout the skin over the whole of the body surface, while the coiled type are found under the hairless skin surface as in the highly developed tactile areas of the fingers. The special sense organs, the eyes, the ears, the nose and the tongue, have end-organs which are adapted to their special tasks. Information on limb movement will come from end-organs which are coiled round some muscle fibres or tendons which respond to changes in the volume of the muscle and so convey messages to the brain about changes which result from muscle activity. The foregoing are examples of some of the many types of end-organs which exist in the body and are representative of the senses which play the most important part in receiving messages from the external environment and from bodily movement.

The Connector Neurones

Messages from the sensory neurones are fed into the central mechanism. Here they are taken up by connector neurones which form an intricate inter-woven pattern throughout the length of the spinal cord and up to the brain. These connector neurones can receive messages from a great many sources and can in turn communicate with a great many other neurones so that messages may be relayed to the different parts of the central mechanism, some being dealt with in the brain and others at a lower level. The lowest level is the *reflex* action in which the message is transmitted through the spinal cord directly to the appropriate motor neurone. This complex organization of inter-woven neurones enables a great variety of activity to be carried on simultaneously in a way which would be quite impossible if all messages had to be dealt with in the cerebral cortex, which, as will be shown presently, behaves as a single channel mechanism.

The Motor Neurones

Once a message has been received in the central mechanism and has been processed there, it is sent down through the motor pathways to the muscle or muscles which have to be activated to produce the required result. The cells of the motor neurones lie in the central mechanism and their axons travel together in bundles to each voluntary muscle, each bundle being known as a *motor nerve*. This is in a sense a slight misnomer since the motor nerves also contain a number of sensory fibres which feed back messages to the brain about the action of the voluntary muscles. When a motor nerve reaches a muscle, it breaks up into a number of individual nerve fibres which are connected to the muscle fibres. The sensitivity with with which muscle activity is controlled is dependent upon the number of muscle fibres controlled by each axon. In the legs an axon may control many hundreds of nerve fibres; in the arms and hands, where a higher degree of accuracy is found, a much smaller number of fibres are under the control of one axon; while in the muscles controlling the movement of the eyeball, which require a high degree of accuracy, each axon may control a single muscle fibre.

To sum up, therefore, there is a system of sensory neurones which together are known as the *receptors*, which passes information to the central mechanism and this in turn passes orders down through the motor neurones to the motor units in the muscles which are together known as the *effectors*. It now remains to be seen how these messages are passed and to discuss some of the mechanism in the central nervous system involved in the processing of the messages.

The Transmission of Neural Impulses

As has already been mentioned, the sensory end organs act as transducers, but it is not known how these specialized receptor cells convert the external stimulus energy into nerve impulses. It does seem clear, however, that the stimulus energy produces a change in the electrical state of a nerve fibre which, when it reaches a certain intensity will fire off an impulse up the nerve. This minimum intensity is known as the *threshold* of the fibre, and if this is exceeded nerve impulses will continue to pass up the fibre. Increase of intensity of the stimulus energy will increase the frequency with which the impulses occur but not their strength—and impulses can either occur or not occur. The rate at which impulses will travel along a nerve fibre depends on the thickness of the sheath of myelin which surrounds the fibre. When the sheath is thin the rate of travel may be as high as 120 m./sec, while if the sheath is very thick the rate of travel may be only 1/10 m./sec. The passage of a nerve impulse can be detected with suitable apparatus, and takes the form of a wave of depolarization travelling along the fibre surface which is due to a temporary leaking of ions outward from the centre. When the impulse has passed, these ions return again, thereby restoring the fibre to a condition ready for the passage of another impulse. Until this has occurred the fibre is inhibited from passing another message and is said to be refractory. A message is thus passed up a nerve fibre as a series of impulses. A similar refractory period is to be found in the nerve end-organs which can only fire off messages after a very brief recovery period. In those nerve fibres where the message travels very rapidly the recovery is also very rapid and impulses can be carried up to a rate of about 1,000 per sec.

When an end-organ is under continuous stimulation, it will respond rapidly at first, but after a time its activity will die down and it is said to adapt itself to constant stimulation. If, however, the intensity of the stimulus changes, the end-organ will become active again. In this way the body becomes unconscious of many of the events occurring in the immediate environment. In addition, many external events will continue to be transmitted by the sense organs but the messages which they send will be filtered out in the central mechanism as currently irrelevant. But this process differs from the state of constant stimulation in that they can be received by the cerebral cortex should they become relevant.

It is virtually impossible in the live body for an external stimulus to activate a single nerve fibre, so that the response to the stimulus will be conveyed to the central mechanism by a number of routes, along some of which the messages may travel rapidly and along others more slowly. Thus the central mechanism will receive a pattern of information dispersed both in time and in space and it appears that it is this type of pattern which enables the brain to distinguish between different forms of stimuli. How the brain does this is not known. It is clear, however, that the ability to differentiate between the impulses from the different senses is due to their travelling to different parts of the brain, and it may be that something similar happens to the information conveyed within each sense, so that it is possible to distinguish between one type of sensation and another, and for instance to feel the difference between fur and wood. This distinction is perhaps made by different patterns of stimulation of the tactile nerve endings in the tips of the fingers.

THE CENTRAL MECHANISM

The central mechanism consists of two parts, the *spinal cord* and the *brain*, and the brain in turn is usually conveniently divided into three main parts, the *hind-brain*, the *mid-brain* and the *fore-brain*. The hind-brain contains the *medulla* (the adjectival form of which is bulbar), and the *cerebellum*. The mid-brain is relatively small and serves as a connector between the parts of the hind-brain and the fore-brain. The fore-brain contains the *thalamus*, and the *hypothalamus* with its associated *pituitary* gland, and the *cerebrum*. In man the fore-brain is the largest of the three parts and it is this that distinguishes man from the other vertebrates. This list is by no means complete and anatomists and neurologists sub-divide these organs in order to describe more particularly the functions of the different parts, but it is hardly necessary to consider these sub-divisions in order to understand how man behaves in his working environment.

The Spinal Cord

The spinal cord may be rather inelegantly described as being like a stick of rock with an 'H'-shaped core of grey matter running down the centre. This grey matter, which is similar to the grey matter in the brain, contains the cell bodies of many of the motor neurones. It plays an important part in forming a *reflex arc* in which the signal coming up a sensory neurone is connected directly to a motor neurone without reference to the brain. In this function, the grey matter of the spinal cord may be considered as a lower order brain such as is found in the more primitive forms of life. The spinal cord also acts as the channel through which sensory messages pass from the body to the brain. On the sensory side there are connector neurones running only short distances along the cord between one vertebra and another, but in the motor system there is no inter-connection between the different levels of the spinal column. The cell bodies of the motor nerves all lie in the grey matter of the brain or of the spinal cord; the axons of the neurones in the brain pass down the main motor nerves until the point is reached at which they leave the spinal column to be distributed over the body, while those which are concerned in reflex activity come straight from the spinal cord to the part of the body concerned.

The Brain

The parts of the brain in which we are interested are, working forward from the spinal column, the medulla and cerebellum of the hind-brain, the mid-brain, and the cerebrum.

The medulla joins the spinal cord to the higher centres of the brain and is the channel through which the majority of the cranial nerves will pass. It also contains the nuclei of the motor neurones of the autonomic system, which control respiration, heart-beat, blood pressure, vasco-constriction, and so on. It therefore forms a very vital centre upon which the continued existence of the whole organism depends. The cerebellum which lies at the base of the skull resembles the cerebrum in that it consists of two hemispheres each with an outer layer of grey matter, its interior being made up of white matter. It smoothes and coordinates the messages which are to be relayed to the voluntary muscles and it may therefore be thought of as the centre for motor coordination. It seems likely that one of its functions is to maintain a repetitive movement which has been initiated in the forebrain and that it must therefore play an important part in the learning and maintenance of repetitive skills.

All that need be said of the mid-brain is that it acts as a channel of communication between the hind-brain and the fore-brain.

The major part of the fore-brain consists of the cerebrum which is made up of two hemispheres, the outer layers of which are of grey matter known as the *cerebral cortex*, the inner part being of white matter. The cerebral cortex is the major part of the brain which enables a man to understand the meaning of the signals from the various senses which tell him what is going on both around him and within his body. This process is known as *perception* and it will be dealt with in more detail later. In addition the high development of the cortex in man enables him to reason and to make decisions. Exactly how this process is carried out is not known, but many of the limitations which are imposed on the process have been measured and these will be discussed later. Finally the cortex is the area of the brain in which information is stored in the memory. Again, it is not known exactly how this information is stored, but the process has definite limitations.

The fore-brain also contains the thalamus which acts, as it were, as the relay station of the brain. It contains an immense number of nuclei of neurones which connect with the other parts of the brain and with the spinal cord. Its main function is to sort out messages which it receives from the other parts of the brain and to direct them into the correct motor channels in order to produce the total effect which is desired by the 'thinking' part of the cortex. It thus acts as a coordinating and integrating organism for the voluntary activity of the body. The hypothalamus carries out a similar function for the autonomic activities of the body. These are controlled individually by the medulla, but the integration of the activity of each part of the autonomic system to maintain balance within the organism is the function of the hypothalamus. For instance, if an individual is subjected to heat, the heat regulating mechanism of the hypothalamus will initiate vaso-dilation and increase heart-rate and respiration and sweat gland activity. The exact levels at which each of these activities is to be maintained is decided by the hypothalamus and these levels are maintained by the medulla.

To sum up, therefore, the decisions to take action are made in the cerebral cortex. These decisions are sorted out into their correct channels in the thalamus and they are controlled in so far as muscular activity is concerned by the cerebellum, which can continue to control muscular activity which is repetitive in the absence of further instructions from the cortex. The autonomic system is controlled in much the same way with the hypothalamus acting as the integrator of the various bodily functions, while the medulla controls those functions at a level determined by the hypothalamus. At a lower level the spinal cord receives and despatches messages involved in reflex activity.

THE SENSES

Information is fed into the sensory system through end-organs or receptors, which are divided into three main groups, the *exteroceptors*, the *proprioceptors* and the *interoceptors*. The interoceptors convey information about the internal state of the body, and produce a feeling of hunger, or fullness of the bladder for instance; they are of no great concern in this context. They are the main source of information for the maintenance of the autonomic functions. The other two groups are of great importance. Exteroception includes sensitivity to light, to chemicals, to touch, to pain, to heat and to cold. Proprioception is concerned with motor function and involves two kinds of receptors, the kinaesthetic and the vestibular.

It is necessary at this point to make clear the difference between sensation and perception. External forces falling upon the body will produce definite patterns of stimulation in the sensory parts of the nervous system and the resulting messages will be conveyed to the brain. This is the sensory function and is definite and measurable in relation to the external stimuli. The nature of the picture, however, which results from all these stimuli is a synthesis which gives an impression of the size, position, distance, colour, shape, sound, smell and so on of the surrounding environment. This picture may or may not be a true representation of the environment and may be modified in the mind of the subject by previous experience or memory. Perception may, therefore, be loosely defined as subjective whereas sensation is objective. Sensation will be discussed in this section and perception will be dealt with later.

Chemical Receptors

Of the exteroceptive senses already listed, the chemical senses are of relatively small importance in the present context. Chemical receptors are to be found in the nose, the tongue and the skin, the first two being responsible for smell and taste and the latter for the feeling produced when a chemical comes in contact with the skin, which is the least sensitive of the receptor functions of the body. While a sense of taste and smell can be important in some occupations such as wine-tasting, cookery or cheese blending, the sensory processes involved are so complex and so little understood that it seems unprofitable to discuss them in detail. Chemical sensitivity of the skin would also seem to play little or no part in working activity. In properly designed processes chemicals should not come in contact with the skin, and it is perhaps only when something goes wrong that this sense may act as a warning device.

The Cutaneous Receptors

The cutaneous receptors are responsible for sensations of touch (or pressure), pain, heat and cold, but before proceeding to discuss touch and pain, it is necessary to clear up a point of terminology. Touch is due to pressure on the surface of the skin transmitted to the nerve endings, but the term 'pressure' may be used to describe the force applied by a limb to a lever, such as on the brake pedal of a car. It is highly unlikely that the skin surface of the sole of the foot, which is relatively insensitive to pressure distributed over the whole of its area, will give very much information about the force which is being applied by the leg. It is much more likely that information will come primarily through the kinaesthetic sense. It must be made clear, therefore, that when the word pressure is used in the discussion which immediately follows, it is not used in the sense of application of force by any part of the body.

Touch and Pain

Opinions differ as to whether the sensations of pain, touch, heat and cold are due to the stimulation of special encapsulated end-organs or whether they are felt through free nerve endings which can serve as receptors for any or all of the cutaneous senses. Whichever is the case, it seems that free nerve endings are responsible for sensations of pain and that encapsulated nerve endings, such as the Meissner corpuscle, are to some extent responsible for the sense of touch, and that the hairs on the skin surface are sensitive to touch through nerve endings in the hair bulbs. There are various theories on the mechanism of heat and cold sensation, but it seems likely that it is related to changes of temperature gradient in the skin surface which in turn cause constriction or dilation, at different depths beneath the skin's surface, of the minute blood vessels with which nerve endings are associated. It seems unlikely that any thermochemical process is involved.

There is no doubt that touch and pain are subjectively two entirely different sensations. There is some controversy as to whether overstimulation of the sense of touch will produce pain, and such evidence as there is (for instance, the overstimulation of the tactile nerve endings associated with the hair roots will not result in pain) suggests that it probably does not. Moreover, it is possible to map areas of the body which are particularly sensitive or insensitive to pain and these do not coincide with areas which are particularly sensitive or insensitive to pressure. On the other hand, excessive cold and excessive heat will produce pain, but it is an open question whether the same sensory nerve endings are involved in the sensation of heat and cold and in the sensation of touch.

The dendrites of the sensory neurones involved in touch break up into a number of branches near the surface of the skin. These branches will join and re-divide forming what is known as a *plexus* from which arise the sensory nerve endings. Each plexus may cover as much as one square centimetre of the skin's surface, and in any given area many plexes will overlap each other. As a result the tactile sense is not very sensitive to shape unless the shape can be sampled by several independent channels. Thus if the tip of a finger is pressed on the end of a round pencil it will be difficult to distinguish between the sensation received and that which would be received from a hexagonal pencil. On the other hand if the top of the pencil is felt with the tips of three or four fingers, the shape is being sampled through several independent channels and a much clearer picture of its form is transmitted to the cortex. This important point should be borne in mind when designing tasks whose successful execution depends on touch alone. On the other hand, it seems to be a great deal easier to distinguish texture, even with the tip of a single finger, and this would seem to be due to very slight variations of pressure over the area which is in contact with the textured surface. These variations in pressure have to convey information to the cortex which is interpreted in the light of previous experience, and would appear to differ in quality from the information conveyed when small shaped areas are touched by a single tactile area.

Referred Pain

An important instance of a connection between excessive pressure and pain, which is not related to pressure on the receptors, is pressure applied to a nerve trunk. A *nerve block* is set up which will temporarily prevent impulses travelling along it. If the pressure is applied for long enough it will render the area served by the nerve trunk insensible to pain – the area 'goes to sleep'; the same effect may be obtained by the application of cold. If the pressure is maintained for too long, permanent paralysis may result, the risk of this being increased by the lack of sensitivity in the area as a result of the pressure. When the pressure or the cold is removed, the area of the nerve which has been affected will usually be small and is the locus of the pain sensation. But the sensation of 'pins and needles' which accompanies the removal appears to come from the whole of the affected limb. This sensation is known as *referred pain*, and it appears that the brain is unable to distinguish between impulses which arise from the stimulation of pain receptors in the skin and those which arise through the release of pressure on a nerve trunk.

Referred pain is also experienced from other causes. For instance internal disorders or pressure on a nerve trunk caused by spinal damage will be referred to the skin surface, with the result that pain is felt in an area which is not actually affected. This mechanism makes it very difficult to diagnose the exact location of many pains. Referred pain can also be caused by the reflex spasm of the blood vessels in the muscles of a painful area. This will cut off blood supply to the muscle which will result in the accumulation of the products of combustion which in turn will cause pain in the muscles over an area greater than that originally affected.

It seems likely that mechanisms of these kinds account for quite a number of aches and pains which are experienced by industrial workers, and in searching for design faults which might cause such pains it must be remembered that the location of the pain may be somewhat remote from its actual cause. But with this proviso in mind, it should be possible to make a reasonably intelligent diagnosis of the probable cause of the pain and make the necessary alterations in the equipment.

THE SPECIAL SENSES – VISION

The Eye

The eye in many ways resembles a camera: it has a lens, a stop (the iris) and a photo-sensitive area upon which the image falls. The eye differs from the camera, however, in several important respects. First, the lens is fixed in relation to the photo-sensitive area and, unlike the camera, is not moved to and from it in order to focus on objects at different distances away, change of focus being accomplished by changing the curvature of part of the lens. Secondly, the photo-sensitive layer of the eye is curved and varies in sensitivity from the centre outwards, whereas a camera has a flat film equally sensitive over the whole area. Thus the image which is recorded by the eye covers a much wider visual angle than that which can be obtained by most camera lenses, but the image is sharply in focus only at the very centre of the picture. Thirdly, the range of intensity of illumination under which the eye can work is far greater than that under which a camera will work using a given type of sensitive emulsion. Both the eye and the camera have a 'stop' which can be opened or closed according to the prevailing brightness. This very wide range of sensitivity is one of the unique characteristics of the eye. Apart from all this the same optical principles which govern the operation of the camera will also govern the operation of the eye.

Only a very small portion of the visual image is sharply in focus and this drawback is overcome by the scanning action of the eye, which is accomplished by a set of six muscles attached to each side and above and below the eyeball. Using the two eyes together *binocular vision* is obtained which enables special relationships between objects and the environment

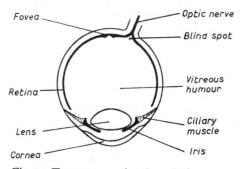

Fig. 27. Transverse section through the eye.

to be perceived. This binocular vision is the result of precise coordination between the muscles which turn both eyes in the right direction and those which alter the curvature of the lens so that the object viewed is sharply in focus.

The eye is very nearly spherical, but with a clear portion protruding from the front on a rather smaller curvature than that of the remainder of the eye (Fig. 27). This transparent portion, the *cornea*, is a lens of fixed focus. Behind the cornea is the *iris*, which regulates the amount of light which enters the eye. The iris is controlled by a ring of small muscles which act reflexly, not being under voluntary control. Immediately behind the iris is the *crystalline lens* which will alter in shape to bring objects at different distances into focus. The whole of the eyeball, except for the cornea consists of hard, white tissue which together with the semi-fluid which it contains, serves to keep the eyeball in its correct shape. On the side opposite to the cornea, but slightly offset from the centre towards the nose, the eyeball is penetrated by the *optic nerve*. Here the nerve fibres pass into the eye and spread out all round the lateral half of the interior surface of the eyeball to form the photosensitive coating, the *retina*. In the centre of the point where the optic nerve enters the eye there is a small area, approximately 1/10 inch in diameter, which is known as the *blind spot*. It is as if a bundle of threads was parted outwards in all directions leaving a small hole in the middle. The nerve fibres form the inner surface of the retina nearest to the lens; beneath them is a highly complex assembly of cells overlying the photosensitive receptors, the *rods* and *cones*, which form the deepest layer. The cones function under normal conditions of illumination but are insensitive to low levels. The rods give night vision. The outer layers of the retina are nearly transparent and light has to pass through them in order to reach the sensitive layer underneath, except in one area at the focal point of the lens where they are absent, so that incident light can fall directly upon the cones. This area, the *fovea centralis*, is approximately 500 μ in diameter. It contains cones only, there being between ten and twelve thousand in this tiny area. This is the only area of the eye in which the optical image is sharply in focus and is used when we look 'directly at' something. The fovea is surrounded by a small ring known as the macula where vision is also acute. Acuteness of vision in the foveal region is facilitated by the attachment of each cone to a separate nerve fibre so that the stimulation of minute areas of the fovea will enable very fine detail to be discriminated. Beyond the foveal region, progressively more cones are associated with each nerve fibre. This, together with the fact that the light has to pass through the retinal layers before falling upon the cones, reduces the ability to discriminate detail towards the edges of the visual field. As we shall see later there is a limit to the amount of information which the brain can receive at any one time and this arrangement in the eye prevents too much primary information being sent to the brain at once. Nevertheless, the remainder of the retina enables a man to be conscious of a large area of his surroundings, and should any event occur within the visual field which mertis his attention he can turn his eye towards it and look at it with foveal vision. It must be remembered, however, that the degree of attention which is possible at the extreme edges of the retina is far less than that which is possible in the region of the fovea, and for this reason stimuli at the extremes of the visual field must be more intense than those in the foveal region. This point is of great importance in the design of instrument panels.

The second receptor elements of the eye, the rods, are sensitive only to very low levels of illumination. They are completely absent from the fovea and for this reason sharp vision in conditions of near darkness is impossible and it is easier to see an object in the dark if it is not looked at directly. Unlike the cones, the rods are not colour sensitive and for this reason night vision is in monochrome.

The interior of the eyeball is filled with a semi-liquid transparent material, the *vitreous humour*, which scatters incident light to some extent; it plays a part in producing glare and influences sharpness of vision.

Accommodation

In the activity of seeing, the single eye displays two separately definable characteristics. The first of these is the ability to focus on objects at different distances from the eye. This is known as *accommodation* and is achieved by changing the shape of the crystalline lens which becomes thicker and more curved the nearer the object is to the eye. For practical purposes all objects further than 6 metres (20 ft) away from the normal eye are sharply in focus and the nearer an object is to the eye from that distance the greater the amount of muscular effort which is required to maintain the lens at the correct curvature. Eventually a point is reached where nearer objects can no longer be focused and this near point gets further away throughout life due to a gradual hardening of the lens. At the age of 16 the normal eye should be able to accommodate down to about 8 cm, but at age 45 this distance will have increased to about 25 cm, and by age 60 will have increased fourfold to about 100 cm. This loss of accommodation with age, known as *presbyopia*, is a perfectly normal ageing process but must clearly be taken into account when designing industrial tasks. It can, however, be corrected by the proper use of supplementary lenses, either as spectacles or as optical aids in the work.

Visual Acuity

The second function of a single eye is its ability to discriminate fine detail, which is known as *visual acuity*. For this to take place it is important that the focal point of the unaccommodated lens should coincide with the retina. If, due either to incorrect curvature of the cornea or to slight elongation of the eyeball, the focal point lies in front of the retina, a condition exists which is popularly termed 'long-sightedness'. Alternatively, the eyeball may be somewhat flattened so that the focal point lies behind the retina, a condition which is commonly described as 'short-sightedness'. In fact both these terms are misnomers, but they do represent conditions in which the musculature of the lens is constantly being activated when the lens should be at rest in an effort to correct for deficiencies of structure. Moreover, in the normal eye the adjustment of the lens to the focal distance of the object to be perceived is reflex, but in the short-sighted and long-sighted eye a conscious effort has to be made to maintain the eye in focus. This produces symptoms of eye-strain which are evident in the later years of life, particularly in the long-sighted individual.

Visual acuity depends not only on the size of the object which is being viewed, but also on its brightness. Thus an object which cannot be seen at all in low illumination may become clearly visible when illumination is high. Over the range of illuminations normally found, acuity varies linearly with logarithmic increase in illumination. There may be several reasons for this relationship. First, it will be remembered that the iris of the eye closes and opens reflexly in response to the amount of light falling upon the retina, thus when the illumination level is at its highest the aperture of the iris will be small. In a camera a sharper image and greater depth of focus are obtained with small apertures than with large ones and this is also true of the eye where a small aperture can reduce aberrations of the lens due to imperfections in curvature. In addition, the internal media of the eye will disperse a portion of the light, and the larger the aperture, the more scattered stray light will be present, which will effect the clarity of the retinal image. Secondly, the sight receptors must receive a minimum threshold of stimulation for which, it has been suggested, the product of area and intensity is constant. As the area gets smaller, so the intensity must become greater.

Any measurement of visual acuity must be related to illumination, and for the results to be meaningful the illumination under which the tests are carried out must be stated. A further complication arises from the existence of several different forms of acuity which are not strictly comparable. The three types of acuity most commonly recognized are *line acuity*, sometimes called *minimum visible acuity*, *space acuity*, sometimes called *minimum separable acuity*, and *vernier acuity*. A further form of acuity sometimes used is *spot acuity*. Line acuity is the ability to see a very fine line of known thickness, and spot acuity is a variation of this. Space acuity is the ability to see two spots or two lines as separates. In effect it is the ability to see a space between two solid areas and as such may perhaps be thought of as the ability to see a white line on a dark ground. Vernier acuity is the ability to detect a discontinuity in a line when one part of the line is displaced slightly. Of these three measures, line acuity gives the smallest value, followed by space acuity and vernier acuity (Hecht and Mintz, 1939; Weston, 1949).

In order to standardize a measurement which is uninfluenced by distance, acuity is usually expressed in terms of the angle subtended at the nodal point of the eye by the object in view; since light rays passing through the nodal point are undeflected, this is also the angle subtended at the nodal point by the image of the object on the retina. Visual acuity is sometimes given as the reciprocal of the visual angle in minutes of arc, which has the advantage of increasing in value with increase in acuity. In dealing with thresholds of sensation which are dependent on subjective measurements it is not surprising that different experimenters get somewhat different results, and this is true of visual acuity which is one measure of visual threshold. The differences are, however, not very large and from a practical point of view of not very great importance.

It might be supposed that visual acuity would bear some relationship to the size of a foveal cone. In fact no direct relationship is clearly established. This may be partly due to variation in the size of cones in the fovea. According to Polyak (1941) there are in the very centre of the fovea a few cones as small as $1\ \mu$ to $1.5\ \mu$ ($12''$ to $18''$ of arc) while in the central area of about $100\ \mu$ diameter there is a density of about 500 cones to the millimetre which gives a cone diameter of $2\ \mu$ ($24''$ of arc). In the next area, extending to about $100\ \mu$ beyond this, the cone density falls to about 350 to the millimetre and in the outer $100\ \mu$, density becomes 300 to the millimetre, which gives cone diameters of about $2.9\ \mu$ ($34''$ of arc) and $3.3\ \mu$ ($40''$ of arc) respectively. Thus in the quite small distance of only $250\ \mu$ the cones may nearly treble their size from about $15''$ to $40''$. It is hardly surprising therefore that various authorities may quote cone sizes in relation to acuity which may differ by this amount or more.

From the results obtained by various workers it would seem that a white spot can be seen when it subtends an angle of about $10''$, but a black spot on a white ground will have to subtend an angle of about $30''$ to be seen. Now if this black spot is extended so that it becomes a fine line, the line is visible when it subtends a visual angle of between $0.5''$ and $1''$, that is when it is only between 1/25th and 1/50th of the diameter of a foveal cone. In the central area, minimum separable acuity lies between $24''$ and $30''$ of arc and the displacement of lines for vernier acuity is about $4''$ to $6''$ of arc. The explanation for this apparent anomaly may well be that the image of a dot or a line on the retina has a penumbra due to its passage through the vitreous humour. Thus while a single cone may receive the major part of the image of a single dot, the cones surrounding it will also receive a small part of the stimulation. This would account for the visual angle of the smallest discernible white dot being less than that for a single black dot. At the same time the extension of a dot into a thin line will mean that a row of adjacent cones are stimulated, and this will trigger off a message to the brain when the area of each individual cone which is stimulated is far less than it would have to be if a single dot was being observed. Thus it appears that for the perception of very small objects to be possible, more than one cone has to be stimulated and that there must be a threshold difference between the intensity of light received by adjacent cones for a visual image to be recorded.

The limit of ability to see two objects as separate is given by different workers as between 24″ and 60″ of arc. If the diameter of the smallest cones is between 12″ and 24″ the space between the cones stimulated by the two images must be at least two or three cones wide. Again it would seem that slight scattering of the image by the vitreous humour will cause the cones adjacent to those receiving the image also to be stimulated so that it is only when the images are sufficiently far apart for a line of cones to be free from the penumbra that two separates are perceived. The image will be sharpened and the scattering reduced when the aperture of the iris is very small so that minimum separable acuity will be increased by increasing the brightness of the objects viewed.

From this it follows that in order to see fine details the contrast must be as high as possible and the brightness of the object must be as intense as possible. Both these variables can be manipulated by the designer and their relationships to good seeing will be dealt with later.

The figures of visual acuity quoted above represent what can be seen by an individual with perfect eyesight under the best possible conditions. Although there are a number of exacting industrial tasks, it is unlikely that many of them will be of a degree of fineness corresponding to the most extreme conditions under which visual acuity is measured. Nor is it likely that under normal conditions of factory lighting the illumination available will approximate to that under which the laboratory measurements were made. For this reason normal visual acuity is often taken as an ability to see as separate two objects subtending a visual angle of 1 minute of arc at the nodal point in normal room lighting, that is at a brightness of between 10 and 20 ftL.

So far we have been dealing with the acuity of vision in normal lighting. When, however, the brightness reaches the region of 0·005 ftL, the cones cease to function and the task of vision is taken up by the rods. Since these are very much less sensitive than the cones and are absent from the foveal region, minimum separable acuity for rod vision is unlikely to be better than 10 minutes of arc.

Convergence

Coordinated directing of both eyes at the object of regard so that its image falls on the foveal region of each eye is known as *convergence*. It is this function which makes clear binocular vision possible. The near point for convergence is approximately $3\frac{1}{2}$ in. from the eye and this does not appear to be affected by age. In practice, however, the nearest point to which the eyes will normally converge is that at which the lens will focus the image sharply on the retina. At distances greater than 6 metres no convergent activity is required and the convergent muscles are at rest. At all other positions between this and the near point, convergence requires muscular activity, whether the eyes are looking at a fixed object or are scanning. As with other muscles of the body static work is fatiguing to the muscles of the eye. For this reason gazing fixedly at an object for a period of time is more fatiguing than scanning which allows the muscles to expand and contract.

Night Vision

Dark adaptation, which is the basis of an ability to see in near darkness, can be of great importance in some industries, such as transport. Dark adaptation appears to take place in two phases. In the first phase the adaptation is rapid at first and then levels off. This is followed in the second phase by an increase in the speed of adaptation which finally levels off to its minimum threshold. There is evidence that during the first phase the cones are undergoing adaptation and that the rods take over from the cones at the beginning of the second phase and are in turn adapted. But it also appears that the rods themselves have been undergoing adaptation during the first phase. The length of time for dark adaptation to be completed may vary from 10 to more than 40 minutes, and the length of the first phase depends on the intensity of the light which was experienced before dark adaptation began, which may also influence the illumination level at which the change-over from cones to rods takes place; it seems, however, that the cones will not function at brightnesses of less than about 0·005 ftL. At this level of illumination the acuity of both the rods and the cones is very nearly the same, so that the change from one to the other has very little practical significance except for those people who have poor rod vision. This deficiency in rod vision, which is usually spoken of as night blindness, may sometimes be accompanied by deficiencies in cone vision.

Dark adaptation is very rapidly destroyed by exposure to a bright light, the time required for this light adaptation being very much shorter than the time required for dark adaptation; it is not, however, by any means instantaneous. It would appear that exposures of under 15 sec are unlikely to effect dark adaptation very much. As with dark adaptation the speed of light adaptation will depend on the brightness of the light to which the eye is being adapted. Deep red and near ultra-violet appear to have far less effect on dark adaptation than does white light or yellow light of similar intensity; the dark adapted eye is least sensitive to the red and violet and this may account for this effect.

Colour Vision

The retina of the eye is able to distinguish between light of different wavelengths thus providing man with colour vision. The retina is not, however, equally sensitive to light of all wavelengths, thus different colours of equal intensity will appear either brighter or less bright to the eye according to their wavelength. When the eye is light-adapted, the brightest spectral colour is at about 555 mμ which gives an impression of yellowish-

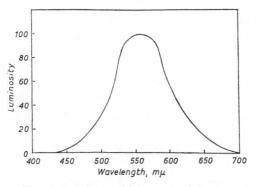

Fig. 28. Relative sensitivity curve of the eye.

green and the brightness progressively diminishes as the wavelength approaches either 700 or 400 mμ (Fig. 28). This illustrates the difference between luminance and luminosity (or brightness) which was referred to in Chapter 1. Owing to the differential sensitivity of the retina to, for instance, yellow and red light, the luminosity of the yellow seems to be greater than that of the red even though both are of equal luminance.

The dark adapted eye has a luminosity curve similar in shape to that of the light-adapted eye when the maxima of the two curves has been adjusted to allow for the differences in sensitivity between them. However, this curve shows a shift of the peak towards the blue end of the spectrum of approximately 50 mμ. That the retina of the dark adapted eye is differentially sensitive to light of different wavelengths does not imply that the dark adapted eye can see colour. It appears that the messages which are sent to the brain by the dark adapted eye do not give information about the colour of the seen objects and to all intents and purposes the dark adapted eye is colour blind.

It will be remembered that the Young-Helmholtz tri-chromatic theory postulates that all perceived colours can be obtained by mixing various combinations of three primary colours. On this basis tri-chromatism postulates the existence of three types of receptors in the retina, red, green and blue sensitive. Thus the perception of any particular colour would be the result of combined signals sent from these three types of receptor, the differences in hue being due to differences in the degree of stimulation of the three receptors. This theory has held the field for nearly 150 years but recent research has given reason to doubt its correctness and there is some physiological evidence that there may be as many as seven distinct types of colour receptors. There are several reasons for this doubt. In the first place the theory postulates that yellow is perceived by means of the red and green receptors. If this were so then it should be seen binocularly by stimulating one eye with red and the other with green. This, however, is not normally the case though it can be shown with a good deal of difficulty Secondly, the equal parts of the retina are not equally sensitive to different colours. The area surrounding the fovea appears to be sensitive to all colours, and beyond this progressively fewer colours can be perceived. The two smallest areas are those responding to green and red, the red being slightly larger than the green and extending roughly from 20° to 30° from the foveal region. Yellow response is obtained in the next largest area being between 40° and 50° and blue in the largest area between 50° and 70°. These areas are roughly oval in shape, being offset towards the side away from the nose. Beyond the limits of the blue area only grey can be perceived. There is evidence that in the fovea itself only red and green receptors are present and that the colour which can be perceived by the fovea is obtained by a di-chromatic mixture. This can be demonstrated by looking at a pattern of patches of blue and green which, if taken to a distance at which the pattern is seen only in the foveal region, will appear to be similar in colour. As a result of this differential sensitivity, there are parts of the retina where a yellow stimulus can be received but which are blind to red and green, and it seems unlikely that the sensation of yellow in this area can be due to the stimulation of red and green receptors. Thirdly, since red and green and blue and yellow are complementary colours it seems more likely, in the absence of other evidence, that there will be primary receptors in the eye for these four colours.

Evidence for the existence of four or even more receptors in the eye has been reported by Granit who obtained readings from single cones in the retina. This evidence is very fully reviewed by Morgan and Stellar (1950) to whom the interested reader should refer. All that it is necessary to say of his work here is that he has obtained luminosity curves for various cones which have peaks corresponding to the four primary colours, green, red, blue and yellow. Yellow of 580 mμ corresponds to the peak of Granit's yellow receptor curve, while primary green of a wavelength of 510 mμ corresponds to the peak of the green curve. Primary blue lies in the region of 440 to 480 mμ while primary red is seen above 630 mμ. When light of any one of these four wavelengths falls upon the retina, only one of the four receptors is excited to any degree and the others very little. Thus green at 510 mμ excites a green receptor maximally and the blue and yellow receptors practically not at all. The red and green receptors appear to be far more sensitive than the blue and yellow and it appears also that each of these pairs of receptors may be linked in the ganglion cells of the retina. It may be that the stimulation of one of the linked pairs of receptors for the complementary colours may cause the inhibition of the other and that their activity is mutually exclusive. If this is so, it would explain the phenomenon of complementary colours.

The fact that a reasonable sensation of most colours can be obtained by mixtures of three primaries as postulated by the tri-chromatic theory is not necessarily an argument against the existence of four or more receptors in the eye, since it is possible to explain perception of tri-chromatic images on the basis of four receptors. It is possible to get quite adequate three colour reproduction with quite a wide range of three additive primaries, and on the whole the wavelengths of these colours do not seem to be critical provided they are in the red, green and blue bands. In practice colours which are usually chosen as being most satisfactory will have a red in the region of 680 mμ, a green in the region of 545 mμ and a blue in the region of 430 mμ. The blue and the red will stimulate only the blue or red receptors, but the green falls halfway between the peak of the green and yellow receptors, and must excite both. If the suggestion that a receptor be inhibited by its complementary is correct, the addition of some red to the green may inhibit response of the green receptors and cause a 'yellowing' of the response. In the same way, the presence of blue may tend to inhibit some of the response of the yellow receptors and this will cause the change of hue towards blue-green. Thus a combination of these three primary additive colours can produce most of the colours which can be perceived, even although four or more receptors take part in the process of perception.

The matter is, however, not by any means settled. Rushton (1961) reports the identification of three chemical receptors in the retina; two of these, the green sensitive and the red sensitive can be fairly readily demonstrated by comparing the normal eye with the red blind eye of a protanope by using an intense bleaching light and measuring the density of the visual pigment. The identification of the blue sensitive pigment is very difficult because there is very little of it and it cannot be bleached without at the same time bleaching the other visual pigments. Rushton, however, has recently succeeded in identifying this pigment, which appears to absorb light at 450 mμ. Rushton says: 'If there are four pigments they cannot be independent but must be connected by a linear equation which exactly removes all the advantage for colour discrimination that an extra dimension of colour could have achieved. This seems to me so perverse an evolutionary step that I unhesitatingly refuse to entertain it until at least one fact is adduced in support.' It will thus be seen that a good deal more work still needs to be done before it can be said with any degree of certainty just what is the mechanism of colour vision and the number of receptors in the eye which are involved. It will be appreciated however that the work of Granit on the sensitivity of cones to light of different colour and the work of Rushton and others on the chemical basis of colour vision is not as incompatible as might appear at first sight.

THE SPECIAL SENSES – HEARING

The Ear

The function of the ear is to convert the pressure waves of sound into neural signals which can be transmitted to the brain. This is not accomplished directly but through a series of transformations, first into mechanical movement, secondly into hydrostatic pressure and thirdly, probably, again to mechanical movement. The whole mechanism is, therefore, complex and as will be seen later is liable to damage if abused.

These transformations take place in three different parts of the ear, usually called the outer, middle and inner ears (Fig. 29). The outer ear consists of the part attached to the head externally (the part called 'the ear' in general parlance), and a tube leading into the interior of the head, which is closed at its further end by the tympanic membrane (the ear-drum). The purpose of the external appendix and the tube is to channel the sound waves to this ear-drum, and in some animals this is aided by the erection of the external appendage so that it points in the direction of the sound.

The Middle Ear

The ear-drum separates the outer ear and the middle ear and to its centre is attached a small bone which is one of three, the malleus, the incus and the stapes, which make up a mechanical lever linkage which transmits movement from the ear-drum on one side of the middle ear to an oval window on the other side, and it is important for the proper functioning of this part of the ear that the air pressure in the middle ear should be the same as that in the environment. Thus the middle ear is connected to the

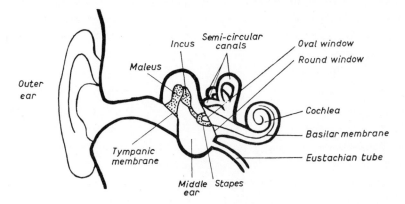

Fig. 29. Medial section through the ear.

back of the throat by the eustachian tube which is opened by the process of swallowing. When the atmospheric pressure changes as when travelling in a lift or in an aeroplane or when 'snorting' in a submarine, it is necessary to 'clear the ears' by swallowing, and this is one of the reasons why air-passengers are sometimes offered sweets before take-off or landing. Individuals who find it difficult to 'clear the ears' may suffer discomfort or pain due to inequalities of pressure on either side of the tympanic membrane.

The Inner Ear

The most important part of the inner ear is the cochlea, which is a coiled structure similar to the shell of a snail from which it derives its name. In man it comprises rather more than two and a half turns which taper towards their end. The wide end of the cochlea abuts the middle ear from which it is separated by a partition in which there are two apertures covered with membranes known as the oval and round windows respectively. The membrane of the oval window is attached to the stapes. Stretching across the cochlea and running its full length are two membranes dividing it into three canals, one above the other, all of which are filled with fluid (Fig. 30). These three canals are known as the vestibular, cochlear and tympanic canals respectively. The vestibular canal and the tympanic canal are connected at the apex of the cochlea, but the cochlear canal has no connection with the other two. The oval window is in contact, on one of its sides, with the vestibular canal and the round window with the tympanic canal and this latter window acts as a kind of safety valve to react to the compression waves in the fluid filling the cochlea which originate in the oval window. The lower of the two membranes is known as the basilar membrane which contrary to what might be expected is at

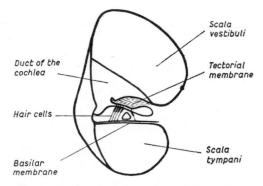

Fig. 30. Section through one turn of the cochlea.

its broadest at the apex and gets narrower towards its point of attachment near the middle ear. This basilar membrane might be likened to the string of a violin and it appears that the vibrations transmitted to it by the fluid produce the sensation of hearing. The cochlear canal which lies above the basilar membrane contains the nerve cells which are concerned with the transduction of pressure waves to neural impulses. They lie on the inner and upper side of the basilar membrane and form the organ of corti, which consists of rows of hair cells which are embedded in a membrane known as the tectorial membrane; these cells are connected through various neural channels to the auditory nerve.

Opinions differ somewhat as to the exact mechanism of hearing but it appears that sound pressure waves cause the ear-drum to vibrate, and these vibrations are transmitted mechanically to the oval window which acts like a plunger in the end of the vestibular canal. This movement sets up hydrostatic pressure waves in the fluid of the canal which in turn cause the basilar membrane to vibrate, different parts of the membrane being sensitive to different frequencies. The movement of the basilar membrane in relation to the tectorial membrane either compresses or pulls on the hair cells, the sense of pitch of the sound appearing to depend on which of the hair cells are stimulated in this way.

It will be remembered that the louder the sound the greater the amplitude of the sound waves, and this in turn would seem to cause a greater amplitude in the hydrostatic vibrations in the cochlea and in the vibration of the basilar membrane, which must in turn affect the hair cells to a greater extent. There is evidence that when too great an amplitude persists over a long period some permanent damage is caused to the basilar membrane and to the hair cells. The basilar membrane may become distorted and this will result in temporary or permanent deafness. It appears that this deafness will extend first to sounds which have a frequency of about 4,000 c/s which corresponds to the part of the auditory system which is most sensitive to sound. This deafness caused by noise will be dealt with in greater detail in Chapter 13.

THE PROPRIOCEPTORS

The Vestibular Receptors

The vestibular apparatus is part of the assembly of the ear but has nothing to do with hearing. It consists of two cavities, the utricle and saccule, and three semi-circular canals. Both groups of organs are filled with fluid and they contain hair cells which give information about the attitude of the body. The hair cells in the utricle and saccule have, as it were, tiny weights upon their ends which are acted upon by gravity, and they convey information to the brain about the position of the head in space. They are thus static or positional receptors. The hair cells in the semi-circular canals are arranged in a rather different way. They form part of a small receptor organ known as the crista which has a blade of gelatinous material to which the hair cells are attached, movement of the fluid in the canals will bend the crista and stimulate the hair cells. Thus the crista is sensitive to acceleration and deceleration and is the dynamic organ of posture. The three canals are arranged at right angles to each other in three planes, so they will give information about movement in all directions.

The vestibular receptors enable man to maintain his upright posture and determine his position in space. Without the utricle and saccule you would not know whether you were standing upright or standing on your head and without the semi-circular canals it would be impossible to move without falling over. Inability to remain upright is one of the symptoms of disease of the inner ear.

The Kinaesthetic Receptors

Kinaesthesis is of great importance in work and is involved in every movement we make. It is the sense which enables us to know about the position of the body and the limbs in space and to make movements without monitoring them by vision. It is probably the sense most deeply involved when trained operators make habitual movements without vision. In addition to the sensory pathways from the kinaesthetic receptors to the cortex, there are direct connexions from these pathways to the cerebellum and to the grey matter of the spinal cord. It would seem therefore that when a repetitive movement has been learned, control of these movements is taken from the cortex and is monitored by the cerebellum, and in some cases at an even lower level in the reflex pathways of the spinal cord. If this were not so and all movement had to be monitored in the cortex, the learning of rapid repetitive movement would probably be impossible.

The kinaesthetic receptors are of three main types. The first of these has nerve endings in the muscle itself which can either be spirally wound round the muscle fibres or can have spray-like endings distributed over the muscle fibre. These end-organs are stimulated when the muscle stretches but not when the muscle contracts.

The second type of receptor is located at the junction between the muscle fibres and the tendons and have spray-like nerve endings distributed over this area. They are stimulated either by a decrease or an increase in the tension at the junction but it is not known whether they can distinguish between them. This however does not appear to be important since the nerve endings in the muscle itself can tell when the muscle is stretched so that the two acting together will give information when the muscles stretch, but the tendon receptor will give information alone when the muscles contract. Thus the pattern of information from these two receptors will tell the brain what is happening in the muscles.

The third type of receptor are larger versions of the encapsulated end-organs found on the surface of the skin and they are located in the tendon and muscle sheaths and also in other parts of the body. They give information about deep pressure and are sensitive to any deformation of the tissue in which they are embedded.

These kinaesthetic receptors play an important part in the controlling of muscular activity. They form as it were the feed-back loop which tells the brain the extent to which its instructions have been obeyed by the muscles.

PERCEPTION

Perception in its widest sense can be taken to cover all those processes which lie between the input and the output of an organism. Physical stimuli will produce sensation in the sense organs; and the physical effect of these stimuli on the receptors in the neural pathways can be measured by physical methods. In this respect, measurement is similar to that in a machine, where a known input will produce a predicted effect. But in man while it may be possible to estimate what his output will be from a given input, the degree of certainty with which this prediction is made is very much lower than in the machine. For example, two individuals looking at the same visual pattern may give two entirely different and unpredictable verbal outputs. On the other hand, the output of two individuals who are punched on the nose will be likely to be very similar and predictable. It can be argued that the perceptual process ends when the incoming data have been organized in order that a decision on the response output can be taken, and that the organization of the output response is part of the motor function, and therefore is not perceptual. While this may be logically true, the nature of the perceptual processes can be inferred only by studying the output which results from a particular input, so that it is convenient to describe perception as the relationship between the input and the output, even if this is not strictly accurate.

It is not intended to discuss perceptual theory in detail. Excellent summary papers (e.g. Bevan, 1958; Graham, 1958) and books have been written on this subject (e.g. Broadbent, 1958). It is, however, necessary to call attention to the fact that the process of perception will differ from that of the reception and transmission of neural messages and it must not, therefore, be assumed that because the nature of a stimulus is known, its perception and the resulting output can be predicted. Although the theory of perception will not be dealt with here it has some features which are important in the study of ergonomics. The first and probably the most important is that of detection. We have seen that man is provided with a number of receptors which enable him to receive information about what is going on around him. But in order to learn anything about the environment from these sensations he must be able to appreciate qualitative and quantitative differences in stimulation. Thus if everything in a room were pure white and the lighting was such that it was completely shadowless it would be impossible to distinguish one object from another. Similarly, if voices were of one frequency and amplitude only, meaningful speech would be impossible. The level of stimulation at which a difference can just be detected is the threshold (also called the *just noticeable difference* (j.n.d.), or the *limen*) between one stimulus and another, while the level at which a stimulus can just be detected is known as the *absolute threshold*.

A great many experiments have been conducted on the thresholds for different sensations from which it has emerged that the j.n.d. for a particular sense of modality is a relatively constant fraction of the standard stimulus which is used. This is often called Weber's Law and is expressed by:—

$$\frac{\Delta I}{I} = \text{constant} \qquad (3)$$

The ratio of the j.n.d. (ΔI) to the stimulus standard (I) is known as the Weber Ratio or Weber Fraction. For example, let us assume that a standard stimulus of 10 units requires an increase of 1/10th to obtain a j.n.d.: the value of a second stimulus which can be discriminated from the first will be 11 units. With this as the standard stimulus, for the next j.n.d. on an increasing scale, an increase of 1/10th of this would produce a third value in the series of 12.1 units – and so on. The values for ΔI for the first ten steps of this series are given in Table 7.

Table 7.   Values of ΔI for a Series of Just Noticeable Differences Starting from 10 Units

ΔI	$Log\ \Delta I$
10·00	1·000
11·00	1·041
12·10	1·083
13·31	1·124
14·64	1·166
16·11	1·207
17·72	1·248
19·49	1·290
21·44	1·331
23·58	1·372

It will be noticed that ΔI does not increase arithmetically with equal steps of j.n.d. On the other hand, log ΔI does increase arithmetically and it is generally accepted that over a limited range the stimulus must increase logarithmically to obtain equal increases in perception (Fechner's Law). Examples of this relationship are (*a*) that just noticeable differences in pitch vary as the log of the frequency; (*b*) that visual acuity measured on a linear scale varies as the log of the intensity of illumination; and (*c*) that the decibel scale for measuring intensity of sound is logarithmic.

The Weber Fraction will also indicate the relative quality of a discrimination since the smaller the fraction the better the discrimination will be. It has been shown by experiment that the ratios for visual brightness, tactual pressure and the pitch of sound, are smaller than those for other sense modalities' sensations, and from this it may be deduced that these modalities are capable of conveying more information. Unfortunately the logarithmic relationship between some stimulus magnitudes and their corresponding sensations is approximately linear over only part of the range of the stimulus and ceases to hold good at the extremes where large changes in the stimulus magnitude produce only small changes in sensation. Gregory and his co-workers (1959) have proposed that at very low intensities this effect may be due to the presence of 'noise' in the subject himself which may be accounted for by random nerve cell activity. Gregory has, therefore, suggested that the Weber Function should be written as:—

$$\frac{\Delta I}{I + k} = \text{constant} \tag{4}$$

where k is the 'noise' expressed in the same terms as I. This concept of noise in the neural pathways of the human being is of importance in ageing.

A more drastic proposal than this has come from Stevens (1961 and 1962) who suggests that, in spite of its general acceptance for 100 years since it was put forward by Fechner, the 'logarithmic law' is false and that the relationship between stimulus (ϕ) and sensation (ψ) magnitudes can be expressed by a power function of the form:—

$$\psi = K(\phi - \phi_0)^n \tag{5}$$

where K is a constant dependent on the choice of units, n is dependent upon the sense modality and is likely to be between 0·3 and 3·5, and ϕ_0 is determined by the effective 'threshold' from which measurement of the stimulus must start. On a log–log plot this function will give a straight line whose slope is determined by the value of n. Thus Stevens would replace a relationship which equates equal stimulus *ratios* to equal sensation *differences* with one which equates *equal ratios* for both stimulus and sensation. To support his argument he shows a number of relationships including loudness/s.p.l., luminosity/luminance and cross modality matching of loudness/vibration. When plotted on log–log coordinates these functions give straight lines.

At the time this book was being written, Stevens' proposals were so recent that they had not been discussed or accepted. As a result many relationships between stimulus and sensation magnitudes are given unaltered in the literature as being linear/log, and it is in this form that they will have to be reported in the present text. The acceptance of Stevens' thesis will mean that a statement that 'x varies linearly with log y' may have to be changed to 'log x varies with log y' and this should be borne in mind when these relationships are encountered.

It will be noted that we have been dealing with the detection of differences in a continuum and man is quite good at making these relative judgements, but is not anything like as competent at making absolute judgements in which comparisons are made, as it were, against subjective standards which have been built up by experience. Thus while many people with relatively little practice can make accurate judgements of the direction and extent of differences of pitch, relatively few people can reproduce a tone on a command. Within this limitation man can be a very good sensing device. He can sense properties which different stimuli have in common which may not be identical with their physical dimensions, he can

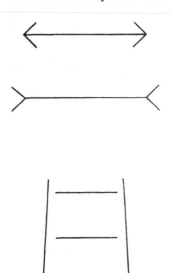

Fig. 31. Two illusions.

recognize the direction and amount of changes in a stimulus, he can discriminate between stimuli and make similarity judgements. These may lead to identification or to recognition when definite standards are laid down against which comparisons can be made. Moreover, he can make these judgements of similarity or difference over a wide range of attributes or against a wide range of standards.

The success with which a man carries out these functions will depend upon the context in which the stimuli occur. Many stimuli will be simple to recognize but the recognition of others is made difficult by design faults which provide an inappropriate context to the stimulus. For instance, after a railway accident in Luton in 1955, the driver of the train involved stated that he had difficulty in identifying the yellow distance signal against the background of yellow lights on a nearby factory. In another collision in Newcastle in 1960 the driver appears to have read and accepted the signal for an adjoining track because of spatial misalignment due to the curvature of the track. In both these instances the stimulus to which the driver had to respond was inappropriate to the context of the task. As a result, although it might have been possible to measure the receipt of a physical signal on the retina from the correct stimulus, the meaning of the stimuli was wrongly perceived. This is a factor of the utmost importance in equipment design since lack of thought on the need for perception can lead at least to difficulty in interpreting the meaning of signals and at the worst to a gross mis-interpretation. That very simple factors can influence perception is illustrated by the well known Müller-Lyer illusion shown in Fig. 31. The two lines in the top diagram are of the same length. Illusions of one sort and another can be created relatively easily by unsatisfactory design features. They can also occur through the presence of after-images which result from looking fixedly at intense stimuli. It has even been suggested that mis-reading of traffic signals can be due to the production of a green after-image after looking too fixedly at an over-bright red signal.

Perception of depth and of movement in depth can be good provided that an adequate frame of reference is available. It is quite difficult to judge the position and speed of approach of an on-coming car at night on a straight road when there is nothing to which the car lights may be related, whereas in the day time this task is relatively easy. The need to provide a frame of reference when depth detection is involved must always be borne in mind when designing equipment. The author recently saw a crane cab in which the older type of half window was replaced by glass running right down to and under the floor, so as to give the crane driver an unobstructed view while comfortably seated. It had, however, been found necessary to place a bar at approximately the position of the bottom of the old window to act as a reference point, because when working in very large open bays the driver appeared to find it difficult to judge the height and position of his hook, and he had been hitting obstacles more frequently than before.

In addition to the various external factors which can affect correct perceptual judgement, there are factors in the man himself which are also of importance. The first of these is the amount of learning or previous experience, which can influence the interpretation of what is seen or heard. Individuals will tend to refer the stimuli received to stored criteria which have been built up through learning or experience and if these criteria are inappropriate to the situation at a particular time, then incorrect perceptual judgements may be made. In this way any individual may develop a *set*. The concept of set recognizes that a stimulus does not arrive 'out of the blue' on a completely unprepared subject. On the contrary an individual may through learning, previous experience, instructions, prejudice, the influence of others and so on, be anticipating a certain kind of input and what happens to the actual input may to a greater or lesser extent be influenced by the set which has developed. For instance, in an experiment on noise, Mech (1953) was able to make his subjects perform in a predetermined manner by making them believe in advance that a certain performance in noise or in quiet was normal. Thus one group gave a better performance under noisy conditions whereas a comparable group gave a better performance under quiet conditions. It can be seen, therefore, that even when equipment is well designed, trouble may result if the operator is prepared for inappropriate data or is inappropriately prepared for the data actually received. For example, norms in inspection can be based on expectancy and defects which are in accordance with the norms will be seen more readily than those which are not. It is clear that allowance should be made in equipment or work design for sets which have been established or will be established by the operators, so that the data given by the operator's task shall not be contrary to that for which the operator is prepared. Further, adequate training should also be given so that the operator is correctly prepared for the data which will be given. This aspect of perception has received a vast amount of attention and is the basis of a number of fundamental theories of psychology. The interested reader who wishes to delve further into these matters should refer to some of the numerous texts which are available.

CHAPTER SIX

Man as a System Component

In the introduction to this book the man/machine unit was referred to and it is necessary now to look more closely at this unit and at the part which is played by the man. In general man's role will depend not only on the requirements of the unit, but also on his ability to meet these requirements. The requirements of the engineering, electrical or electronic parts of the unit, that is the machine, are already extensively studied by the appropriate scientists. In the same way, the functions of the man are the subject of studies by biological scientists, and their results indicate the limitations imposed by his capabilities, the conditions under which he can function at his best, the activities he can perform well, and the system characteristics which will enable the best use to be made of his capabilities.

THE NATURE OF MAN-MACHINE UNITS

When a man is said to be functioning as part of a man-machine unit, the word 'machine' is used to imply any piece of equipment with which an individual accomplishes some purpose. The pencil with which we write, the racket with which we play squash or the spade with which we dig the garden are, in this sense, just as much 'machines' as the car we drive or the lathe on which we may be working. A man-machine unit has three basic functions: (1) an input function which conveys information to the man's senses, (2) a control function carried out by man in the central mechanism, and (3) an output function which will usually, though not invariably, be achieved by the activation of the man's motor system and the application of muscular force. When there is no direct link between the output and the input, the unit is an open loop, but when the output may have some influence on the input, the unit functions as a closed loop in which the man is acting as a control element.

In pressing the start button on a machine in response to a decision that the time has now come for the machine to be started up, the operator is acting as part of an open loop (Fig. 40), but should the machine be a lathe on which a cut is to be taken by hand, the rate at which the crank which drives the saddle is turned will depend on information coming through the eyes from the nature or colour of the swarfe or through the ear from the sound of the tool cutting. The speed of turning will be modified through this feedback to attain the optimum rate of cutting and the most satisfactory cut. The man is then acting as part of a closed loop (Fig. 37) and thus, in its simplest form, the closed loop is giving him immediate information on the effect of his action. Another kind of closed loop is that in which an operator has to control steam pressure continuously against a varying load by opening or closing a valve in order to restore the pointer on the pressure gauge to the desired position. If this response is continuous, the man is exercising a function which is generally known as *tracking*. Tracking takes two forms, compensatory tracking in which an index has to be maintained at a pre-determined position as in the example above, or pursuit tracking in which a control index is kept in alignment with an index which may be moving in a random fashion. Because tracking is a task which is easy to study and because it is possible to build various types of circuit characteristics, including delays, into the function of the controls it is one of our main sources for understanding the functioning of the human operator as a controller. The literature on the subject is vast and the development of the concept of the human controller in non-psychological terms has been developed by Craik (1947, 1948), Hick and Bates (1950), Tustin (1947) and North (1956) and others. The description of the function of the human operator in these terms has been a particular interest of the members of the Psychology Branch of the Naval Research Laboratory in Washington, in particular Birmingham and Taylor (1954), Taylor (1959), and Taylor and Garvey (1959), and their work has been drawn on heavily in the present account.

Closed loops do not always give immediate information in this way. Suppose that an operator is manually controlling a function in a chemical plant; a change of reading on an instrument may require a change of setting of a control, but owing to the nature of the chemical process the effects of the change of setting may not be evident on the display until some time has elapsed, that is the feedback is delayed. Under these circumstances, there may well be a tendency to overcorrect so that the process hunts on either side of the ideal situation. If the information given to the operator by his dials is not ideal for the function which he has to carry out, or if he has to use information from more than one dial, he may be carrying out the functions of a differentiator, an amplifier or an integrator, or he may be required to act as a computer of geometric or logarithmic functions. When this is borne in mind, it is not very surprising that units which call upon him to act in this way are not always very efficient and the hunting may be accentuated. Methods of simplifying the controller's task by 'quickening' will be discussed more fully later in this chapter.

MAN-MACHINE SYSTEMS

So far, we have been discussing a single man-machine unit. These units can be combined into a system which can be defined as one or more man-machine units that convert an input (which can be information and/or material) into an appropriate output in accordance with a planned purpose. In such a system, the several man-machine units may interact with each other, the functions of one unit depending upon the efficient functioning of another. Alternatively, a system may consist of one machine and several men, for example, a rolling mill which is controlled by two operators functioning simultaneously, or one man may form the link between several machines. The study of systems of this kind has come to be known as *systems engineering* – an activity which embraces study of both the human and physical components. It is a major purpose of research into systems that the functions of man and equipment should be carefully balanced in order to obtain the optimum result, and for this purpose the various capacities and limitations of people should be taken into consideration.

MAN AS A CONTROLLING COMPONENT

The sequence of events which leads from the receipt of the external stimulus to the completion of a response to the stimulus, is an immensely complicated one, involving as has been shown the transduction of the external or internal stimulae, the perception of the meaning of the signals from the transducers, the making of judgements, the placing of information into temporary or permanent storage in the memory and the recall as necessary, leading up to the making of a decision. Since decisions can be made only at a limited rate, it is necessary to have a substantial amount of coding so that a single cue for recognition may convey the optimum amount of data. After the decision has been made, the various cues for muscular action must also be coded and sent to the parts of the body at the right time and in the right order. Although it is convenient to discuss all this activity separately, the sensory, central and motor activities are so closely related that in most instances it is almost impossible to draw a dividing line between them. Thus, while the sensory processes will be discussed first, followed by considerations primarily related to the central processes, it must be understood that it is not intended to imply that these activities fall into three water-tight compartments.

When acting as a controller, man has a number of qualities, some of which make him outstandingly useful, while others put limitations on his performance. These qualities must be considered in relation to functions which are done well or not so well by a machine. These are listed in Table 8.

Of man's useful qualities perhaps the most characteristic is his flexibility. Unlike a machine, which is designed for a specific purpose, a person can change his rôle rapidly and frequently, moreover he has a multiplicity of channels which he can select as required and his homeostatic processes

Table 8.

Functions which can usually be done better by a man than a machine, unless it is highly complex
Sensing minimum stimulae
Amplification
Improvisation and flexibility
Switching
Long term storage (memory) of great capacity
Perception of space, depth and pattern
Interpolation
Extrapolation and prediction
Translation
Inductive reasoning
Making judgements or decisions
Homeostasis

Functions which it is usually best to allocate to a machine
Computing
Differentiation
Integration
Response at great speed
Application of massive force smoothly and precisely
Precise repetition
Short term information storage (memory)
Deductive reasoning
Complex simultaneous functions
Simple YES-NO decisions

A man in a system will introduce:
Delays
'Noise'
Limits on data transmission which will be related to channel capacity

enable him to do this effectively under a variety of conditions both favourable and inimical. He can continue to function even when some channels have been rendered inoperative, either temporarily or permanently. Functions which man seems to be able to carry out quite well are amplification, interpolation, switching, storing information and translation, although the latter two are affected by age. He can also extrapolate and predict and, particularly, he can make decisions. Activities which he is not so good at carrying out (although his degrees of success will depend very largely on circumstances) are computing, differentiation and integration. It is a

measure of man's flexibility that he can carry out these functions more or
less successfully either singly or in a number of combinations at one time.
His success in a task will depend on the nature of the simultaneous demand
for the exercise of these functions and on the rate at which he may be
called upon to carry them out.

The performance of a system can be influenced by several human
limitations. Men are variable in performance between one moment and the
next, and one individual may differ substantially from another This can
cause imprecisions in carrying out the various functions and can introduce
'noise' into the system. Men will also introduce delays, and will also have
a definitely limited capacity both in terms of amount of data and speed of
data transmission.

The Sensing Functions

We will now look more closely at some of man's attributes and see how
they may influence decisions on system design. Of first, and primary
importance, is that man can act as a *sensing device* to minimal stimulae
from a wide range of sources simultaneously and bring these signals
together to produce a complete picture of an event. In carrying out this
sensing he may respond to very dim lights or quiet sounds in an all-or-none
function, and in this a machine may be able to act equally well if it is
sufficiently sensitive; but what is more important, a man can sense changes
in these functions which may be very small indeed and can relate these
changes to changes in other functions so as to arrive at a 'view' of their
meaning. Since combinations of changes may be of almost infinite variety
it might be very difficult, if not impossible, to programme a machine to
cover the wide range of eventualities which can be covered by a man. The
complexity of some of the cues used by experienced people is often not
realized until attempts are made to determine exactly how a man arrives
at his conclusions and on the basis of what information. This may be
necessary either when planning training programmes or designing machines
to replace man.

An example of the discrimination of minimal differences in sensation is
the job of 'Fish Taster' in the port of Hull (Smith, 1961). This title is
somewhat of a misnomer since smell, touch and vision all play a part in
deciding the quality of the fish. For instance, a fish smelling of seaweed
is superior to one that is 'mousey', 'musty' or 'peppery' and a 'firm elastic
fish' will be of good quality whereas one that is 'soft and flabby' will not.
Other qualities judged are eyes by shape and colour, gills by colour and
quality of skin and slime. Qualities of this kind can be subjectively defined
and passed on by example (most people would recognize the smell of

mouse!), but although smells can be measured by gas chromatography it is not easy to determine in measurable form the difference between the smells of seaweed and mouse and to build a machine which would not only distinguish between them, but the degrees of difference between them.

One of the convenient concepts which psychology has borrowed from the communications engineer is the notion of a channel which is used to describe the means by which data are conveyed to the central mechanism. The term is also used in discussing the central mechanism itself which at its highest level is believed to act as a single channel. Of the sensory organs which act as transducers to the external or internal stimulae, the visual channel is one of the most important. The efficiency of this channel will depend not only on the nature of the visual information which has to be transmitted but also upon the way in which it is displayed. In industrial situations it is often more important to ensure that the operator gets the kind of information he needs than to establish the optimum design of the instruments which give this information, although both must be right if the optimum result is to be obtained. Limitations of the visual channel, such as visual acuity, colour discrimination and so on have already been discussed.

Another important sensory channel in industrial activity is the kinaesthetic channel. Since this would appear to be more important in industry than for military purposes, relatively less research seems to have been done upon it. It seems that it must play an important part in all tasks involving positioning movements without vision and in those tasks which involve the 'feel' of a control. Quite a lot is known in an empirical way from the studies of methods of training in complex skills, and Gibbs (1953) produced evidence to suggest that kinaesthetic feedback, which becomes available when using a pressure control, is greater in amount and more rapidly conducted than the feedback available from a displacement control. In both types of control kinaesthetic information is available together with visual feedback, the difference between the two being that in one case the information comes from the force which is being applied by a limb whereas in the latter the information comes from the amount and perhaps the rate of displacement. It may be that two different types of receptor are involved, but if this is so, evidence has not yet been adduced to demonstrate the matter clearly.

In addition to the receptor channels involved in a particular activity, there is also a substantial amount of data transduced and transmitted through other sensory channels of the body at any one time. This will produce a great deal of redundant information which has to be filtered before the messages reach the highest level in the cerebral cortex. The same is true of a single sensory channel and since the capacity of channels is probably small, the efficiency of each must be related to the extent to which unwanted information is presented through the channel. Display design must take this point into consideration and ensure that only the data required are presented to the operator.

Another way of looking at the process of decision making, which has again been borrowed from communication engineers, is the concept of information theory developed by Shannon. It has been shown that the time taken to respond to a signal is related to the number of choices involved in the response (Hick, 1952). The measure of information gain has been called a *bit* (which is a contraction of binary digit) and the bit is the amount of information which is contained in a single choice between two

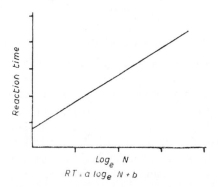

Fig. 32. Reaction time as a function of information.

alternatives. Until recently it has been generally accepted that the time to respond to a signal forming one of a number of alternatives will be directly related to the logarithm to base two of the number of alternative signals (Fig. 32) but this has now been shewn to be due to a lack of practice (Mowbray and Rhoades, 1959). When subjects are practiced over a long period differences in R.I.T. can disappear (Fig. 33). It is too early yet to be able to say what the effect of these findings will be upon the theories which have been developed from experiments with naïve people. (For reviews of this subject see Welford, 1960, Gore, 1960 or Edwards, 1964.)

Before leaving the sensory channels, it is perhaps appropriate to recall the limits of ability to make sensory discrimination. It will be remembered that over a wide range of sensations which covers the major area of industrial activity, the ability to discriminate varies approximately as the logarithm of the intensity of the stimulus. This fact must be taken into consideration when means of giving data to an operator are being planned.

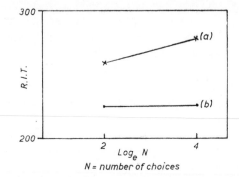

Fig. 33. Reaction time (*a*) without and (*b*) with practice (after Mowbray and Rhoades, 1959).

The Central Processes

When man has received some cue, minimal or otherwise, he may take some action upon what he has perceived. In doing this he is acting as an *amplifier*, and on the basis of these cues he can set a complex train of consequences under way. His range of activity in this field is vast but perhaps the example of piano playing will illustrate the point. A single sheet of music can be amplified by the pianist into a complex series of physical movements which will produce for the audience a sense of aesthetic pleasure which, so far, is not attainable by purely mechanical means. A player-piano using a punched tape can play the notes in the right sequence, but it is not difficult to distinguish between a work played by a live pianist and a work played on a player-piano.

When given two pieces of data a man can often accurately arrive at a result somewhere between the two: that is, he can *interpolate*. As a simple illustration of this, it has been shown that a satisfactory way of reading scaled industrial measuring instruments is to interpolate into fifths rather than to mark each of the five intervals on the scale. Again given two items of data he can also *extrapolate* to give data in extension which can be in the form of a prediction. Prediction can also be carried out from single pieces of data.* For instance, the appearance of the road sign 'SLOW, MAJOR

*There are semantic difficulties in describing information on which someone may or may not have to act. The obvious word 'information' has been appropriated by information theorists and given a special meaning so that its use in the ordinary sense could lead to confusion if not qualified. To avoid misunderstanding the term 'piece of data' is used to convey that an operator is receiving one or more items in a 'chunk'.

ROAD AHEAD' will cause the driver to decrease speed even though the road junction may be round a bend so that he cannot see it. With the right data an operator can be fairly accurate in his estimate of when an event or a state are likely to occur. However, his predictions from rate of change, which involve integration, are not usually carried out entirely successfully.

Memory may be either permanent or transitory. In the permanent memory are stored data relevant to the operation of the system which cumulatively become 'experience'. In the transitory memory are stored data which are required for immediate use and which, once used, can then be discarded. Without getting involved in learning theory it can, in a simple form, be suggested that learning and experience may both involve a form of permanent memory, in that they enable a series of pathways to be set up (almost as they might be in making a complex telephone call) so that a particular cue for action will elicit the required response without the decision process being deeply involved. In any event, systems which have the maximum amount of memory built into them – that is, systems which rely as little as possible on the memory of the operator – are liable to be learned more quickly and operated more efficiently, particularly if the operator is becoming old, since one of the characteristics of older people is loss of transitory memory. Another function which appears to be affected by age is *translation*, that is an ability to go from A to C via B. Ability to translate varying functions is one of the flexible characteristics of the human operator, but since it is, to an extent, age-related, it is a function which should be called upon only when absolutely necessary.

We come now to the *decision-making* part of man's activities. The word 'decision' is one which may be misunderstood since it has a very wide range of meaning. This does not imply that the processes involved are necessarily very different. The decisions made by an executive, which have formed the subject of extensive study in our management schools, are perhaps the most readily recognized, involving as they do the weighing of a number of factors and the prediction of the eventual outcome of the decision which is to be made. It is much less easy to recognize the same process in the almost automatic decisions which are made from moment to moment when a motor response or a series of motor responses are made as a result of a signal for action. The exact boundary between decision-making and reflex action is hazy, but many activities which might appear to be automatic do, in fact, involve the making of decisions and these decisions do not differ very greatly in their procedure from the processes of decision-making by the executive. It is largely a question of the time-interval involved. Thus the operator will be receiving data from his environment or his equipment; these data must be compared with data stored in the immediate memory from which he will arrive at a number of alternative outcomes from which he must select the one most likely to give the optimum result. In selecting the optimum outcome, the operator may be assisted by previous experience which has enabled him to set up a series of rules to assist in reducing the uncertainty of the outcome and increase the speed of decision. The simpler the set of rules the quicker they can be established; the smaller the number of alternative outcomes, the more efficient the process of decision-making will be.

In carrying out the decision-making process, a man is probably performing the function for which he is best suited. There are, however, definite limitations to his performance and he will be able to carry out his task as a controller only if these limitations are taken into account. There is now evidence to show that at the highest decision-making level in the cerebral cortex man acts as a single-channel mechanism (Welford, 1959a), and that the rate at which decisions can be made is strictly limited. When once a piece of data enters the channel a period which has been described as the psychological refractory period (Welford, 1952) inhibits the receipt of further data and the taking of the appropriate decision. It appears that the refractory period is about 0·5 sec, so that about two pieces of data can be passed through the channel per second. If two pieces of data arrive almost simultaneously it follows that one of them will be passed through the channel but the other may be delayed or lost altogether; if the data arrive absolutely simultaneously they can be coded and pass through the channel together as one piece. Designers should therefore accept the general principle that under no circumstances should equipment be designed or a machine cycle be evolved which requires the operator to make two simultaneous decisions, whether through one or more sense modality. This limited transmission rate has been described by Birmingham and Taylor (1954) as the 'band width' of the human operator and they have specified this human band width as 'the region between zero and three radians per second'.

A rate of data transmission of about two pieces of data per second will give a minimum time between successive responses of about 500 msec, but only part of this time may be occupied by the central processes. This part will be used by the operator to make the decision and to organize the response which, once the neural networks have been established, will usually be carried out without any further voluntary control on the part of the operator. The effect of this 'reaction time', therefore, is to introduce *delays* into the system which must be allowed for in design.

In functioning as a single channel mechanism with a delay, the human operator will behave in an intermittent fashion in his responses. The situation has been described by Birmingham and Taylor (*loc. cit.*): 'It would seem that if any type of servo motor could be taken as an analogue of human behaviour, it would have to be an intermittently sampling servo instead of a continuous follower' and it is in this activity of intermittent sampling that we have an explanation of commonly observed behaviour which might seem to be contrary to the limitations just described. For instance, it is often pointed out by students that an accomplished pianist will be playing notes at a greater rate than two a second. It has been suggested that it is a characteristic of an accomplished pianist that he is able to recognize groups of notes in music which will elicit a series of response movements which have been learned as a whole, so his perception of one group will trigger off a whole series of finger movements which once they have been initiated will continue to finality without any further voluntary control. It is in this way that the accomplished pianist differs from the learner who has to make a discreet decision for every note played. Sampling can also account for some apparent aberrations in behaviour which might otherwise seem inexplicable. For instance, a car driver who is continuously sampling data from a dog which may run across his path, may at the same time fail to sample from an important road sign or from the trafficator of a vehicle ahead, even although both should have been clearly visible to him. Light rays from the signals may have fallen on his retina but because at that particular moment he is sampling the activity of

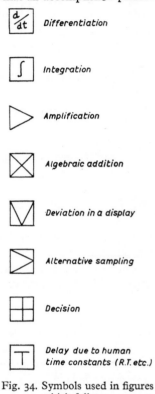

Fig. 34. Symbols used in figures which follow.

the dog they have not reached the level of the cerebral cortex. Thus a person can, in colloquial language, look at something and not see it.

If a man has to spread his sampling over a wide range of data sources the amount of data which he is receiving from any one source will be reduced in proportion to the sampling pattern. This means that the more functions he has to perform, the less data he will have available for the performance of any one of them, and this will introduce *imprecision* into his performance. It seems possible that even when he has to sample only one source of information he is likely to be less precise than a specialist mechanical or electronic device designed to carry out this function. A decision whether a man should be used or whether a machine should be designed to replace man must therefore depend, to some extent, on the precision of response required. Where it is greater than can be expected from a man with the limitations already discussed it may be economically worth while to replace a man by a machine. When, however, the precision required is substantially less, it may not be worth while replacing the man. Precision, however, is not the only factor which has to be taken into account. Even when carrying out the simplest functions, man is *variable* in his performance and this is probably one of his most serious limitations. He will vary from moment to moment, from one period of the day to the next and from day to day. It is one of the penalties which have to be paid for the greater adaptability of the human organism. Variability is one of the major causes of the introduction of *noise* into a close loop system by the human component.

Some of the delays and imprecision in a system may become less as the operator builds up expectancies based on previous experience in operating the system. Frequent occurrences may have statistic properties which an operator can detect. Relationships between sequential events may also develop which can be learned and these together can reduce the randomness of situations with which he is presented and so lead to more effective performance. The building up of expectancies is part of the process of becoming a competent operator and the more the design of the equipment helps in this, the greater will be the rate of performance improvement.

Motor Activity

The output function of man is usually carried out by muscular activity. Muscular activity, as we have seen, will have three main characteristics: the force which can be applied and the speed and accuracy with which a movement can be made. The effectiveness with which force can be applied is to a very large extent dependent on control design. Once it has been decided what part man is to play in a system, it is usually possible to design the characteristics of controls so that they are within the physical capacity of a man to operate them; but before a final decision is made on the part to be played by a man in a system a check should be carried out to ensure that the many physical factors relating to man's efficiency as an operator of controls, which are discussed elsewhere in this book, have been taken into account.

The other two factors, speed and accuracy, can together be partly responsible for delays and/or imprecision in the system. The effect of the information content on response initiation time has already been discussed as also have other factors which may influence this period such as the sense modality used, the length of the fore period (if any), age, sex, the limb used, the response required and so on. Due allowance must be made for these factors, particularly when speed is of importance in the operation of the system. Probably of less account, but still not negligible, is the actual time which it takes to make a control movement. Free movements between one control and another or between a rest position and a control will vary with the distance travelled: but not proportionately. For instance, a horizontal movement of 10 cm will take roughly 550 msec, whereas a movement of 40 cm will take about 750 msec. Other factors which have to be taken into account when determining the time to be allowed for the human operator are the direction in which the movement has to be made, whether it can be monitored by vision or has to be a 'blind' positioning movement, and the nature of the manipulation which is to be carried out at the end of the movement. The time to carry out the manipulation itself will vary so much and will be so closely related to specific design characteristics of the system that very few general principles can be laid down, except to say that the greater the accuracy with which the control movement has to be made, the longer it is likely to take; moreover, accuracy itself will be influenced by a number of factors in the control design, including gear ratios, inertia, friction and load.

The preceding pages have dealt with the kind of information and the kind of output which will enable a man to carry out his designated part in a system. The design requirements which will enable him to do this efficiently are the subject of the chapters which follow, and it is now possible to discuss some of the ways in which man may be fitted into various systems. In doing so notations will be suggested which may be useful in the theoretical exercise of sorting out which functions should be undertaken by an operator.

MAN'S ROLE IN A SYSTEM

In many systems, the effect of man's output is to produce an immediate adjustment of the display. In other systems, notably those associated with the control of ships, aircraft, or chemical processes, the dynamics of the process may include a complex series of delays between the making of a control movement and the ensuing change in the display. In systems with lags of this kind, there can be a tendency to hunt, which must be prevented by the operator learning to anticipate the course of events and to make corrections before an indication appears that these are needed. This means that he may have to *differentiate* and/or *integrate* several times over. The theoretical considerations behind a situation of this kind have been very fully investigated by Birmingham, Taylor, and their collaborators at the Naval Research Laboratories in Washington, on whose work the exposition which follows is based.

When operating as a link in a control loop, man's basic output is force, and when this is applied to controls with different characteristics they will respond in ways which require different mathematical representations. A simple control might be a spring-centred joystick which can be represented as in Fig 35(a). If force is applied, the resulting displacement is directly proportional to its magnitude, but may vary with the spring loading; in other words, the joystick functions as an amplifier of adjustable gain. If instead of being spring-loaded the joystick is viscously damped, it will move with an angular velocity, or rate, proportional to the force applied, and the displacement is proportional to the first time integral of the force as shown in Fig. 35(b). If now force is applied to a joystick with high inertia, it will move with the angular acceleration which is proportional to the applied force: displacement is then proportional to the second time integral of the force as shown in Fig. 35(c). Thus depending on the physical characteristics of the control mechanism, the control output in terms of time/displacement is zero, one or two integrations of the force applied.

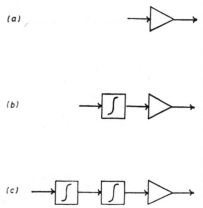

Fig. 35. Dynamics of (*a*) a spring load control, (*b*) a control with viscous friction, and (*c*) a control with inertia.

The simplest practical tracking system which may be made to follow precisely a constant velocity input may be represented in a number of ways, one of which is shown in Fig. 36. In this system, the output of the first integrator is the rate component. This rate is differentiated in order to obtain the position component. When a human operator is introduced into this system as, for instance, when he is expected to respond to a deviation by the operation of a damped joystick, a system such as that shown in Fig. 37 might be envisaged. In this system the mechanism acts as a simple amplifier while the damped joystick will provide one integration. As a result, in order to complete the loop, man will have to act as if he were a differentiator, an amplifier, an integrator and two algebraic adders all

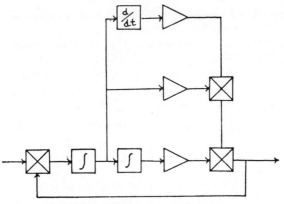

Fig. 36. A perfect tracking system.

functioning at the same time. This he will not do very well. There is ample evidence to show that a man will act best if his task is that of a simple amplifier only, so that in order that he may track with perfect aiding, the mechanism should be required to carry out the integrations, differentiations and adding, as shown in Fig. 38. In other words, the functions of the man and the mechanism are reversed. This is an idealized concept and the nearest practical system to it is one which is used in gunnery control under the title of 'rate-aiding' which is similar to Fig. 38, except that the derivative term is not included. Birmingham and Taylor (1954) point out that similar idealized systems can be described for tracking with both inertial and pressure joysticks. In practice, systems in which a human operator may be found are not, as a rule, required to follow constant velocity displacements and in practical situations the displacements may contain factors of acceleration, rate of change of acceleration, or even higher derivatives.

Methods of achieving the optimum performance of such systems have been proposed in principle by Birmingham and Taylor under the terms 'unburdening' and 'quickening'. Unburdening implies that the operator is relieved of the need of applying force continuously or in some timed sequence pattern. In the systems of the type quoted above this is done by the integrators in the mechanical portion of the system illustrated. This is a simple example, but Birmingham and Taylor suggest that functions more complex than simple integration may be required and that the principle of unburdening may have general application to many situations. Quickening is designed to give the operator immediate knowledge of the result of any control movement which he may make. This is achieved by providing feed-forward loops which may give information about position, rate, acceleration, or rate of change of acceleration to output integrators which give the operator immediate knowledge of the effect of his control actions. Such a quickened display is shown in Fig. 39. In conditions such as this it is possible for the display to be a dial with two pointers, one of which gives the input while the other is controlled by the quickened feedback. The task of the operator then is to keep the two pointers in alignment at all times, and Birmingham and Taylor report that under these circumstances, tests have shown that hunting in a system can be completely eliminated by quickening.

Taylor (1957) reports an experiment in which subjects were required to track under four conditions, three of which incorporated delays. In one of these, tracking was in one coordinate, in two others in two coordinates, all without quickening; in the fourth condition subjects tracked in four coordinates simultaneously with quickening. Results showed that some subjects were able to cope with the first condition after eight days' practice, that the second and third conditions were impossible and that in the fourth, quickened condition practically all the subjects achieved perfect performance on the second day. From this, Taylor concludes that quickening the display to give immediate knowledge of results will increase the number of control loops which one operator can handle (in the experiment from one to four) and that learning time will be substantially reduced.

So far this principle has been applied mainly to flight instruments, one of which is the Sperry Zero Reader.

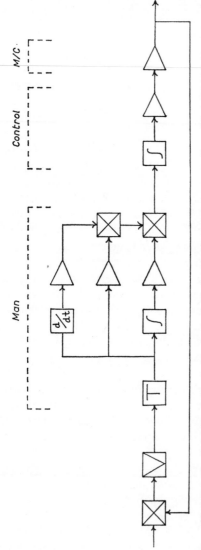

Fig. 37. An 'unaided' tracking system.

It has been applied with particular success to the operation of helicopters, in which with conventional instruments it is almost impossible for the pilot to maintain a stationary hover under blind flying conditions. Sweeney *et al.* (1957) have shown that by quickening the system, hovering performance can be improved by over 500%.

The above discussion has implied continuous control of a function which is varying. There are other situations, in which the controlling may be discontinuous, and this is particularly true of operations in the chemical industry where continuous controlling has virtually disappeared. A situation may arise in which an operator has to make a decision based on a change in an indicated variable such as temperature. He may be given a thermometer and he may be required to decide, should the temperature start to deviate from its designated value, whether this deviation is likely to be temporary or whether the process is going out of control. In the former circumstance he should 'ride it out', but in the latter he will have to stop the plant. With the information he is given he is having to estimate the rate at which the temperature accelerates (Fig. 40) and this we have seen he is not very good at doing. If, however, the display is quickened to give him the rate of change of temperature, or better still the rate of rate of change, the chances of making an error will be

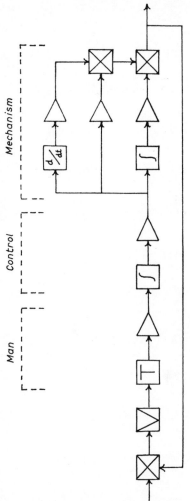

Fig. 38. An 'aided' tracking system.

substantially reduced (Fig. 41). It is therefore of great importance if an operator is retained in a system in order to make decisions of this kind, that he should be given the best available information on which to act. If he has to estimate the rate of rate of change from the rather slow moving

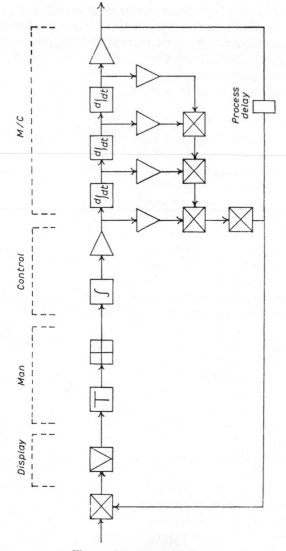

Fig. 39. A 'quickened' system.

indication of temperature change, he might make the right decision with
a frequency which will not differ much from chance. If, however, he is
given directly the rate of rate of change, the chances of him making the
wrong decision become very small indeed. Furthermore, he is likely to
be able to make the correct decision with very little experience, a situa-

tion which is probably not true of the conventional methods of display in common use at the present time.

The need for expensive and often costly training to enable operators to carry out some of the more complex functions which we have discussed has been pointed out by Taylor and Garvey (1959). They called the con-

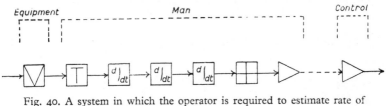

Fig. 40. A system in which the operator is required to estimate rate of rate of change.

ventional method of training individuals to operate sub-optimal systems a 'Procrustean' approach, after Procrustes, the legendary highwayman, who adapted his victims to an iron bed either by stretching them to the required length or by shortening their legs. To support their view that this may be an inefficient or even dangerous approach they quote an experiment (Garvey and Taylor, 1959) in which two matched groups of subjects were

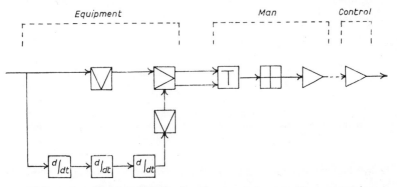

Fig. 41. A system in which the functions carried out by the operator in Fig. 40 are transferred to the equipment.

practised on a tracking task with two forms of control, position control and acceleration control. With position control there was no transformation between the operation of the joystick and the movement of the marker on the display, but with acceleration control the output was transformed by two integrators, so that the acceleration of the marker depended on the amount of displacement of the control from the neutral position. To operate the acceleration control satisfactorily the subjects had, therefore, to act as an integrator and a differentiator, as well as an amplifier. One way of looking at the situation is to say that acceleration control is an unquickened system; another way is to say that with position control the control/display relationship is compatible (see Chapter 10), whereas with acceleration control the relationship is incompatible at least part of the time. The learning curve for these methods of control is shown in Fig. 42,

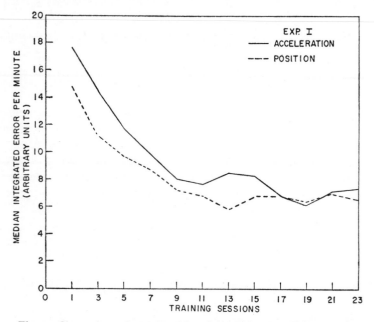

Fig. 42. Change in performance with training with two different types of control systems. Each point is based upon results with eight subjects (from Taylor and Garvey, 1959).

from which it will be seen that during the first days of training the acceleration control system was inferior to the position control system. This difference, however, became less as training proceeded, until on the 17th day the two systems produced equivalent performances. We can see from this that the initial inferiority of the acceleration control was nullified by training. When training was completed both groups of subjects were given one trial lasting one hour and also shorter trials with six different types of 'task induced stress'. These stresses were (1) reversed display-control relationship, (2) left-handed tracking, (3) two-handed tracking in which the subjects were required to track two dots simultaneously, (4) two-coordinate tracking in which a target was tracked in two coordinates with a joystick, (5) a secondary visual task in which subjects had to detect and

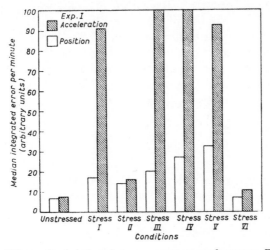

Fig. 43. Effects of task-induced stress on system performance. Each bar is based upon results with eight subjects (from Taylor & Garvey, 1959).

report the range and bearing of targets on a simulated radarscope and (6) a secondary arithmetic task in which subjects were required to subtract mentally a one digit number from a two digit number.

The results of the one hour continuous experiment is shown in Fig. 43. It will be seen that both time and stress have affected performance adversely, but that the performance with the acceleration control has been more adversely affected than that with the position control. During the one hour run, performances on both were equivalent for the first 25 minutes, but thereafter performance with the acceleration control deteriorated more rapidly than did performance with the position control. In the unstressed situation, which was reached at the end of practice, there was no significant difference between the two types of control, nor was there any significant difference when the operators were called upon to track with the untrained hand (stress condition 2), but with the other five stress conditions significant differences between acceleration and position control developed, and in all the tasks the deterioration was very severe indeed. These experiments show, and there are others which demonstrate the same kind of effects, that performance on an inefficient system may be brought to the same level as that on an efficient system if sufficient training is given, but the superiority of the efficient system will become very marked when a stressful condition of work obtains. This can be highly dangerous under circumstances where extreme stress or emergency may possibly arise, since the expected performance of the operator, however well trained he may be, may deteriorate to a point where breakdown will occur.

At this point it may well be asked why, if man should act only as an amplifier, he should not be superseded in this function as well. Logically of course, the argument does lead to this point, and in certain instances systems may work more efficiently if man's amplification functions are given to the machine and the man is eliminated altogether. On the other hand the removal of the man may require his replacement by highly complex sensing equipment which might or might not have the versatility of the human operator or be able to incorporate the same safety factor which is present when an operator is employed. It is clear that in any practice of systems research, data must be obtained which will enable a decision to be made on whether a man is to be present in the system, and if so, what his rôle should be and what information he is going to need if he is going to carry out this rôle efficiently.

PART TWO

Practical Ergonomics

CHAPTER SEVEN

Design Factors I.
Layout of Equipment

Ideally, equipment design should start with the operator, who should have his equipment laid out around him in positions which will ensure that his posture is adequate, that he can see what he has to do and can operate his controls in the most effective manner. This course of action is not always possible, but at the very least the main areas for vision and for the placement of controls can be determined and the design can be arranged so that the relevant components are placed within these areas. Unfortunately, even these modest requirements are not always met; it would seem that too often a piece of equipment is first designed and then the operator is added as a kind of afterthought leaving him to cope with unsatisfactory arrangements of display and control as best he can. This situation may arise either because the importance of designing equipment round the operator has not been realized, or because a new design is a modification of an older design. Too often it is accepted that because a machine has always been the way it is there is no good reason to change it. 'What was good enough for our grandfathers is good enough for us.' Thus the unquestioning acceptance of the correctness of an existing design may mean that when a new design is being produced it will not be 'thought-through' from first principles.

It is almost impossible to walk round a factory without seeing the effects of tradition acting in this way. By way of an example we can consider the centre lathe, which historically seems to go back to a simple wood turning device in which the bed was a baulk of timber and the chair-leg, or what-have-you, was turned by a cord attached to a bent sapling and operated by the foot. Basically, centre lathes have not changed since that time. The bed is still horizontal, the bent sapling has been replaced by an electric motor, the tool is now mounted in a cross-slide, but otherwise the arrangement is unchanged. It will only take a moment's thought to realize that while the controls may come in an adequate position for the hand, the work is too far away from the eye and as a result the turner has to operate in a bent posture which is unsatisfactory and often heartily disliked.

Now, if a lathe had never been made before and a designer was starting from scratch, he would first define the area in which the work should be placed so that vision would be easy with good posture. Then he would define the areas in which controls should be placed for ease of operation. When this has been done, it will be evident that instead of the bed being horizontal so that the centre line of the machine is beyond the tool post, the bed should be tilted at an angle of about 50° to 60° and raised so that the tool and the work can be seen by a standing operator. If the machine is not too large, the operator will probably be able to sit, but whether he sits or stands there will not any longer be a need for him to bend in order to see what is going on. An additional advantage which comes from this arrangement is a greater ease of swarfe removal, since it will no longer fall all over the bed of the lathe, so that wear will be reduced. An experimental attempt to validate this idea using an old lathe has been reported by Alcock (1961), who built a special bracket on which the lathe was mounted. As can be seen in Fig. 44, the operator can now sit comfortably instead of standing, he can clearly see the tool, and the control comes easily to hand. When its advantages are stated in this way one wonders why this kind of arrangement has not been built into, at any rate, the smaller centre lathes many years ago: it seems so obvious when looked at from the point of view of the operator.

Another instance of tradition apparently influencing current design is the control layout for diesel and electric trains. These, historically, would seem to owe their origin to the early tram-car where the electrical control was a large vertical contactor operated by a crank on the top. These large contactors are now no longer necessary, but the principle of the operator driving a train by means of a large crank placed in front of him still persists. If an arrangement of this kind also incorporates a 'dead-man's handle', the strain on the driver can be substantial, since it is very difficult to maintain pressure downwards at arm's length for any length of time. In practice the drivers often do not use the back-rest of the seats provided because they have to lean forward in order to bring pressure to bear on the 'dead-man's handle'. Again if one were starting to design from scratch, it would seem that the logical place to put the controls would be on either side of the seated operator. There is a rotary movement about the shoulder which could be made to coincide with the rotary movement of a crank-operated controller and should a 'dead-man' action be required it can be achieved by using the weight of the forearm; should the driver collapse, he would fall forward off the controls rather than on to them as in the present arrangement.

Similar arrangements of controls are to be found in many crane cabs

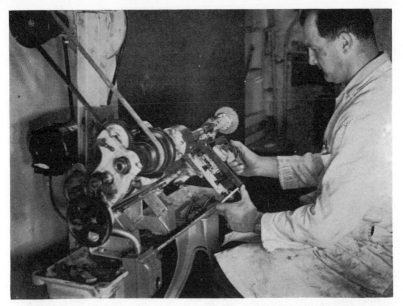

Fig. 44. Centre lathe tilted and raised so that the operator can see without stooping. He can also sit down (by courtesy of Rural Industries Bureau).

Fig. 45. Working position in an old type of crane cab (from Laner, 1961; by courtesy of the British Iron and Steel Research Association).

Fig. 46. Vision from an old type of crane cab (from Laner, 1961; by courtesy of the British Iron and Steel Research Association).

Fig. 47. Unrestricted vision from a re-designed crane cab (from Laner, 1961; by courtesy of the British Iron and Steel Research Association).

Facing page 129

with the old type rotary cranks bearing no relationship either to the movement of the crane or of the hook and having little or no relationship to the natural movement of the limbs (Figs. 45 and 46). The design of crane cabs has been tackled from first principles by ergonomists working at BISRA (Laner, 1961). They have produced a design in which the controls, in the form of a joystick, are at the side of a seated operator, Fig. 47. It is obvious that the improved vision achieved in the new cab and the optimum placement of the controls must put much less strain on the driver and, as a result, he should be more efficient in his work.

THE ARRANGEMENT OF MACHINE CONTROLS OR OF ASSEMBLY WORK

The arrangement of machine controls or of assembly work will be dealt with together because many of the principles which apply to one will apply equally to the other. In either case, if the designer starts with the operator by laying out the areas for vision, for controls, for sitting, for leg room and so on, the body dimensions which have been given in Chapter 4 should be used. It cannot be too strongly emphasized that if a machine is liable to be used by a woman or will be exclusively used by women, the anthropometric data for women should be used. It may be noted in passing that the areas of reach which are given in many text books are applicable to men and are not appropriate if women will be using the equipment.

A basic point which will have to be settled at the outset is whether the operator will stand or sit or should be able to do both. It has long been one of the tenets of method study that high benches should be used so that an operator can stand or sit at will. This 'principle' does not appear to be based on any definite evidence that high benches are better than lower ones, provided that work is planned in such a way that operators leave their work places at regular intervals, in order to obtain a change of posture. The reasons for planning work in this way are given later in Chapter 17. The standard of much factory seating is deplorably low (a common type of factory chair is used as an illustration in the next Chapter on seating), so it is not very surprising that workers have found it necessary to stand up from time to time; the seats are probably too uncomfortable to be sat upon for more than a limited period. However, if good seating is provided, there seems to be no reason to expect that operators will want to stand up if they can sit down. At any rate this has been the experience of the author on production lines with which he has been involved: great care was taken to ensure that the seating was the best that could be provided, and the operators were never seen standing under these conditions.

An experiment with seating was conducted by Farmer (1921) during his investigations into metal polishing. During one period a seat was provided, while during a second period the operator had to stand all day. Farmer noted that when the seat was provided, the operator started the day standing but soon sat down and remained seated for the rest of the day. Output was found to be 32·6 units/hour when standing and 36·2 units/hour when seated. Other evidence on this point is provided by Vernon (1924) who studied the number of changes of posture made voluntarily by girls who could either sit or stand at will. The findings of this investigation are summarized in Table 9. From this it will be seen that the amount of time spent standing was quite a small part of the working day. It must be a matter for individual decision whether this amount of standing (or anything like it) needs to be catered for.

Table 9.   Proportion of Time Spent Sitting or Standing

Operation	Number of Girls	Percentage	
		Standing	Sitting
1	23	18	82
2	10	11	89
3	4	—	100
4	23	4	96
5	8	4	96
6	8	26	74
Total	76	11	89

If high benches are used, foot-rests become essential, and while in theory these have the advantage of being adjustable, in practice adjustments are often not made. A foot-rest on a high bench is satisfactory only if it is a flat surface on which the feet can be moved. The more common type of bar used as a foot-rest is not satisfactory since it will force the user to remain static in one position for long periods of time. Clearly it is more difficult to fit an adequate adjustable platform foot-rest than a bar type foot-rest and it is unfortunate that the line of least resistance is so often taken. Under the optimum conditions, with a well-designed foot-rest, it should be possible to ensure that the relationships between seat top and work and between seat top and foot rest are correct for a wide variety of sizes of operator. On these grounds of adjustability, therefore, a case can be made for a high bench, but in practice this adjustment is often ignored and operators are left to get along as best they can. It is pertinent to note that no suggestion is made that the same principle should be applied in the office so that a person working at a desk could stand if he or she wishes, in fact the old type counting house desk at which this was possible seems to have disappeared many years ago. It seems unreasonable, therefore, to suggest that people on assembly work should need to stand any more than should typists, and, in any event, low work benches are quite customary in a number of industries, such as the shoe industry, where no difficulty seems to be experienced.

Following this line of argument, a Birmingham firm changed to a floor-level production line which has been reported as being highly successful, Mann (1961). A standard chair was adapted and low benches were built to suit the chair height (Fig. 48). For the women, footstools were provided as necessary. Throughout the whole of the operation, the people who were going to be affected were consulted and the seating finally adopted was

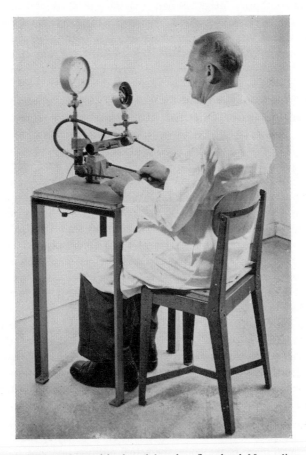

Fig. 48. Section of an unitized work bench at floor level. Note adjustment of test gauge at eye level (from Mann, 1961; by courtesy of S. Smith & Sons (England) Ltd, Industrial Division).

chosen by them. Mann states that the result of the change, which was made in 1951, was greater productivity, a saving of working space and a greater satisfaction on the part of the labour force. This is a practical example of breaking away from tradition.

There are many advantages in working at floor level. There will be a saving in the cost of benches, and mechanical complication due to having to build foot-rests will be avoided, although footstools on the floor may have to be provided for smaller workers. Operators will have greater freedom to move their legs and bodies while working and when products have to be moved by the operators the tote pans can be picked up and put down at a more appropriate height. When everything is taken into consideration there seems to be no reason why the alternative postures of sitting and standing should be required under modern conditions provided that good seating is installed, that the relationship between the floor, the seat, and the task is correct and that operators' work is planned so that they will leave their work places at regular intervals.

When benches are laid out for assembly work or when controls are positioned, the general principles of motion economy as laid down by Gilbreth, and re-stated by a number of authors, have proved on the whole to be well founded. There is additional knowledge now available which makes it possible to understand the principles more clearly and to extend them. We know now that reducing movement too much can cause local muscular fatigue which will not be present if the actions are spread over more muscles in the body. The accuracy with which hand movements can be made falls off rapidly as the hand moves away from the body or across the body. Accurate positioning or manipulative movements should, therefore, be made near the body and controls which need fine adjustment should be placed close to, and opposite, the hand which will operate them. The best area for the operation of controls has been shown to be in a wedge, base down, with a base of approximately 1 ft square and a height of 1 ft 9 in. The base will be located approximately at symphesis height and the top of the wedge will thus come at about shoulder height. The vertical centre line will be located opposite the shoulder of the hand which may be involved in making control movements. These dimensions have been determined essentially for a man, but there is no reason to suppose that they will be very much different for a woman. While this has been suggested as being the best position, we have already noted that, provided controls are in a position where they can be comfortably operated and close to the frontal plane of the body, the actual position does not appear to be of enormous importance. If a control is to be operated by both hands its axis should be as near as possible to the centre line of the body. This is particularly important if the control is a lever or a joystick. An asymmetric arrangement of two-handed controls can cause errors in control movements which are related to the degree of displacement from the centre line of the body. Reciprocating controls, whether they are levers or pedals, should be placed opposite the limb which is to operate them and the direction of movement should be in line with the shoulder or hip as the case may be. Levers which have to be operated by movement to the side of the body or pedals which have to be operated by moving the leg across the body produce unnecessary strain.

When the operator is standing, foot pedals, whether for bench work or on machines, should be avoided if possible. If it is essential that a pedal should be used by a standing opera-tor, and especially if the use is frequent, the pedal should be of a rocking type, pivoted in line with the tibia so that the operator's body weight can remain securely on both feet and the pedal can be operated by extending or flexing the ankle. The usual alternative to this kind of pedal requires the operator to stand on one leg for a good deal of the working day while he operates the pedal with the other. This is a thoroughly unsatisfactory arrangement since in most designs the movement of the pedal does not follow the natural arc of the foot and the operator must complete the movement in an unbalanced position. This effect is shown in Fig. 49. Whether they are operated sitting or standing,

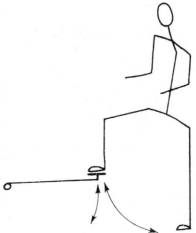

Fig. 49. Pedals which travel in an arc different from that of the operating foot can cause an unbalanced and tiring movement.

the return spring pressure on a pedal should be sufficient to support the weight of the foot or of the leg, as the case may be, otherwise very tiring static work will have to be done to maintain either a foot or a leg in the 'off' position. In extreme cases actual muscular damage can be caused if pedals with insufficient return spring pressure are operated for a long period of time. Where pedals have to be operated by a seated operator, they should be placed so that both feet are at the same level most of the working time. If possible they should be placed so that they can be operated by either foot at will. For precision, a treadle is superior to either a rocking or reciprocating pedal, but should be so placed that the feet can change position when the operator feels the need.

Information should be given to the operator in the simplest manner possible. If of 'yes'-'no' type, alternative lights may be used, or simple pointer instruments with appropriate areas clearly marked. Where adjustments have to be made to bring a reading to a standard, it is better to have the standard position clearly marked with a plus or minus on either side, rather than to have a graduated instrument on which the operator has to

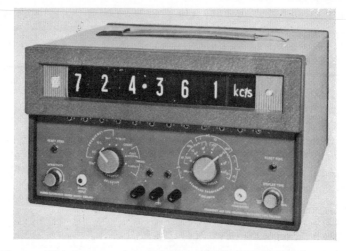

Fig. 50. Digital presentation of frequency and time measurement (by courtesy of Venner Electronics Ltd).

adjust to a given reading. Where quantitative information has to be given, digital presentation should be used wherever possible in preference to the reading of graduated scales. A recent application of this principle to the measurement of frequency and time has been made by Venner Electronics Ltd, in their equipment TSA 3336 illustrated in Fig. 50. The clarity of this presentation, as compared with the use of a graduated scale, is very obvious. A very unusual use of digits is the feature of the Decimicrobalance (Fig. 51) manufactured by L. Oertling Ltd on which the adjustment is made by means of a control at the side of the case. Since this can be turned quite rapidly the two final digits are obtained from a drum graduated and numbered in 5s. The same problem of rapid counting is found in the Indicator of Machine Tool Travel (described in Chapter 20) only in this device the drum is numbered in 2s, the first counting wheel being for the 100s as it is on Oertling's balance.

Control Panels

Control panels can be found in all kinds of situations, from petrol refineries to aircraft. In almost all instances the human function is basically that of a monitor, but the importance of the monitoring function will depend on the extent to which automatic controllers have been installed to look after the moment to moment fluctuations in the physical characteristics being measured. In the petroleum or chemical industries, for instance, this

Fig. 51. Partial digital presentation on a Decimicrobalance (by courtesy of L. Oertling Ltd).

automation will have been carried nearly to the extreme and the operator will only have to intervene when something goes radically wrong. The operator will have a greater degree of control in a power station where the load may fluctuate, and the pilot of an aircraft may depend almost entirely on monitoring his instruments in order to ensure that the aircraft is functioning correctly. There are so many possible variations between these two extremes that it is not possible to deal with them all, therefore a number of general principles will be laid down for the design of control panels which may, or may not, be applicable to a particular situation in the light of the amount of automatic control which has been installed.

The operator's function as a monitor will be discussed in greater detail in Chapter 17. It is sufficient to say now that it is not possible to rely on an operator always being alert to changes in a function over a long period of watch, if this information is presented only visually. The physical conditions under which the operator works may well affect his efficiency as a monitor. Both temperature and noise are environmental conditions which could affect the efficiency of an operator (Fraser, 1957a). There is a difference of opinion as to whether it is possible to make an operator too comfortable. One view is that a well-designed and comfortable seat may cause him to go to sleep, whereas another view is that his efficiency may be gravely impaired if he is physically rather uncomfortable, at least sufficiently uncomfortable to ensure that he stays awake. Unfortunately there is no real evidence to support either view but if some auxiliary tasks can be built into the operator's job which will compel him to leave his seat at regular intervals it is unlikely that he will go to sleep even if provided with a comfortable seat. In any event the design of many panels is such that it would not matter if he did!

The type of instrument which is fitted in a control panel will be dictated by the process. Many of them will be automatic controllers keeping records on paper and the trend towards miniaturization may mean that in some instances these traces may be difficult to see. It is important that when a control panel is being laid out the information which is vital to the operator should be determined. We have already seen that the speed with which an operator can take in information will depend on the number of choices which have to be made and if there is a large number of redundant instruments on the panel, the operator may find it more difficult to control than if the redundant instruments are banished to a part of the panel which is not strictly operational. Senders (1952) has shown that there is a linear relationship between the time required to identify a deviating instrument and the number of instruments to be checked. Thus 16 instruments took about 2·8 sec to check, while 32 took 5·6 sec. This means that there should be a departure from the present tendency to group all instruments related to a particular function on one part of a panel whether they are all required for normal operation or not. The exact relationship between one instrument and another will also require a good deal of thought. There are two ways in which instruments may be related, one will be for normal operation and the second for the most probable emergency conditions. In many instances when the equipment is looked after by automatic controllers it is the latter function which may well receive priority. In arranging these instruments on a panel it should be remembered that there is a strong tendency to scan from left to right, that there appears to be a tendency to notice more quickly the instruments which are in the top left or bottom right of a small panel and that as time goes on there is a tendency to ignore instruments which are away from the centre of vision.

Important instruments should therefore be concentrated as far as possible in a central area in relation to the operator. This should subtend an angular width of not more than 65° at the eye and the panel should not extend more than 10° above or 45° below eye level. The upper dimension is rather critical since there is evidence that displays much above eye level are not only very fatiguing to watch, but indications on them are frequently missed. Other less important functions can be placed on panels at the side. All function's of importance should be guarded by automatic warning systems which will tell the operator when a function is going out of limits. These should be a combination of both an auditory and a visual warning, and care should be taken that the visual warning should be quite evident and not misleading. If visual warning alone is used there is evidence that lights which are 30° from the centre of vision will be noticed more slowly than lights which are central, particularly if they are rather high up (Elliot and Howard, 1956).

If the efficient operation of a panel depends on accurate check reading, the limits within which the pointer must be maintained should be clearly indicated on the instrument; the memory of the operator should not be depended upon. The time required to notice a deviation on an unmarked instrument will be about 35% longer than when a 'safe area' is clearly marked (Kurke, 1956). Several research workers have suggested that by the choice of suitable scale maxima instruments can be orientated so that when reading normally the pointers will lie in a cardinal position on circular instruments. In the same way pointers of vertical straight instruments could form a line across a group when reading normally; a substantial saving in horizontal width can often be achieved by the use of this type of straight instrument which can be used with confidence instead of a circular instrument (Murrell, 1960a). This recommendation is based on the finding that a deviation on one of a group of dials patterned in this way is noticed more quickly than when the group is unpatterned (White et al., 1953; Johnsgard, 1953). The magnitude of the difference is related to the number of dials involved; a deviation on one of 16 unpatterned dials took 2·8 sec to notice compared with 0·25 sec when patterned and on 32 dials, 5·6 sec compared with 0·5 sec (Senders, 1952). Although in theory this kind of arrangement is desirable, it seems that in practice it would be difficult to achieve except under special circumstances, such as when a number of similar instruments indicating the same function in different plants are fitted in one panel. Where this is so, it is desirable that the instruments should be fitted side by side and close together (Fitts and Simon, 1952). An example of this would be the four instruments showing boost in a four-engined aircraft.

There seems to be a marked difference between the ability to discriminate a number of objects above and below six; at six or below, the number present is 'grasped by intuition' or *subitized*, above six the number has to be estimated (Kaufman *et al.*, 1949). This suggests that where a number of instruments are closely related in a panel, it could be best to

Fig. 52. A bezel which is too heavy can 'kill' the pointer and scale (after Murrell, 1957).

divide them into groups of either five or seven instruments. To aid identification of controls associated with these instruments, particularly if the controls are in a remote console, they should be patterned in the same way as are the instruments. In addition to coding by position, controls may also be coded by shape and/or colour (Jenkins, 1947; Hunt, 1953).

Speed of noticing deviations may be assisted by the choice of a good pointer design which will make it stand out prominently against the surroundings, Fig. 52. The prominence of the pointer can be substantially reduced by instrument cases or by black or chromium bezels and these should be avoided. Self-coloured sand-blasted brass or anodized aluminium bezels or cases can be used and it is possible for them to be colour-coded to correspond with, for instance, the colour coding of pipes. This will be an added aid to the operator in appreciating the nature of a function which is displayed on an instrument.

In all situations where the operator is not required to be dark adapted, white dial faces with black markings should be used and the background colour of the panel should have a reflection factor about half way between that of the dial face and the general surroundings. The actual colour of the panel is probably immaterial, some people may prefer a warm colour, whereas others may feel a cool green is more restful. This is purely a matter of personal opinion and there is no evidence that any one colour is likely to be superior to any other. What, however, is quite certain is that panel backgrounds should never be black and that they should never be shiny. Lighting should be restful, should not throw sharp shadows and should be so placed that the reflections from the cover glasses of the instruments is avoided. A luminance of between 30 and 50 ftL is recommended.

When controls are fitted in the panel itself, it will usually not be difficult to identify which control relates to which display, but whether controls are fitted in panels or in a remote console, it is important that the movement of the control and of the instrument should be compatible. This is particularly so when electrical and pressure instruments are mixed in the same panel and a standard relationship, either compatible or incompatible, should be adopted. If coding of controls by position is not thought to be adequate, coding by shape can also be used.

When an arrangement of either displays or controls or both has been decided upon, this layout should be maintained on all equipment to which it is applicable, especially if an operator has to move from one equipment to another. To do otherwise is to invite, at the best, inefficiency, and at the worst, disaster. It is known that aircraft mishaps were caused during the war by pilots pulling the wrong control or reading the wrong instrument in a cockpit with perhaps an unfamiliar layout (Fitts and Jones, 1947a and b). In America a gas main had to be 'blown down' so that repairs could be carried out. The main trap gates were not operated correctly and, as a result, gas was cut off from a small town (a highly dangerous situation). It was found that the layout of the gates was not standard and that the wrong valve had been opened by the operator (Pickens, 1961). Many other examples could be quoted to emphasize the importance of uniformity of arrangement between related equipment.

Unless special circumstances make them particularly desirable, graphic panels are not likely to be very satisfactory, since the instruments will usually have to be spread over a much larger area than would otherwise be the case, and important and unimportant instruments will be mixed up together, causing a good deal of redundancy, to which will be added the information given on the graphic panel. No evidence has been traced to show any great advantages in graphic panels under normal conditions. However, under special circumstances – for instance in signal cabins displaying train routes or in power stations where alternative routes for power under conditions of breakdown must be displayed – graphic presentation will be necessary in order to achieve clarity. Under these circumstances the graphic information is essential to the operator's task and it will usually be given in a separate display panel.

MAINTENANCE AND REPAIR

Maintenance is usually taken to mean action to ensure that breakdown will not occur, whereas repair involves putting right faults which have developed. In the former activity operators will, or should, follow prescribed routines which should be planned when the equipment is being designed. Failure to do this, so that points which will require regular maintenance are placed in inaccessible places or require extensive dismantling, will lead not only to excessive cost, but also to neglect. It is not possible to make specific recommendations on this point since every piece of equipment will be a problem in itself, but there is no reason to suppose that the optimum position for application of force on a spanner is any different from that on a lever, so that the many points on design which are given throughout this book should be intelligently applied to each specific situation remembering always that a man has only two hands and is not usually double-jointed. Ideally, if a prototype is built it should be given user trials for maintenance efficiency as well as for operational efficiency.

The process of repair will usually fall into two parts, diagnosis and rectification. Diagnosis, or 'trouble-shooting' as it is often called, may be simple; often it will be quite obvious what is wrong, but at other times the process may be lengthy and complicated, particularly with electronic equipment. The kind of individual who can do this and the way he should be trained is an important field of research for psychologists, and some useful theoretical and practical work is now being done (e.g. Stolurow *et al.*, 1955; Dale, 1958). Most of this work is related to military equipment but is of equal or greater importance to industry as plant becomes more complex and expensive.

Diagnosis can frequently be assisted by careful design. The location of faults can be indicated to the operator or units can be laid out so that the cause of failure can be isolated. This method has been adopted by an American motor manufacturer (Murrell, 1958, p. 67) on engine machining assemblies. Electrical faults could be identified by the use of probes on special panels and the location of mechanical faults, such as a broken tool, would be indicated on another panel. In this way the task of the operator has been greatly simplified. The converse may be true if insufficient information is given to the operator; with air-operated valves the position of the valve may be indicated only by the air pressure in the operating cylinders, it will not necessarily tell the operator when a linkage has broken so that the valve is not in the indicated position.

As with maintenance, rectification can be expedited by forethought in early stages of development. It may be feasible to predict the possible location of components which may require rectification and to ensure that they are as accessible as is practicable. The added cost of ensuring access to potential trouble spots must be weighed against the cost of having plant shut down while a rectification which, for instance, should take 2 hours, takes 20 hours, because the location of the trouble is completely inaccessible. The application of the relevant ergonomic data should help to ensure that time lost through breakdown is minimal.

AN ERGONOMIC CHECK LIST

The various factors which have been discussed can be summarized in the form of a check list, which could be used by designers whatever the nature of the equipment which is being designed.

(1) What rôle is the operator expected to play? Will the optimum use be made of his inherent capacities or will he be called upon to undertake functions which he cannot do very well? Can these be transferred to the equipment?

(2) How will the equipment fit the operator? Is the existing equipment the result of tradition or has it been planned from the start with the operator in mind?

(3) Can the operator sit or must he stand? In either case will his posture be satisfactory?

(4) Is the equipment likely to be operated partly or exclusively by women? If so, what population of women must be provided for?

(5) What information does the operator need in order to do the task? In what form must he receive this information and how can it best be displayed? Should it be a visual, auditory or tactual display?

If the information is to be given by a display, which type will give information most quickly and with the minimum of ambiguity? Which instruments are essential for the efficient operation of the equipment and which can be relegated to a minor position?

(6) What controls will be needed and what type should they be? If the operator is standing, can foot control be dispensed with? What force will the operator be called upon to exert and will some form of servo assistance be required?

(7) What form of communication will have to take place between operators? Must this communication be verbal and if so, will there be interference from noise? If interference from noise is expected, can information between operators be transmitted by means of instruments?

(8) What physical work will the operator be required to do? Will it be within his or her physical capacity or will some form of mechanical assistance be required?

(9) What are the ambient conditions likely to be? Will there be noise? Will there be heat? What illumination is likely to be required?

(10) Will the physical or mental demands of the task be such as to overload the operator, if so, what steps can be taken to reduce the load?

(11) What are the expected maintenance requirements? Has the equipment been designed to make the diagnosis of faults easy? Has the equipment been planned so that probable repairs can be carried out with the minimum of delay? Are all locations for regular maintenance accessible?

The use of such a check list will ensure that most of the points which have been discussed in relation to the design of equipment will not be overlooked. This will enable the best use to be made of the ergonomic data which will be given in detail in the chapters which immediately follow. A similar check list for organizational factors will be found at the end of Chapter 17.

CHAPTER EIGHT

Design Factors II.
Design of Seating

Modern man spends so much of his time either sitting or lying down that it might be said that *homo sapiens* has ceased to be an upright animal! This being so, it is not surprising that a great deal of research has been done into the design of good seats. What is surprising is that, in spite of this, most people at work are still sitting on seats which are badly designed and generally too high. For some special purposes such as typing, adjustable seats are often provided, but since there is no clear understanding of the best height for a particular individual, and since sitting height is so often determined by the height of the desk, these seats are often not used in their best position. Alternatively, most chairs of fixed height for clerical work in offices, in canteens, dining rooms and so on are a conventional 18 in. which is too high for comfort for a majority of the population.

It is the requirement of good seating that the person sitting in a seat should be able to maintain a good posture which will not cause overstrain of any particular group of muscles. It has already been explained that continual strain on one particular group of muscles can cause fatigue and therefore it would seem that a good seat should enable the user to change posture at intervals so that different muscle groups may be called into play. At the same time the use of a well designed and positioned back-rest may relieve the back muscles of a good deal of postural work. Any seat which maintains the body in a rigid position is likely to be unsatisfactory, especially if it causes bad postural habits which may result in the development of postural abnormalities. Some writers have suggested that a satisfactory posture is achieved only when there is a lumbar concavity but this may well be too rigid an interpretation.

A second important factor is that the seat should not press unduly on the tissue of the thigh which is not designed to withstand pressure as is the tissue of the buttocks. Thus if the thighs are pressed into the front edge of a hard seat of the type so often found in many factories, they may be compressed by a quarter or even a third of their full thickness before the weight is taken by the femur (Fig. 53). This pressure will restrict the blood flow and may cause pressure on the nerve trunk which runs on the underside of the thigh and will cause discomfort and may cause the limb to 'go to sleep'. The parts of the body which are intended to be sat upon are two bony protuberances of the pelvis known as the *ischial tuberosities* and the tissue in the vicinity is adapted to withstand pressure without restriction of the blood supply, in fact after the feet and the hands it is probably the part of the body best adapted to the bearing of weight. A well designed seat should therefore bear the weight of the body in a good posture on the buttocks and not on the thighs, and when achieving this objective there are anatomical, physiological and anthropometric principles which have to be taken into consideration and these may impose a few restrictions on the

Table 10.   Anthropometric data relevant to seating

	Males Mean	Percentile				Females Mean	Percentile			
		1st	5th	95th	99th		1st	5th	95th	99th
Stature (without shoes)	68·0	60·0	64·0	72·0	76·0	63·0	56·5	58·5	67·0	69·5
Sitting height	36·0	31·5	33·5	38·0	40·0	33·5	28·5	29·5	34·5	35·5
Elbow height above seat	9·5	7·0	7·5	10·5	11·0	9·0	8·0	8·5	10·0	10·0
Buttock – back of knee	19·0	16·5	17·0	20·5	21·5	18·0	15·5	16·0	19·5	20·5
Lower leg length (popliteal height)	16·5	14·5	15·5	17·5	18·5	15·5	14·0	14·5	16·5	17·0
Elbow width	17·5	14·5	15·0	19·5	21·5	16·0	13·5	14·0	18·5	19·5
Buttock width (bi-trochanteric width)	14·0	12·0	13·0	15·5	16·5	14·5	12·0	12·5	16·0	17·0
Shoulder height above seat	23·0	20·5	21·5	25·0	26·0	21·0	19·0	19·5	22·5	23·5

Note. These data are for nude subjects or subjects in foundation garments. Allowances should be made for the bulk of ordinary clothing. Dimensions of length will be relatively little influenced but dimensions of bulk may change appreciably if thick clothing is worn and a dimension such as buttock width or buttock – back of knee would have to be increased by about 2 in.

design of seats for specific purposes. The elements of good seating will depend on the length, width and shape of the seat; to a limited extent on the material of which the seat is made; on the shape and height of the back-rest and the height of the seat above the floor.

The relevant anthropometric data have been given in Chapter 4, but for convenience they are repeated here in Table 10.

Seat Height

If pressure is to be avoided on the under side of the thigh, the seat height must clearly be related to the length of the lower leg from the under side

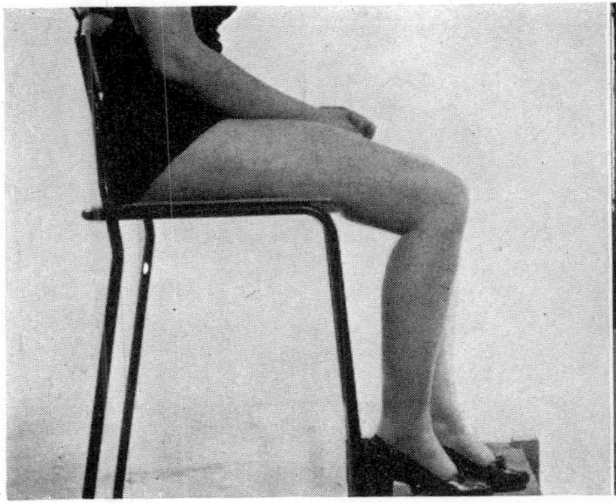

Fig. 53. (a) A girl sitting on a 19-in. chair.

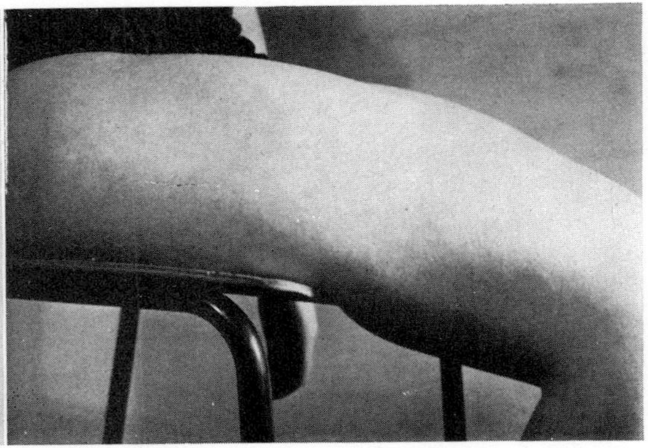

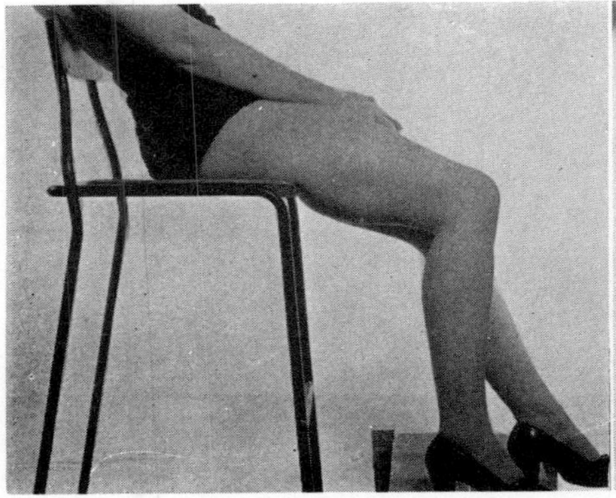

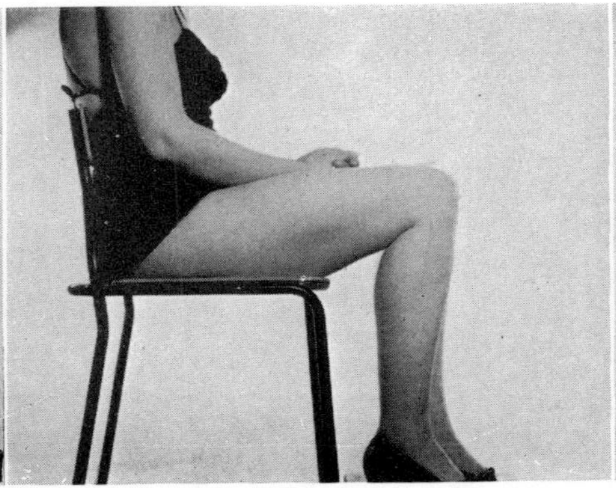

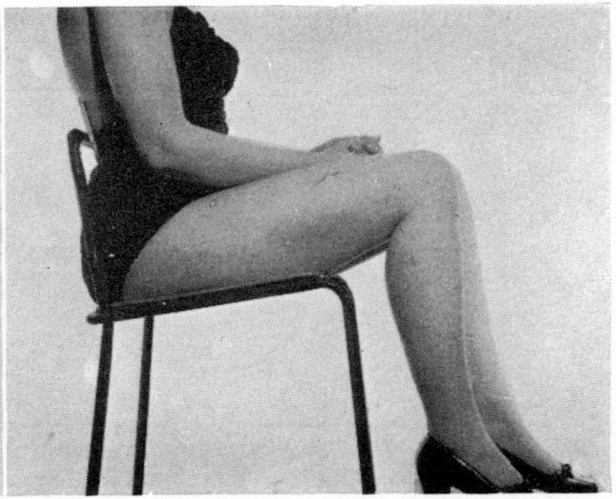

(b) Compression of thigh on too high a chair.

Fig. 54. An unsatisfactory posture often adopted on too high a chair.

Fig. 55. Sitting on a 15-in. chair.

Fig. 56. A 15-in. chair with a sloping seat.

(b)

(a)

56

55

54

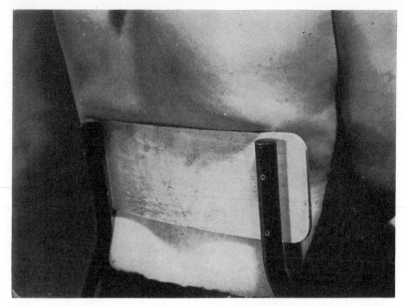

Fig. 57. (*a*) A fixed back-rest cutting into the flesh.

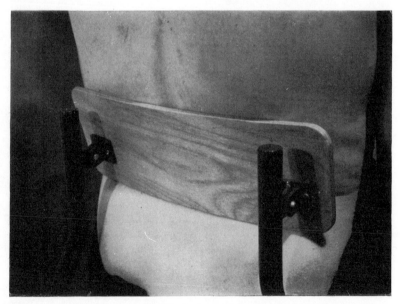

Fig. 57. (*b*) An adjustable back-rest fitting the back snugly.

of the knee to the heel (the popliteal height) and to the curvature of the thigh. Due to variations in the plumpness of the thighs of different individuals there may be a difference of as much as 3 in. between the lower leg length and the height of the unsupported thigh above the floor. In practice, when the sitting individual is clothed, the lower leg length will be influenced by the height of the heels worn and it would seem that the increased plumpness of the thighs in the female may be to some extent offset by the higher heels which are normally worn. It must be remembered, however, that in most factories the wearing of very high heels is discouraged by management because of the accidents due to tripping.

Most of the recommendations which are made on seat height are based on the assumption that the sitter will have the feet flat on the floor and the lower leg vertical (e.g. B.S. 3079:1959). This may not necessarily be the posture most often adopted by people who are sitting for any length of time. It was suggested earlier that the most relaxed position for the leg is when the knee is bent at about 45°, and in order to do this or to tuck the legs under the seat, the seat height might be somewhat lower than the height which would be adopted if only the dimension of the lower leg is considered. Moving the foot forward its own length is equivalent to increasing the seat height by 2 in. If one seat has to accommodate all ranges of people from the smallest woman to the tallest man, or even the smallest woman to the tallest woman or the smallest man to the tallest man, opinions differ as to whether the smaller people should sit on a seat which is too high or the taller people should sit on a seat which is too low. It has been suggested that sitting on a seat which is too low, thereby causing the angle between the trunk and the thigh to become acute, will cause low back and stomach pain. On the other hand it might be argued that it is much easier for a tall person to stretch the legs forward, thereby bringing the thighs into a horizontal position, than it is for a short person to sit with the feet barely touching the ground. Another factor which must be taken into consideration is the possibility that when working on a desk or bench the trunk may be inclined forward from time to time, thereby reducing the angle between the trunk and the thigh. This could of course be taken as an argument for making seats rather higher and less deep, so that only the buttocks are on the seat with the thigh sloping slightly forward. Although this would seem to be a theoretically sensible solution, it seems in general that the population of sitters do not approve of shallow seats. The exact importance of maintaining the angle between the trunk and thigh has been emphasized by several writers who suggest that when this angle is not maintained it may lead to a convexity of the spine which over a long period may lead to damage to the ligaments of the vertebral column.

On the whole, therefore, it would seem that seat height should be at or near the mean of the population for which it was intended if a range of adjustment is not to be made available.

If a chair is too high and pressure is felt on the underside of the thigh so that the legs become numb, relief will probably be sought in one of two ways. In both instances the buttocks will be slid forward on the seat and the individual may either sit upright without using a back-rest or may slump back against the back-rest using it in an unsatisfactory position with a convex curve to the back (Fig. 54). A good indication that seats are too high is if girls, say in a typing pool, are seen sitting on the front edges of their seats and rarely using the back-rests. This bad posture may be found even when there are adjustable seats due either to the desk being too high or to the fact that in everyday life most people accustomed to sitting in canteens and homes and elsewhere on 18 in. chairs, feel that a chair of the 'right height' is too low. For this reason it is not usually a good idea to leave people who have adjustable seats to adjust them themselves without some instruction. A good test for height is whether it is possible to slip the fingers readily between the front edge of the seat and the thighs with the legs vertical. If this cannot be done the seat is too high.

A decision on the height of a seat will depend on whether it is to be adjustable or non-adjustable, whether it will be used by men or women or both, and whether it will be used in the office or in the factory. It must also take into account the height of the heel likely to be worn and the character of the thigh. As mentioned earlier, owing to its plumpness, the underside of the thigh half way along its length will be nearer to the floor than the bend of the knee, and it seems not unreasonable to assume that the plumpness of the thigh and the heel height of low heel shoes would be approximately equal, so that the height of the lower leg may be taken as the basis for the height of the seat. However, if high heels of about 2 in. are worn by women this will have the effect of reducing the difference between the leg length of men and women by 1 in. Thus in offices it may be possible to use the same chair height for both sexes, but in factories, where high heels are usually frowned upon, it will be better to have a different height. Ideally, of course, seats should be adjustable to the individual and this would also imply that the height of the working surface should be adjustable. If an adjustable seat is intended for women, it should have a range between about 14 and $17\frac{1}{2}$ in., but if it is intended for men the range should cover 15 to $18\frac{1}{2}$ in. to cater for 90% of the population. A fixed seat height for factory women should be between 15 and $15\frac{1}{2}$ in., and this will accommodate rather over half the female population (Fig. 55). A fixed seat for men should be about $16\frac{1}{2}$ in. A seat of fixed height to accommodate both men and women should have a height somewhere in the region of 16 to 16½ in., which will accommodate about 75% of the male population and about 40% of the female population or 60% if high heels are worn. The use of footstools with this seat height for women will be of assistance to the smaller members of the community. In practice it should not be difficult in factories and offices to predict the sex of a user of a particular chair and so to use the correct height. For instance, in a typing pool it is highly unlikely that any men will be employed and on many production lines the same will be equally true. There may be some problems in providing chairs of different heights but this should not prove to be an insuperable manufacturing or economic problem once the importance of having good seating is appreciated.

When high chairs are used at work benches, there will have to be some form of foot-rest and this should be a flat surface rather than a bar, which will cause fatigue by forcing the operator to keep the feet in a fixed position. Wherever possible, adjustable foot-rests should be used but when this cannot be done, a foot-rest should be set at a distance below the seat in accordance with the seat heights suggested above. When padding is used on the seat, the seat height should be calculated allowing for a degree of compression by the thighs of the sitter.

The extent to which the heights which are recommended above will be adopted will depend on the degree to which the public is prepared to accept and the manufacturers to make chairs which, to the unaccustomed eye, may appear to be rather too low, and it goes without saying that the height of desks, benches and so forth must follow the height of the seat. In fact the relationships between the seat and the bench height are in many instances more important than the relationship between the seat and the floor.

The Size and Shape of the Seat

The depth of a seat should be sufficient to allow the buttocks to move to permit changes of posture but should not be so great that the seat cuts into the back of the knee. Hooton (1945) found that for vehicles the 90% range of seat length for females was 16·8 to 20 in. and for males 17·4 to 20·2 in., which agrees fairly well with the above data but does not allow any clearance between the back of the knee and the front edge of the seat. For working seats, a clearance of about 2 in. is desirable, which means that the maximum seat length to accommodate 90% of females should be in the region of 15 in. or 16 in. at the most, and except under certain circumstances which will be mentioned later, there seems to be no reason why the depth of the seat should be made any less than this. If no padding is fitted, the front edge of the seat should be curved to avoid having a sharp edge to cut into the underside of the thigh.

The seat width should also be sufficient to allow a certain amount of movement of the buttocks and so should be over rather than under the bi-trochanteric width. Hooton in his survey found that the 95th percentile for hip breadth, both male and female, was in the region of $17\frac{1}{2}$ in.: this is higher than the above figures, but his subjects were fully clothed. From this it would seem that a seat breadth of about 17 in. would be acceptable to a majority of the population. If arms are to be fitted, they must be wider apart than the seat and a minimum of an additional inch each side of the seat would seem to be reasonable, which would give a width between the arms of 19 in. The arms should be $8\frac{1}{2}$ to 9 in. above the top of the seat, and should project 10 to 12 in. forward from the back of the seat.

Many industrial seats are made of smooth wood. If such a seat is horizontal and the sitter leans firmly back against the back rest, an opposing force in the direction from the back to the front edge of the seat will be set up and the buttocks will slide forward on the seat. This effect is not by any means confined to the factory, since board room chairs with a firm padding made of good quality leather will show this effect also. The solution of this difficulty is to slope the seat backwards at an angle of about 3° to 5° which is sufficient to prevent the sliding force from operating (Fig. 56). In passing it may be mentioned that a suitable chair for an unwelcome visitor would be one with the front legs shorter than the back legs and a highly polished seat!

Some padding on seats is to be desired, but for seats which are intended for use at work it should not be too soft, otherwise the buttocks will sink in too far and the thighs will be elevated. Padding of horsehair or similar material which produces a convex surface to the seat top is highly undesirable since it results in a contour which is exactly contrary to that of the buttocks. Shaped seats are also undesirable since they involve fixing the pelvis in one position and do not allow for any movement.

The Back-rest

The back-rest in most industrial seats will consist solely of a support in the lumbar region. It should be curved to a radius of about 16 in. and should be sufficiently high to allow the buttocks to protrude beyond the back-rest when a person is sitting erect. It should, therefore, be not less than about 8 in. above the back of the seat. Its depth should be between 4 and 8 in. and it should not be more than about 13 in. wide to avoid hitting the back-rest with the elbows when the hands are drawn towards the hips. Owing to the differences in the shape of the lumbar region in different individuals, it is preferable that the back-rest should be hinged on its support so that it can accommodate itself to the different angles on different individuals (Figs. 57a and b). It should be padded if this is practicable. There is probably a good deal to recommend the use of a more continuous back-rest, especially if padding can be used to allow for the wide variability of the shape of the back. The angle of such a back-rest should be between 95° and 110°. Authorities differ on the best shape for this back-rest but one solution which has had a wide acceptance is that which has been proposed by Åkerblom (1948). This back-rest has a backward curve above the lumbar support and will also give support to the upper part of the back. The contour of this back-rest is shown in Fig. 58. If a continuous

Fig. 58. A seat back contour proposed by Åkerblom, 1953.

padded back-rest is used on office furniture its top may be between 16 and 18 in. above the seat top.

The Seat Frame

Since it is desirable that people should be able to tuck their feet under the seat as well as stretch them forward in order to obtain changes of posture, it is preferable not to have a bar connecting the two front legs if it can be avoided. This will also assist in maintaining balance when rising from the seat, otherwise the frame may be any shape which will produce the required strength. A common type of seat frame which can cause acute discomfort is the tubular stacking chair with a canvas seat. The stretch of the canvas frequently makes the sitter rest on the metal chair frame which is usually about 14 in. wide and is too narrow for nearly half the male population. To ensure that this shall not happen to a reasonable proportion of the population, the bars should be at least $15\frac{1}{2}$ to $16\frac{1}{2}$ in. apart.

An interesting development in seating intended to minimize muscular fatigue has been proposed by a British firm who have developed rubber-jointed seating. The frame, seat and legs are separated at their junctions by rubber bushes which enable the seat to move forward and laterally or to twist with the movement of the body while at work. It is thought that the movement of the seat will absorb some of the movements of the body and will thereby reduce slightly the muscular activity involved. No laboratory validation of this principle seems to have been made, but the personal experience of the author and his staff seems to suggest that this is a comfortable seat to use. One of these seats is illustrated in Fig. 59.

The foregoing is a very brief explanation of the principles underlying good seating. They apply to seating of all kinds although the dimensions given are intended primarily for industrial and office usage. Readers who wish to study the matter more closely are recommended to read papers by Floyd and Roberts (1958), by Åkerblom (1948) and Åkerblom (1954).

Fig. 59. A seat with rubber jointed legs (by courtesy of Admel International Ltd).

Seating Where Leg Room is Restricted

In circumstances where leg room is unavoidably restricted, the seat can be made higher provided that it is also made shallower, reducing the seat depth approximately to that of the buttocks only. Suggested dimensions to be used under these circumstances are a seat depth of 9 to 10 in. and a height of 22 in. for a leg room of 28 to 30 in. If the leg room is progressively reduced to the minimum of about 24 in. and the seat raised to the point where it is little lower than the gluteal furrow height (about 29 in. for

women and 32 in. for men), a situation approximating to the use of a
shooting stick is produced. In factories 'standing supports' or 'rump-
rests' on which the operative can support the buttocks may have many
applications in places where there is insufficient room for an ordinary
chair; such a support would normally have only one leg, a shaped seat to
fit the buttocks and a non-slip base so that the leg can be tilted at an angle.
A design has been produced by Lammers (1960), and a British commercial
design is shown in Fig. 60.

Fig. 60. A standing support ('rump-rest') of adjustable height and rake
(by courtesy of Admel International Ltd).

Seating for Specialized Purposes

Apart from offices or factories, seating will be found in many other places – in the executive suite, in the motor vehicle, in places of entertainment, for relaxation in the home and so on. In general the dimensions which have been set out above will provide appropriate sizing limits. Seats which are too deep from back to front will be just as uncomfortable whether they are found in the factory or in a motor car, and they will be just as uncomfortable if they are too high.

The principal difference between factory seating and the various varieties of seating with which we are now concerned is the padding or upholstery which will usually be provided. For this to produce a com-

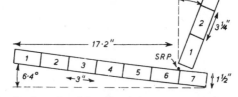

Section	Average deflection ins		Total load lbs	
	Cushion	Back	Cushion	Back
1	–·21	·33	4·9	6·5
2	·10	·20	11·2	6·5
3	·50	·10	18·2	6·0
4	1·05	·09	22·8	5·4
5	1·23	–·22	24·7	3·9
6	·97	–·80	20·2	1·8
7	·37	–1·38	13·9	0
Total			115·7	30·1

7 springs per section spaced at 3" centres

Fig. 61. Distribution of weight of a sitting person on the cushion and back of a seat (after Lay and Fisher, 1940).

fortable posture without undue pressure the support provided by different parts of the seat should be varied to suit the expected loading. This aspect of seating has been studied by Lay and Fisher (1940) who used an intricate apparatus called the Universal Test Seat in which almost everything was adjustable including the supporting pressure in various parts of the seat to suit the feeling of comfort on the part of the subjects. Their results, in so far as they relate to the cushions, are shown in Fig. 61. Fig. 61 also gives an indication of the spring characteristics which should be developed if the pressure is to be evenly distributed in relation to the parts of the body which are in contact with the seat. This factor is probably most important in seating used in transport.

The length of the seat and the angles of the seat and cushion, which are also given, are the means of the values selected by the subjects. The 90% range of these values is not very great, being 101° to 108° for the angle between the seat and the back, 4° to 8° for the seat angle, and 15½ to 18¾ in. for seat depth. (The mean seat height arrived at in this study was 17·1 in. but the histogram is substantially level between 16½ and 18 in. with a 90% range of 15¼ to 18¾ in.)

Seats which will be used for relaxation can be somewhat lower than office and factory chairs (14 to 16 in. has been recommended), width between arms 22 in. and the height of the back 20 in. above the seat if the shoulders only are to be supported or up to 30 in. if the head is to be supported also. Seat depth can be up to 20 in. but it must be remembered that if a seat is made too deep most women and many men will be unable to sit upright, but this is not to be expected in an 'easy chair'. There are so many variations in this kind of seating that only these very general indications of some of the limits can be given. For more detailed recommendations see Schroeder (1948).

When seating is used in conjunction with other equipment, as in crane cabs, transport vehicles or rolling mill pulpits, its size and height may have to be modified to permit efficient operation of associated controls. Each situation will have to be considered on its own merits in the light of the general principles laid down and taking account of the anthropometric characteristics of the population for which the equipment is intended.

CHAPTER NINE

Design Factors III.
Design of Instrumental Displays

Strictly speaking, a display is any device or event which gives information about a situation which is occurring or has occurred. In this sense the movement of the bed of a grinding machine is a display, since it gives information to the machinist about the progress of his job, while the sound of the grinding wheel and the nature of the stream of sparks coming from it may act as auditory and visual displays, giving him information on whether the amount being ground is correct.

Although in some contexts it may be necessary to use the term 'display' in this wider sense, in this chapter we shall be dealing only with devices which give information about an event or situation, where this information is not (or cannot be) obtained by seeing or hearing the event or situation directly. For instance, gauges may be needed to register the steam pressure in a boiler cr the amount of petrol in the tank of a car, since we cannot usually tell directly through our senses what this pressure or quantity may be. But even this definition is somewhat wide, since it could be taken to include the flag signals used by the Navy, or even the hand signals used by tick-tack men on the racecourse. Therefore, for the present the subject matter will be restricted to displays which give information about conditions in a piece of equipment or a system.

Instrumental displays are mainly of two types, visual displays and auditory displays. Visual displays are the most commonly used and are also the most versatile; they include, *inter alia*, instruments in which the information is given by means of an index, by means of lights, by means of shutters and those which give data directly in numerals.

The kind of display to be used and the nature of the information which is to be given by it must, to a large extent, depend on the importance of the data which the display will give. In some situations it may not matter very much what kind of display is used, but when accuracy of information is vital, the choice and design of the display can be of great importance. For instance, the misreading of a multi-pointer altimeter in an aircraft, or failure to notice increase of temperature in an automatic process, can lead to disaster. Fitts and Jones (1947b) give details of an analysis of 227 accidents in aircraft which were due to misreading a display. These incidents were classed as 'pilot's error', but it is equally likely that they were 'designer's error' in giving the pilots displays on which it was probable that they would make mistakes under stress. Unless a careful check of reading accuracy is made, it is often not realized just what the performance is likely to be on a particular type of instrument. For instance, an experiment carried out under the supervision of the author, with a graduated micrometer and skilled journeymen as subjects, showed that about $3\frac{1}{2}\%$ of the readings were in error. This finding is contrary to the generally held belief that skilled men read micrometers with zero error. On the other hand, a digital micrometer included in the experiment was read with $\frac{1}{3}\%$ error by the same men. These two examples emphasize the importance of not accepting that a particular presentation is necessarily the best for the purpose just because it is in common use.

VISUAL DISPLAYS

Probably the most common type of visual display is the graduated scale on which the indication of a value is given by means of a pointer. This type will be called a *dial*. Other pointer-type instruments may have no graduated scale and may show the state of a system: for instance, whether a valve is open or shut – information which may also be given by means of shutters or lights – and these devices will be called *indicators*.* A third type of display will call the attention of the operator to a change in a system which may require him to take some action. This is a warning function, and these displays will be described as *warning devices*. Finally, a device which gives information directly in numerals will be called a *counter*.

Uses of Visual Displays

The uses of visual displays can be grouped into two main classes, according to whether or not they are associated with controls. The first class consists of those which have no controls associated with them, but which will give the operator information which he may either have to record or on which he may base executive action. Examples of this class are the ordinary clock, or the dials fitted in the machinery control room of a ship, from which orders are relayed by the officer of the watch to the boiler or engine room. The second class consists of displays which have controls associated with them which enable the operator to make any necessary adjustments himself. Examples of this class are the dial on a radio receiver which is adjusted

*Not to be confused with indicating instruments as defined in B.S.2643:1955.

by the tuning knob, or a tachometer which is adjusted by a motor speed control to give a required speed. These broad classes can be further sub-divided according to the way the information will be used. These uses will be briefly defined (Murrell, 1951b).

Displays Used Without Controls

Dials and counters may be used:

    (a) *For Quantitative Reading*. The operator has to read the value shown by the display, either for information or to be reported or recorded. A clock or a voltmeter are examples of this use. It should be noted that the reading of a value in order to discover whether the correct value is being maintained, is not quantitative reading but check reading.

    (b) *For Check Reading*. The operator has to notice deviations from a given 'normal' value, or values, and then take executive action where necessary. This is similar to check controlling but differs from it in that the operator instead of making an adjustment himself gives instructions to somebody else to take any necessary action.

    (c) *For Qualitative Reading*. The operator has to understand the meaning of a change in value, e.g. a compass can indicate the rate of change of bearing.

    (d) *For Comparison*. The operator may have to compare a reading on one dial with that on another, for instance, in order to maintain two pieces of equipment running at the same speed or at the same output.

Warning devices may be used:

    (e) *For Warning*. A display showing the state of a system when it is not under the operator's control, e.g. the warning light on the dashboard of a car which shows whether the oil pump is functioning correctly. It is a characteristic of many warning displays that some urgent action is required to avoid disaster.

Displays Used With Controls

Dials and counters may be used:

    (f) *For Check Controlling*. Using a control to restore the display to a given value whenever a deviation from this value is observed, e.g. adjusting the speed of a pump to maintain the discharge pressure when the load varies.

    (g) *For Setting*. Using a control in order to make a display read a desired value, e.g. the setting of a generator to give a desired voltage output.

(*h*) *For Tracking.* Using a control to follow continuously or to compensate for a changing indication of a target or a pointer. Examples of this are rarely found in industry though they are common in military equipment, for instance in gun laying or in following radar echoes with a strobe.

Indicators may be used:

(*i*) *For Indicating.* Showing non-quantitatively the state of a system when under human control, e.g. for showing whether a valve is open or shut.

There may appear to be some overlapping between the 'check' and the 'warning' functions when a dial is used with the 'normal' reading marked upon it. If such an instrument is ungraduated it could be a form of warning device. But check reading implies frequent deliberate action on the part of the operator in order to determine whether all is well; warning on the other hand comes into action only when a function has gone out of limits and its purpose will often be to alert the operator even if he is asleep.

This division into such a large number of uses may at first sight seem somewhat pedantic. Most text books on human engineering divide uses of displays into only three classes; check reading, quantitative reading and qualitative reading. This, however, leads to confusion in interpretation of research results since the requirements for the nine different uses set out above are all different and it is only by recognizing this that apparent anomalies in some research results can be explained.

When deciding upon the type of display to be used and what its characteristics should be, it is important first to be sure what information the operator needs to do his job efficiently and how this can best be given quickly and unambiguously. This decision can be made difficult when displays have to serve more than one purpose. For instance, a pressure gauge may be used to give information on the rate of increase of pressure when starting up (qualitative); it may be used with a control to set the pressure at a desired running value (setting) and may then be watched to ensure that this desired value is maintained (check controlling). From time to time it may be necessary to log the value for record purposes (quantitative). The requirements for these different uses may conflict, and it will then be necessary to decide which function is of the greatest importance, that is, when failure to take the correct action quickly may be expensive or disastrous. If we consider the example given above, it may be that the maintenance of pressure at its correct value is of supreme importance. It may not matter very much how rapidly the pressure is built up or that the

exact value of the pressure is known to the operator and the keeping of a
record may be an 'old Spanish custom'. In this instance, therefore, the
check function is of primary importance and this purpose may best be
served by an un-numbered dial with the correct reading marked clearly in
some distinctive colour. The operator will then simply have to bring the
pointer onto the mark and ensure that it stays there. Logging, if it does
take place, may be in terms of deviation (plus or minus) from the set point.
If standard numbered instruments are to be used, the range within which

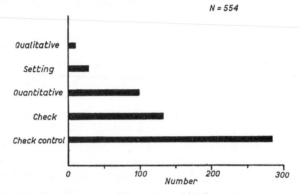

Fig. 62. Results of a survey of the uses to which instruments are put (from
Murrell, 1952b).

the pointer is to be maintained should be clearly marked, for instance with
chinagraph pencil on the cover glass. Whichever method is used, the
operator will no longer have to depend on his memory in order to ensure
that all is well and any method by which 'memory' can be built into the
equipment will lead to increased efficiency and reduced errors.

In practice a great many numbered scales seem to be used for check
purposes. Unfortunately, there is no information available on the uses to
which dials and indicators are put in industry, but a survey in H.M. Ships,
McCarthy (1952), on electrical and steam equipment revealed that only
18% were used for quantitative reading, while no less than 75% were used
for some form of checking, see also Fig. 62. From this it would appear that
if testing and measuring instruments are excluded (and these are in a class
of their own) the primary use for the displays is for checking, and observations made in industry would appear to confirm this.

Types of Display

Visual displays may be of a number of different kinds:
   (a) Moving index with a fixed legend. The index may be of several

types such as a pointer or a light spot, while the legend may be a graduated and numbered scale, an indicated safe area or it may be in the form of words. These will be referred to as *moving index displays*.

(b) Fixed index with moving legend. In this type the index is nearly always a pointer or hair line and the legend a graduated and numbered scale the whole of which is visible at one time. These will be referred to as *fixed index displays*.

(c) A special case of (b) in which only part of the legend, which may be curved or straight, is visible through a window. Indicators in which information is given by words may also be of this type, which will be referred to as an *open window display*.

(d) *Counters* in which the value of a quantity is given directly in numerals.

(e) *Shutters* in which information is given by means of a mechanical movement, which may be once and for all as in the bulls-eye of a manual telephone exchange, or by reciprocation as in the flags in a bell system.

(f) *Lights* which may be used either to warn or inform.

On the whole there can be relatively little variation in the shape of types (b) to (f), but moving index displays, type (a), are commonly of several shapes. *Circular*, in which the scale may be complete or may occupy only a part of the circumference of a circle; *eccentric*, in which the scale forms an arc which, if continued, would extend beyond the limit of the instrument; and *linear* (also called edgewise), which may be horizontal or vertical in which the scale is straight in relation to the observer, although it may be inscribed on a flat or slightly curved surface.

The Choice of Displays

Extensive research has been carried out to determine the best form of display for the various uses which have been described. Some of the results are reasonably clear cut, but others have been obtained only by using conditions which appear at first sight to bear little relationship to what occurs in practice. One such study was carried our by Sleight (1948), who compared five different dial shapes (see Fig. 63), allowing his subjects only 0·12 sec to make their readings. The results of his experiment show that, under these conditions, the open window shape gave 0·5% errors compared with 35·5% errors for a vertical straight scale. It is questionable whether this method of experiment, which has been quite widely used, is valid for our present purpose. It is known as the tachistoscopic method (a tachistoscope being an instrument which presents a stimulus for a brief

period of time) and other experimenters who have used it have found that
differences in dial shape will tend to disappear as the length of exposure
increases. There seems to be an explanation for Sleight's results. It will be
noticed that the area to be searched for the pointer in the three best shapes
gets progressively greater as the errors increase, and it can be argued that
the reason why the open window was superior is that on it a minimum of

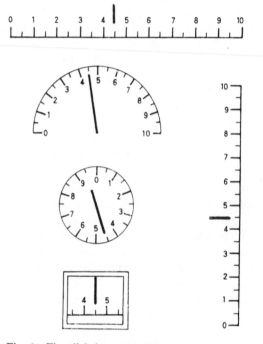

Fig. 63. Five dial shapes tested by Sleight (1948).

search for the pointer was needed. The effect of varying exposure time was
reported by Christensen (1952) whose subjects were required to read
various types of dial at selected exposure times without knowledge of
which one of the various dials would be exposed. Six variables were considered: the subject, exposure time, practice, moving pointer versus
moving scale, clockwise versus counterclockwise and pointer fixation. All
these variables were significant either singly and/or in combination with
other variables. Exposure time was shown to be an exceedingly powerful
factor. Relationships discovered at one exposure time became insignificant
or even reversed at other exposure times. This interaction between dial
type and exposure time casts serious doubt on the practice of inducing

errors in dial reading experiments by the expedient of reducing exposure time. Some of the interactions also changed with practice showing the importance of level of training in experiments of this nature. The tachisto-scopic method may also be criticized on the ground that the time allowed for viewing may bear no relationship to the actual time taken by an observer in practice: for instance, in their study of the check reading of blind flying instruments, Jones *et al.* (1949) show that the fixation time of the eye on a particular instrument varied between about 0·4 and 1·1 sec. One such criticism came from Kappauf *et al.* (1947), who say 'The natural situation in dial reading is for most readings to be paced by the subject himself. Even when readings have to be hurried in accordance with the demands of the task, the dial reader is normally free to check himself by a second glance if he feels it is important to do so'. The method which they developed as an alternative was to expose panels of 12 instruments at one time which the subject read at his own speed, giving both speed and accuracy scores. This method was used by Murrell *et al.* (1958), who found that practised subjects took approximately 1·2 sec to report a value verbally. This time has been confirmed by Jones *et al.* (1965) who repeated Murrell's experiment. Part of this time was occupied by speaking out the value of the indication and it is likely that the subject moved his eyes to the next dial and may even have started to locate the pointer while he was still speaking. The method was criticized by Naylor (1954) on these grounds, and he attempted to overcome the difficulty by making his sub-jects switch on a light in order to observe the display and switch it off again as soon as they had decided what the indicated value was. This was claimed by Naylor to give the actual comprehension time for a particular scale. However, there was probably a slight time-lag between making the decision and actually switching off the light. Unfortunately Naylor does not give an indication in his paper of the range of decision times which he measured in his experiments.

Although the different experiments give somewhat variable results for the time it takes to understand the meaning of an indicated value on a scale, it is quite clear that it must be far greater than the times of about 0·2 or 0·3 sec which are necessary if the tachistoscopic method is to show any difference of importance between the different scale types, so that it can probably be concluded that findings on dial shape obtained by these methods have no very great validity in practice. Decisions upon which type of display to use must, therefore, be based on other evidence.

These experimental methods have been discussed in detail in order to illustrate first, some of the difficulty of interpretation of experimental results obtained by different workers under different conditions and secondly, the paramount importance of not using an experimental finding under conditions totally different from those of the experiment without first validating the results under the new conditions. In the case under discussion, Sleight's results have often been quoted as a recommendation on dial shape without reservation or discussion of the conditions under which they were obtained (e.g. Kellerman *et al.*, 1963). This can lead to unsuspecting users being misled and perhaps designing equipment with displays which are not the best for the purpose.

With these limitations in mind the following is a summary of the research on the best type of display for a particular purpose:

For Quantitative Reading

Several workers have shown quite conclusively that counters, which give the value directly in numerals, are far superior to any other display which uses a graduated scale and an index. Counters, however, should not be used if values are likely to change rapidly. If a graduated scale is used it seems likely that the actual shape of the scale is relatively unimportant in spite of the work of Sleight and that of Graham (1956) who used a similar method and found that circular scales are read more accurately than straight scales. Her results can also be explained in terms of search time.

An exactly opposite finding was made by Naylor (1954) who showed that a small degree of curvature did not effect legibility but that dials with a high degree of curvature were not so easy to read as straighter scales. He also showed that horizontal straight dials are superior to vertical ones. In his paper, Naylor gives only the results of his work without any details, which makes it somewhat difficult for others to evaluate his findings. No difficulty was experienced by subjects in reading circular scales to a high degree of accuracy in either the experiment of Murrell *et al.* (1958) or Jones *et al.* (1965). The design of the actual scale on a dial is likely to have a far greater influence on accurate reading than its shape, and the various factors which must be taken into account when laying out a scale will be described in detail below. Moving index displays are superior to fixed index displays (Christensen, 1952), probably because additional information is given by the position of the index.

For Qualitative Reading

(*a*) Direction of change. Grether and Connell (1948) have suggested that vertical straight dials are better than circular dials because an upward movement of the pointer will always indicate an increase in value. In this respect circular dials are somewhat ambiguous since, when the pointer is in the right hand segment, a downward movement of the pointer indicates an increase in value. This effect may perhaps be over-rated because Morley and Suffield (1951) in their experiment on check reading have shown that there appears to be no greater difficulty in determining the meaning of a deviation from the 3 o'clock position than there was from the 9 o'clock position, provided that no control movement is involved.

(b) Rate of change. Channell and Tolcott (1948), without quoting references, have suggested that a circular dial may be the preferred type because the angle of the pointer will convey information of rate of change quite rapidly. Counters would appear to be unsuitable for indicating rate of change and so would open window dials. It would seem, however, that fixed pointer types where the scale is on a disc, as on a compass card, might be as satisfactory as a moving pointer, but no evidence on this point has been found.

For Comparison

Comparisons of the readings of one display with another have been shown by Murrell (1951b) to be made more quickly on circular dials which are ungraduated than on counters. It would appear that these judgements are made from the angle of the respective pointers, and that an observer can say with certainty that the indications are *not* the same only when the angular orientation of the pointers differs by more than 10°. If counters are used, comparisons are made more slowly but with almost complete accuracy. Whether circular dials would be faster than counters if the dials are graduated is not known, but it seems probable that if the pointers are within 10° of each other it will be necessary for the observer to read the actual values indicated by both and that this might be slower and less accurate than reading the values directly off a pair of counters.

For Check Reading

A number of workers have suggested that moving pointer instruments are superior to moving scale instruments and that circular dials may be better than straight dials, possibly because more of the pointer is visible (e.g. Connell, 1950). However, their experiments have mainly been conducted with single instruments with an alerted subject, who was expecting a deviation from the normal. Whether the same would be true with an operator who is watching a panel for a long period of time and who cannot know when or if a deviation will occur is problematical. Murrell (1960a) conducted an experiment using three panels of 16 dials, two had moving pointer dials and one had open window dials. Subjects were required to watch these panels for a period of 1½ hours and to report occasional deviations from normal. The results confirmed that the moving scale dial was inferior to the moving pointer dial and showed also that the straight dial was noticed slightly more quickly than the circular dial, a difference which was significant. When allowance was made for apparent lack of matching in the groups of subjects used, the validity of the finding is somewhat reduced; in any event the difference was only $\frac{3}{4}$ second, which is of doubtful practical value. It would appear, therefore, that as long as a moving pointer instrument is used it does not matter very much whether it is a straight instrument or a circular instrument. The decision can therefore be made on other grounds.

For Setting and Check Controlling

Graham *et al.* (1951) have shown that horizontal straight dials are set slightly more quickly and more accurately than vertical straight or circular dials. Their experiments, however, were on dials which were mounted singly and the result is probably accounted for by the relationship between the pointer and the control. If this type of dial is mounted in a panel, however, it would be rather difficult to place a control under the centre of each of a series of horizontal dials. Preference may therefore perhaps be given to the horizontal dial when it is used singly, but when mounted in panels with associated controls it is probable that both circular and vertical straight dials may be used with reasonable confidence. Kappauf (1949) has suggested that a setting cannot be made very easily or quickly with counters.

When deviations from normal have to be noted as in check controlling it seems unlikely that horizontal straight dials will be very satisfactory, unless they are used singly, and it also seems that factors influencing speed of observing deviations from normal, already discussed under checking, would apply equally to check controlling.

For Indicating

Pointers with or without ungraduated dials, shutters or lights are usually used for indicating. No research on the relative merits of these three types has been traced. It will be seen, however, that indicators of the pointer type could give some information to the operator by the position of the pointer and at the same time could show the direction of movement of an associated control required to change the indicator settings, provided that the relationship between the two is compatible. It would appear to be difficult to do this with shutters or with lights.

For Warning

Shutters and lights are usually used, and again no research on the relative merits of these devices has been traced, except on the types of warning lights to be used. Although shutters are extensively used, their attention getting value would appear to be small unless they are accompanied by some other warning, such as a noise, so that it would seem that, for visual warning, lights would be preferred.

Summary of Suitable Applications for Types of Display

The foregoing may be summarized in relation to the different display shapes which may be used:

(1) Circular dials are probably best used singly for check reading or for qualitative reading when rate of change is required; in groups for comparison when speed and not accuracy are important, for quantitative reading when the values are likely to change rapidly, or when for mechanical reasons counters cannot be used.

(2) Straight horizontal dials may be used singly and also in small multiple panels for setting and check controlling.

(3) Straight vertical dials are probably best used either singly or in multiple panels for qualitative reading, when direction of change (up-down, increase-decrease) is required or in multiple panels for check reading. It is probable that they are also the best type to use for setting and check controlling in large panels when space is very limited.

(4) Eccentric scales can be used for the same purposes as straight scales; if the scale is on the left of the instrument it may be used in the same way as a vertical dial. If the scale is at the top of the instrument it can be used in the same way as a horizontal dial (Naylor, 1954). Instruments with the scale on the right hand side of the dial with value increasing downward should never be used.

(5) A fixed index instrument with a 360° dial face visible may be used for qualitative reading when rate of change is required.

(6) Counters should be used wherever possible for quantitative reading, except when the value is likely to change rapidly.

(7) Open window dials with moving scales may be used for monitoring when one after another of a number of systems is selected by means of a control for quantitative reading.

(8) As far as possible choose the simplest instrument which will give the information required rapidly and unambiguously, ensuring that the meaning of the indication is apparent and that the observer has to do a minimum of interpretation. Decide exactly what information you want to give with your displays. Choose the best type for your purpose and ensure that it gives this information and this only, all extraneous information being eliminated.

The Design of Quantitative Instruments – Counters

The legibility of counters does not appear to have received much attention. The main factor which will influence this is the design of the numerals, and there is no reason to suppose that work on the legibility of numerals for graduated scales should not be equally applicable to the numerals used on counters. Owing, however, to the lack of a 'continuity effect' it would seem that these numerals should be larger than those recommended for graduated instruments and the height in inches can be obtained by dividing the reading distance in feet by 25. Thus for a commonly found reading distance of 3 ft the numeral height should be 0·12 in. The overall width of the numeral should be between 70 and 80% of the height. Other factors which may influence the efficiency of a counter would be the rate at which the numerals change and whether the numerals change continuously or intermittently. Intermittent movement should be provided on the number wheels for tens and above, since confusion might be caused when numerals are out of alignment.

The number of numerals in a counter may be of importance. Connell (1950) displayed numbers containing from two to seven digits to 20 subjects, and recorded time and errors in detecting differences between verbally and visually presented numbers. He found there was an almost linear increase in time and in errors for each addition of a number beyond two digits in the presentation.

The Design of Quantitative Instruments – Dials

Instruments in which the value of a quantity is indicated by a pointer on a graduated scale are by far the most common type of measuring instrument. They form an important part of the instrumentation of a wide range of equipment, particularly aircraft, and for this reason during and since World War II they have received more attention from research workers than any other type of visual display. Factors such as the distance between scale marks,* the length and thickness of scale marks, the number of marks in a complete scale, the way they are numbered, the size and design of the numerals, the shape and length of pointers, and so on, have all been the subject of extensive experiment. Based on these experiments recommendations on the methods of laying out dial faces have been made by a number of authors (e.g. Woodson, 1954; McCormick, 1957) which are dependent on the relative weight given by each author to differing experimental results.

*Some of the terms used in describing parts of a scale are defined in B.S.3693. A glossary of these and other terms is given in an Appendix.

The Size and Number of Scale Divisions

The fundamental factor in the readability of a scale is the physical width, which is best defined in terms of visual angle, of each of the parts (the *called division*) into which a scale division must be divided by eye in order to achieve the design tolerance. The width of the space between two scale marks or its visual angle (the *scale spacing*) will in turn depend upon the number of parts into which the scale division is to be interpolated. Thus if interpolation into fifths is required, the visual angle of the scale division will be equal to five times that of a called division. If this visual angle is below a critical size, errors will be made in reading which will increase linearly in frequency with decreasing log visual angle, see Fig. 65, p. 169. If, however, the visual angle is increased above this size (at which reading accuracy can be expected to be about 99% with a highly practised observer) no further improvement in reading accuracy will be found. When this optimum visual angle is known, the width of the scale division for a particular tolerance and reading distance can be calculated, and the length of the scale itself can be determined when the number of scale divisions required are known. Thus in starting to set out a scale it is important first to decide the greatest distance from which the scale will be read and from this the width of scale division to be used can be calculated: the number of scale divisions will depend upon the maximum scale value and the required tolerance. This 'law of constant visual angle' has been questioned by Churchill (1959) on the basis of his tachistoscopic results. Murrell *et al.* (1958) suggested that it is possible that the 'law' may not hold when scale divisions are wide, interpolation is in tenths and viewing distances approximate to 30 in. or less, and propose that further research on this point is needed. There seems, however, to be no reasonable doubt that, for distances over 30 in. and with a required tolerance of a fifth or less of a scale division, the determining factor in relating dial size and reading distance is the visual angle (Murrell, 1962a).

The optimum size of the scale space has been investigated by a number of workers in America. This work has been summarized in a number of books, notably Chapanis, Garner and Morgan (1949) and McCormick (1957). All these experiments were conducted with aircraft instruments in mind and reading distances of about 28 to 30 in. were used, which approximate to those found in aircraft cockpits. The results of the experiments are reasonably in agreement and suggest that the separation of the scale marks should be somewhere between 0·2 and 0·5 of an inch. This result is often given in texts without mentioning that this applied specifically to instruments read at this one unique distance of 28 to 30 inches. There is some difficulty in interpreting these results due to lack of uniformity in nomenclature. In some instances the term 'scale marker' is used to refer only to the major mark numbered, while other workers use it to refer to any mark, numbered or un-numbered.

We may consider the applicability of these results to dials other than aircraft dials by considering the work of Kappauf, Smith and Bray (1947). They showed that under speed and accuracy instructions errors were at a minimum when the separation of the major scale marks was 0·88 in. Under

Fig. 64. Three scales compared for legibility by Kappauf, Smith and Bray (1947).

accuracy instructions there was only a small improvement when this was increased from 0·44 to 0·88 in. One of the dials used in this experiment is a 100 unit scale read to the nearest unit (see Fig. 64). Since an observer can interpolate an interval into five parts, this layout may be used for the scale of, say, a 100 lb pressure gauge. With a separation between the major marks of 0·88 in. and ten major intervals, the scale base length of this design will be 8·8 inches. Since a normal pressure gauge has a scale covering 270°, the circumference of the circle of which this scale will form a part will be 11·7 in., with a diameter of 3·75 in.; this would be inscribed on a 4½ in. nominal dial blank.* If the reading distance is increased from 2½ to 10 ft, the diameter of the dial blank must be increased to 18 in., if the same visual angles are to be maintained, or if the smaller figure of 0·44 in. obtained under accuracy instructions is used, which agrees favourably with 0·50 in. from the study by Grether and Williams (1947), we should still have a 9 in. dial blank for a reading distance of 10 ft which is much larger than the dials which are normally met with in practice. These considerations led the author and his co-workers to carry out an extensive study of the relationship between scale spacing, dial size and reading distance (Murrell *et al.*, 1958). Five dial sizes and six reading distances were used, chosen so that different combinations of dial size and reading distance gave a similar visual angle. The results of this experiment are shown in Fig. 65. From this it will be seen that the optimum result is obtained when the angle subtended at the eye by the called division is approximately 2′ of arc. If there are five called divisions in a scale division

*The nominal dial blank is the part of the dial visible inside the spacer ring.

between two adjacent markers, as in the example above, then the angle subtended at the eye by this scale division is 10' of arc. This is rather less than half the usual angle derived from the dimension of 0·44 of an inch. The subjects in the experiment of Murrell *et al.* were able to read scales with a consistent accuracy of about 99% whereas the accuracy achieved by the American subjects under similar instructions was about 96% on the

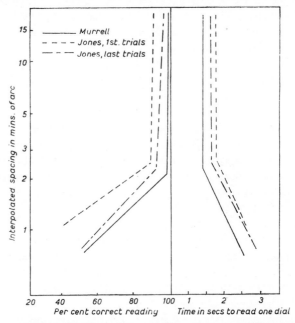

Fig. 65. The relationship between reading accuracy, reading time and angle subtended at the eye by an interpolated spacing (after Murrell, *et al.*, 1958).

largest dial. When reading scale divisions of a size similar to those used in the British experiment they achieved an accuracy of only about 87%. This result obviously needs some explanation and there would appear to be a number of possible reasons for the difference.

First, in dial reading there is a very marked practice effect. So much so that in Murrell's experiment it was necessary to make each subject do some 2,000 readings before scoring was started. As far as can be gathered from the reports of the American work no subjects read any particular dial design more than 150 times. It is obviously a matter of opinion whether dials should be designed for people who use them regularly or for people who use them only occasionally, but it would seem that anybody in industry who has a job which involves reading dials would, in the course of a very short time, become practised. The effect of practice on reading an unfamiliar dial face was demonstrated by Spencer (1961) who used as subjects process workers who were very experienced users of dials. In an extended trial with one man, reading to within $\pm \frac{1}{2}\%$ maximum scale value (m.s.v.), performances improved from 70% within tolerance for the first 30 observations to 97% for observations 71-100. Increasing practice can also be defined in terms of decreasing scale spacing read to a given accuracy. In Murrell's experiment, the larger scale divisions were read with complete accuracy after less than 200 readings, while the very small divisions were not attempted at all. But after more than 1,000 readings these small divisions were being read with acceptable accuracy. It is evident that results from unpractised subjects, who have made less than about 250 readings on any dial layout, will be liable to give larger values for the optimum scale spacing than those which will result from well practised subjects, and this makes experiments with naïve subjects of doubtful value.

Secondly, in the experiment by Murrell *et al.* (1958) the same dial design was used throughout, whereas in the American experiments a number of different dial designs were used at the same time and the subjects switched from one to another. This in itself may introduce errors since it was shown by Nafe (1944) that gun aimers made many errors due to changing from one scale design to another. On the other hand, it is perhaps nearer to 'real life' to have to read a variety of dials, but again practice is of great importance and this is likely to require longer with a mixed array than with an array of one design. The work of Christensen (1952) on this issue has already been quoted.

Finally, the eyesight of the British subjects was carefully tested and in the main experiment only those with eyesight 6/6 Snellen or better were used; the eyesight standard of the American subjects is not known. A subsidiary experiment was carried out with British subjects with eyesight of 6/12 and the results suggest that, for practised operators, the width of the scale divisions should be increased by approximately one-fifth, if people with reduced visual acuity are to be provided for.

The experiment of Murrell *et al.* (1958) has been repeated by Jones *et al.* (1965) who used smaller dials and closer reading distances than Murrell. His results confirm that the optimum called division should subtend a visual angle of 2′ of arc. Practice had relatively little effect on this value, 100 readings giving an improvement of about 5% in accuracy and a decrease in visual angle of about 15″ of arc. What did show a change with practice was the slope of the line describing performance below the optimum visual angle which, with little practice, was shallow but which later approximated to that found by Murrell *et al.* (Fig. 65). Spencer (1961) has demonstrated that the dimension of 2′ of arc is realistic, therefore the recommendations which will be made later are based on the British results.

Fig. 66a.

Fig. 66b.

Fig. 66. Examples of 'open' (*a*) and finely divided scales (*b*) (by courtesy of Sangamo Weston Ltd).

Having considered the optimum width of a scale division it is necessary to decide into how many units this should be required to be interpolated. Some of the experiments referred to above (e.g. Kappauf *et al.*, 1947) have shown quite clearly that there is a marked decrease in accuracy if a scale division has to be interpolated into ten parts. At the same time Leyzorek (1949) has shown that when a scale division has to be interpolated into fifths, it is read more quickly and accurately than when there are five markers each equal to one unit. In the experiment by Kappauf *et al.* (1947) the superiority of the interpolation in fifths over reading to the nearest unit is marked under 'speed and accuracy' instructions when the scale spacing is small (0·44 in. or less); but this superiority is lost when the spacing is large (0·88 in.). Naylor (1954) favours a 'plain dial' without units being shown by scale marks, particularly on circular scales, when he obtained a difference significant at $p > 0.001$. Thus for decimal scales the maximum useful required interpolation is in fifths, but no work has been traced which would indicate whether fifths are superior to fourths or halves. Fig. 66 compares similar instruments which require interpolation in fifths and halves respectively. On the evidence the former arrangement is preferred for rapid reading.

Some instruments may be required to be graduated in the duo-decimal system. Under these conditions interpolation into sixths may be required. No work has been traced which will show whether this will be satisfactory; until evidence to the contrary is forthcoming it would seem that reading in sixths should be acceptable although threes might be easier.

The minimum scale spacing for a given reading distance and the number of units into which a division is required to be interpolated by eye have now been determined, but before deciding how many of these divisions there are to be in the whole scale, the tolerance to which the scale is to be read must be estimated. Most industrial pressure gauges are said to have an intrinsic tolerance not greater than 1% of the scale range, while the tolerance of many electrical instruments may not be even as great as this.

It must be emphasized that the tolerances expressed as a percentage of the scale range must not be confused with reading accuracy, although they can be interrelated. The reading accuracy expresses the expected number of readings which will not deviate by more than a given tolerance from the true pointer position. Even with the best industrial instrument it is unlikely that the accuracy of an average observer will be much better than 97% with a tolerance of $\pm \frac{1}{2}$% m.s.v., though careful and experienced observers may achieve 99%. If, however, the tolerance is reduced to $\pm \frac{1}{4}$% m.s.v. the accuracy will fall to about 85 to 87% if the design demands interpolation into tenths (Kappauf *et al.*, 1947; Murrell *et al.*, 1958). Provided that the called spacing is not less than the minimum width appropriate to the reading distance and the scale division is interpolated into not more than five parts – the reading accuracy will be independent of the tolerance; thus if the scale interval on a 100 unit scale has a value of 0·5 of a unit the instrument *could* be read with 97 to 99% accuracy, with a tolerance of ± 0·05% m.s.v., provided that it is sufficiently large.

Unless we are going to demand that readings are taken to a tolerance which is greater than that inherent in an instrument, the reading tolerance and the inherent tolerance should be the same. It is convenient to classify instruments which have an intrinsic tolerance of 1% (± ½%) scale range or worse and which are read 'at a glance',* as *commercial instruments*, while *test instruments* are those which have a tolerance better than 1% scale range, and are intended to be read slowly and with care. All the experiments described so far have related to instruments which have to be read 'at a glance' so that their results apply to the design of commercial instruments and the discussion which immediately follows is restricted to this class of instruments. Test instruments will be considered separately. The division of instruments in this way is a matter of convenience and must not be taken to imply that instruments used for commercial purposes may not have an intrinsic tolerance which demands that they be read with an accuracy greater than 1% scale range. It is believed, however, that most industrial instruments are neither required nor able to indicate values with greater accuracy.

If a commercial instrument shall be read to 100th of its m.s.v.† and if a scale division is to be read to fifths, the optimum number of scale divisions in a whole scale would be 20. Obviously some scale maxima will make it impossible to use this number of scale divisions, but provided that the scale spacing is not smaller than a minimum associated with a particular reading distance, this is acceptable, although the reading tolerance cannot be exactly 1%. Should a reading tolerance other than 1% be required, the optimum number of scale divisions will be $\dfrac{100}{5x}$ where x is the tolerance required, expressed as a percentage.

Scale Base Length

With a called division of 2′ of arc the minimum called spacing should be 0·08 in. at a reading distance of 12 ft. This figure, however, applies to

*The term 'at a glance' used in connection with *quantitative reading* implies a reading time of 1·0 to 2·0 sec. When *check reading* a 'glance' may be as short as 0·5 sec.

†Or of its scale range if scale numbering does not commence at zero.

practised observers with good eyesight reading under good conditions. To allow for observers with less good eyesight, or for conditions which are not ideal, this value should be increased by about one-fifth to 0·10 in., and with 100 called spaces, the scale base length will be 10 inches. This relationship between reading distance (D) and scale base length (L) can be expressed as:

$$D = 1·2L \qquad (6)^\star$$
$$\text{or } L = 0·833D \qquad (7)^\star$$

when D is expressed in feet and L in inches.

If the same units are used for both dimensions:

$$D = 14·4L \qquad (8)$$

This relationship has been adopted by the British Standards Institution and makes allowance for observers who may have sub-standard eyesight down to 6/12 Snellen (B.S. 3693:1964).

Where it is desired to find the scale base length for a scale which has more or less scale divisions than the standard, a correction factor will have to be applied, and formula 7 will become:

$$L = 0·833D \times \frac{i \times m}{100} \qquad (9)^\star$$

where i = the number of parts into which the scale division is to be interpolated and m = the number of scale divisions.

If it was decided to use a scale with 25 scale divisions on a 0–50 dial it is probable that interpolation will be in quarters to 0·5 unit; formula 9 will give a correction factor of 1 and this instrument will be the same size as one based on the 100 unit standard. On the other hand, a 0–60 instrument interpolated in the same way will have 30 scale divisions and the correction factor will be 1·2, so that this instrument would have to be one fifth larger than standard. If the correction factor is less than unity the instrument in question can be smaller than the standard, but this is unnecessary since no harm can come from an instrument having scale divisions slightly larger than standard. Although the scale base length can be calculated in this way, dials will usually be made with standard case sizes in conformity with B.S.I. or other specifications, which will prescribe the scale base length within fairly narrow limits for any given case size. Knowing this scale base length it should be possible to specify the reading distance for a particular sized dial and this should enable the user, knowing the reading distance, to choose a dial to suit his requirements. If the correction factor is less than unity the instrument in question can be smaller than the standard, but this is unnecessary since no harm can come from an instrument having scale divisions slightly larger than standard.

*These formulae apply when calculations are made in feet and inches. General formulae are given later on pp. 194-96.

Table 11 sets out the relationship between the nominal size of dials, the reading distance and the minimum scale spacing for scales divided into 20 parts. The diameter of the dial blank is given on the assumption that it is six fifths of the diameter of the scale base. That dials based on these dimensions, and on the requirement of interpolation into fifths, will give satisfactory results in practice has been demonstrated by Spencer (1961), whose work has already been quoted. Twenty-four experienced process workers had no difficulty in reading a nominal 3 in. dial (scale base length 5 in.) at 6 ft with the required accuracy, in spite of an almost complete lack of familiarity with the scale designs observed.

The Arrangement of Scale Divisions

We have seen that the optimum number of scale divisions in a scale is 20, but the decision on the number of divisions in a particular scale will depend on its maximum scale value, on the probable interpolation and on the accuracy of reading required, although the two latter will not always be compatible, since the degree of agreement will depend to a large extent on the scale maxima. Thus, an instrument with a scale maximum of 60 units would never be read with a tolerance of 1% since this would involve reading the scale to 0·6 unit while such a scale would in fact be read either to 0·5 or 1·0 unit.

In Table 12, methods of building up scales for common scale maxima

Table 11. Relationship of various scale parameters to reading distance (in general units and in ft/in.)

Reading Distance	Minimum Called Spacing	Minimum Scale Spacing (interpolation in 5ths)	Scale Base Length for 20 spaces	Scale Base Diam. for a 270° Scale (based on column 4)	Diameter of Visible Area of Circular Dial (based on column 5)
1	2	3	4	5	6
x	0·0007 x	0·0035 x	0·0694 x	0·0295 x	0·0354 x
(ft)	(in.)	(in.)	(in.)	(in.)	(in.)
2	0·017	0·08	1·66	0·707	0·869
3	0·025	0·13	2·50	1·061	1·304
4	0·033	0·17	3·33	1·414	1·739
6	0·050	0·25	5·00	2·121	2·608
8	0·067	0·33	6·66	2·826	3·477
10	0·083	0·42	8·33	3·535	4·347
12	0·100	0·50	10·00	4·242	5·216
15	0·125	0·62	12·49	5·303	6·520
20	0·167	0·83	16·66	7·070	8·693
25	0·208	1·04	20·82	8·838	10·867
30	0·250	1·25	24·99	10·605	13·040

Table 12.   Optimum graduation of scales with various m.s.vs related to probable interpolation

Scale Range	No. of Scale Divisions	Value of a Scale Interval (units)	No. of Major Marks Numbered	System of Numbering	No. of Minor Marks	Probable Reading Tolerance (fraction of scale division)	(units)	(%)	Tolerance in 5ths (units)	(%)
0-10	20	0·5	6	0-2-4-	3	$\frac{1}{5}$	0·1	1·0	0·1	1·0
0-15	15	1	4	0-5-10-	4	$\frac{1}{2}$	0·5	3·3	0·2	1·3
0-16	16	1	9	0-2-4-	1	$\frac{1}{2}$	0·5	3·1	0·25	1·25
0-20	20	1	11	0-2-4-	1	$\frac{1}{2}$	0·5	2·5	0·2	1·0
0-20	20	1	5	0-5-10-	4	$\frac{1}{2}$	0·5	2·5	0·2	1·0
0-25	25	1	6	0-5-10-	4	$\frac{1}{2}$	0·5	2·0	0·2	0·8
0-30	30	1	7	0-5-10-	4	$\frac{1}{2}$	0·5	1·7	0·2	0·7
0-40	20	2	5	0-10-20-	4	$\frac{1}{4}$	0·5	1·25	0·4	1·0
0-50	25	2	6	0-10-20-	4	$\frac{1}{4}$	0·5	1·0	0·4	0·8
0-60	12	5	7	0-10-20-	1	$\frac{1}{5}$	1	1·7	1	1·7
0-60	30	2	7	0-10-20-	4	$\frac{1}{2}$	1	1·7	0·4	0·7
0-80	16	5	9	0-10-20-	1	$\frac{1}{5}$	1	1·25	1	1·25
0-100	20	5	6	0-20-40-	3	$\frac{1}{5}$	1	1·0	1	1·0
0-150	15	10	4	0-50-100-	4	$\frac{1}{5}$	2	1·3	2	1·3
0-160	16	10	9	0-20-40	1	$\frac{1}{5}$	2	1·25	2	1·25
0-200	20	10	11	0-20-40-	1	$\frac{1}{5}$	2	1·0	2	1·0
0-200	20	10	5	0-50-100-	4	$\frac{1}{5}$	2	1·0	2	1·0
0-300	30	10	7	0-50-100-	4	$\frac{1}{5}$	2	0·7	2	0·7
0-400	20	20	5	0-100-200-	4	$\frac{1}{4}$	5	1·25	4	1·0
0-500	25	20	6	0-100-200-	4	$\frac{1}{4}$	5	1·0	4	0·8
0-600	12	50	7	0-100-200-	1	$\frac{1}{5}$	10	1·7	10	1·7
0-800	16	50	9	0-100-200-	1	$\frac{1}{5}$	10	1·25	10	1·25

are given, based on the interpolation likely in practice. The numbering is based on straight or sector scales; for circular scales it will probably be necessary to drop alternate numeral groups when the number exceeds nine for instruments with maxima below 100, or seven for instruments with maxima above 100. The 'probable reading tolerance' has been based on the view that for the lower values a reader may think in half units or one unit but not in 0·2, 0·4 or 0·6 units, while in the higher values he will probably read in 1, 2 or 5 units but not in 2·5 and 4 units. The scales given in Table 12 are chosen to give the greatest probable accuracy consistent with not having too many scale divisions. The most unsatisfactory design in this respect is the 0-30 with 30 scale divisions. If, as suggested, inter-polation is in halves, giving a tolerance of 1·7%, this scale with a correction factor of 0·6 can nevertheless be of standard size; if interpolation is into fifths to give a tolerance of 0·7% the correction factor will be 1·5 and a dial half as large again as the standard will be required. The same applies to the 0-60 scale with 30 divisions; a third m.s.v. with more than 20 scale divisions is the 0-50 with 25 scale divisions but since its scale interval of 2 units value will be interpolated into 0·5 units rather than into 0·4 units, it will be of standard size, with a correction factor of 1.

There are quite a number of other ways in which scales might be made up and the most probable of these are given in Table 13. They have been omitted from the main table either because (a) the probable interpolation is 0·2 (15, 16, 20, 40) or is 4·0 (400, 500, 600); (b) they have an excessively

Table 13. Sub-optimum graduation of scales with various m.s.vs related to probable interpolation

Scale Range	No. of Scale Divisions	Value of a Scale Interval (units)	No. of Major Marks Numbered	System of Numbering	No. of Minor Marks	Probable Reading Tolerance (fraction of scale division)	(units)	(%)	Tolerance in 5ths (units)	(%)
0-10	20	0·5	11	0-1-2-	1	$\frac{1}{5}$	0·1	1·0	0·1	1·0
0-15	15	1	4	0-5-10-	4	$\frac{1}{5}$	0·2	1·3	0·2	1·3
0-16	32	0·5	9	0-2-4-	3	$\frac{1}{5}$	0·1	0·6	0·1	0·6
0-16	16	1	9	0-2-4-	1	$\frac{1}{5}$	0·2	1·25	0·2	1·25
0-20	20	1	5	0-5-10-	4	$\frac{1}{5}$	0·2	1	0·2	1·0
0-20	20	1	11	0-2-4-	1	$\frac{1}{5}$	0·2	1	0·2	1·0
0-40	8	5	5	0-10-20-	1	$\frac{1}{5}$	1	2·5	1	2·5
0-40	40	1	9	0-5-10-	4	$\frac{1}{2}$	0·5	1·25	0·2	0·5
0-40	40	1	9	0-5-10-	4	$\frac{1}{5}$	0·2	0·5	0·2	0·5
0-50	10	5	6	0-10-20-	1	$\frac{1}{5}$	1	2	1	2·0
0-50	50	1	11	0-5-10-	4	$\frac{1}{2}$	0·5	1	0·2	0·4
0-50	25	2	6	0-10-20-	4	$\frac{1}{2}$	1	2	0·4	0·8
0-60	30	2	7	0-10-20-	4	$\frac{1}{4}$	0·5	0·8	0·4	0·7
0-80	40	2	9	0-10-20-	4	$\frac{1}{2}$	1	1·25	0·4	0·5
0-100	20	5	11	0-10-20-	1	$\frac{1}{5}$	1	1	1	1·0
0-160	32	5	9	0-20-40-	3	$\frac{1}{5}$	1	0 6	1	0·6
0-400	8	50	5	0-100-200-	1	$\frac{1}{5}$	10	2 5	10	2·5
0-400	20	20	5	0-100-200-	4	$\frac{1}{5}$	4	1	4	1·0
0-500	25	20	6	0-100-200-	4	$\frac{1}{5}$	4	0·8	4	0·8
0-500	10	50	6	0-100-200-	1	$\frac{1}{5}$	10	2	10	2·0
0-600	30	20	7	0-100-200-	4	$\frac{1}{4}$	5	0·8	4	0·7
0-600	30	20	7	0-100-200-	4	$\frac{1}{5}$	4	0·7	4	0·7

large number of scale divisions (16, 40, 50, 80, 160); (c) they are likely to be read with less accuracy (40, 50, 400, 500); or (d) they will tend to be larger than standard (16, 40, 60, 160, 500, 600).

When reading a scale, the observer must locate the pointer with reference to a particular numbered division, and must then decide in which of its constituent scale divisions the indicated value lies. He will then interpolate the position within this division. The numbering system used must therefore be one which is rapidly intelligible without being over-crowded; there is some evidence that, far from being a help, too many numbers will cause confusion. Another factor of importance is the space available on the dial face for the numerals, and the need to restrict the number of different systems to avoid confusion when dials with different m.s.v. are mounted side by side.

The relative merits of different numbering systems have been investigated by several research workers. Vernon (1946) concluded that systems numbered 1, 2, 3 . . . or decimal multiples were best, while next best were systems numbered 2, 4, 6 . . . She also concluded that systems numbered in 4s (i.e. 4, 8, 12, 16 . . .) 8s or 25s were unsatisfactory and that decimals should be avoided. Her work is confirmed by Barbour and Garner (1951) who found that scales numbered in 1s and 2s were superior to those numbered in 5s. Unpublished work by Spencer, however, does not suggest that 5s are markedly inferior to 1s and 2s. Chapanis and Leyzorek (1950) placed first systems numbered in 1s, with 2s and 5s equal second. Remaining systems were ranked in descending order 8s, 4s, 9s, 6s, 3s, 7s, and finally 2·5s which were very bad indeed. The accumulated evidence suggests that any one of the three number systems – 1s, 2s and 5s – may be used without introducing any great chance of error, and that they are greatly superior to any other number system.

There is evidence that one, three or four minor marks between each major mark can be satisfactory. On the face of it, it would appear that scales with one minor mark between each numbered mark are likely to give rise to less uncertainty than scales with four minor marks, since in the former there is only one chance in two that the observer may be reading in the wrong scale division, whereas in the latter there is one chance in five. However, this has not been borne out entirely by experiment, and it would appear that either method of dividing the scale may be adopted provided that the value of the minor marks falls into one of the three numbering systems recommended for the major marks. For example, the major interval 0-20 can be subdivided in three ways, as shown in Fig. 67. When there is one minor mark, its value will be 10 falling in the series 0, 10, 20, and three minor marks will fall in the series 0, 5, 10, 15, 20. But if there are four minor marks, they will have values in the series 0, 4, 8, 12, 16, 20, which is unacceptable. On the other hand, if the interval 0-50 has one or three minor marks, they will fall into the series 0, 25, 50 and

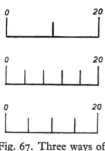

Fig. 67. Three ways of dividing a 20-unit interval.

0, 12·5, 25, 37·5, 50 respectively, both of which are unsatisfactory. Thus it will be seen that the number of minor marks will depend on the series chosen for numbering the major marks: one or four for the series 0, 1, 2 . . ., one or three for the series 0, 2, 4 . . . and four for the series 0, 5, 10 . . .

Scale Design Procedure

We can now set out some principles which will guide us in setting out a decimal scale.

(a) The smaller the number of scale marks (within the limits laid down below), the quicker and the more accurately will the scale be read. It can also be read at a greater distance.

(b) Choose sufficient scale marks so that the space between them will be interpolated into not more than fifths by eye.

Fig. 68. An example of a 200-unit scale 'extended' to 250 units (by courtesy of the English Electric Company Ltd).

(c) Never allow for a reading tolerance greater than the mechanical tolerance of the instrument.

(d) Number the major scale marks in 1s, 2s, or 5s (or decimal multiples or sub-multiples). Avoid 3s, 4s, 8s, or decimals such as 2·5. Use the lowest number system possible without overcrowding, with the proviso that uniformity between one scale and another should be preserved when possible.

(e) There should be either one, three or four minor scale marks between the major marks.

(f) There should be not less than five numbered divisions in a scale including zero.

(g) As far as possible avoid numbers with more than three digits, with the exception of the terminal number. The numbering of a 6,000 lb scale, for instance, in thousands, would lead to a hopelessly over-crowded dial face, and it would be far better to number the major marks in units and put a legend on the dial face to show that the indication is in thousands of pounds.

To see how these principles can be applied to the setting out of a scale, consider a scale with a maximum value of 200 units, to be read to the

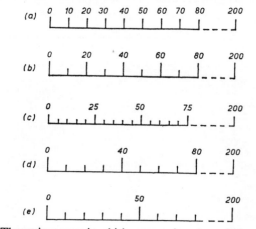

Fig. 69. The various ways in which a 200-unit scale could be numbered (after Murrell, 1952a).

nearest two units, i.e. with a permissible reading error of $\pm$ 1 unit. This scale will have 20 divisions, each equal to ten units, which can be numbered in five different ways, as shown in Fig. 69. Of these, (c) and (d), being numbered in 25s and 40s, are unsatisfactory and should therefore be dis-carded. While (a) has a satisfactory number system, it will, on a circular dial, have too many numbers, while (e) will have too few. The scale most nearly fulfilling the requirements, although it is on the borderline of over-crowding, is (b). If the number of digits is considered too high, (d) *might* be used, in spite of recommendations to the contrary.

In some instances it may be necessary to use an instrument with an m.s.v. which is close to a value to which the foregoing principles can easily be applied. Such a scale should be laid out as if it had the latter value with one or more numbered divisions added or subtracted. The 0-250 unit scale in Fig. 68 has been treated as if it were a 0-200 scale of 20 scale divisions; arrangement (*e*) in Fig. 69 has been used because it is the most appropriate to the added increment. The s.b.l. will have to be longer by a quarter than that of the 0-200 scale for the same reading distance or the distance will have to be reduced if the s.b.l. is kept constant. Other scale layouts will have to be chosen as appropriate, for instance 0-180 or 0-240 scales should be based on (*b*). If the added increment is not a convenient portion of an acceptable scale, either the m.s.v. is adjusted so that it is a convenient portion, or it is numbered independently of the remainder of the scale. For instance, an m.s.v. of 105 might be specified. The normal 0-100 scale on which it will be based may be numbered 0-10-20, etc. and the increment of 5 units does not fit this numbering. An m.s.v. of 110 should, therefore, be used; but if this is *impossible* (and for no other reason) the scale should be numbered 0-10 . . . 90-100-105.

Not all scale maxima will permit these principles to be as readily applied

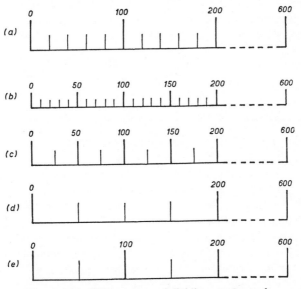

Fig. 70. Five different ways of dividing a 0-600 scale.

as this. One of these is the 600 scale maximum. In theory this should be read to the nearest 6 units, but in practice people do not think in sixes. Five ways of building up this range are shown in Fig. 70. Range (*a*) with 30 scale spaces could be read to 4 units (0·7%) if interpolation is in fifths, or to 5 units if in fourths (0·8%), but this number of scale spaces would require a longer scale base length than 'standard'; (b) with 60 scale spaces and 13 numbers, is too crowded and *could* be read to 2 units (0·3%); (c) has 24 scale spaces but the value of each is 25 units and thus is unacceptable, although it can be read to 0·8% m.s.v. This leaves (d) and (e), both of which can be read to 10 units (1·7%), a reading tolerance which is not likely to differ greatly from that actually obtained on many commercial instruments when they have been in use for a time. Range (d) has too few numbers and so the choice is likely to fall on (e).

Non-linear Scales

The scales so far discussed have been linear, with all the scale divisions of equal physical size. Because of the characteristics of some functions, some scales will have to increase or decrease in size with increasing value. No research on this type of presentation for 'at a glance' reading has been traced, so it will be necessary to decide the best scaling from first principles.

In the first place no scale division should be smaller than x times the called spacing appropriate to the intended reading distance. The value of x will depend upon the interpolation which seems to be the most likely.

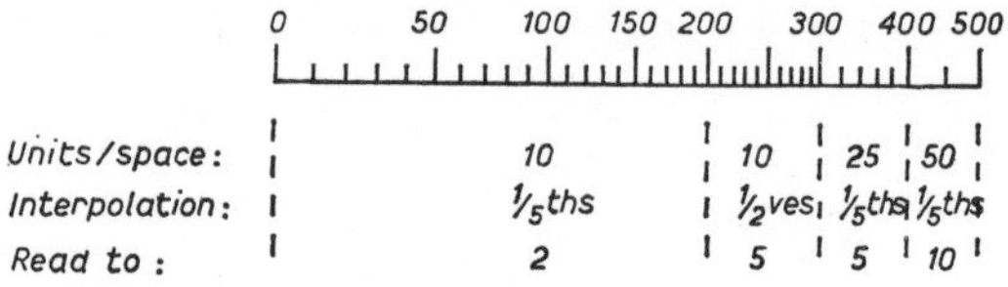

Fig. 71. A layout for a non-linear scale.

On the basis of the finding that errors can occur when making successive readings on dials with scales which are differently laid out, it would probably be best to choose an arrangement which requires the same interpolation throughout.

To demonstrate these two principles a 0-500 scale is shown in Fig. 71, to be read at a distance of 10 ft. From Table 11 the minimum scale spacing will be 0·42 in. when interpolation is in fifths. Since the 200 scale mark is about half-way along the scale, we will take the space immediately below this as being of minimum size and give it a value of 10 units, there will then be 20 minor scale spaces between 0 and 200, and these will be divided by three major scale marks numbered in 5s. This part of the scale will be read with a tolerance of ± 1 unit. The interval between 200 and 300 could be set out in the same way, but interpolation would have to be into halves to 5 units. A superior result can be obtained by using four scale spaces, as shown between 300-400, and maintaining uniform interpolation in fifths.

At 500 the scale spacing would be below the minimum, so that the 400-500 interval is divided into two spaces to maintain interpolation in fifths to 10 units.

In this arrangement both objectives have been achieved. No scale division is less than five times the minimum called spacing and interpolation in fifths is required throughout. The scale length will be about 18 in.

The Direction of Increase of Value and the Position of Zero

The position of zero on horizontal scales and sector scales presents relatively no problem, but this is not so with circular scales. Vertical straight scales and left-hand sector scales should almost certainly read from the bottom up whilst horizontal straight scales, and top and bottom sector scales should normally read from left to right. In certain circumstances, however, if they are used with valves which turn anticlockwise to open, bottom sector scales may read from right to left, i.e. with the pointer moving clockwise; from this it can be argued that it would not be unreasonable for horizontal straight scales also to be read from right to left when necessary; there is some evidence that there is no great difficulty in reading horizontal scales in this direction. Right-hand sector scales are more difficult since it can be argued that either they ought to read from top downwards, that is with the pointer moving in a clockwise direction, or that they should be considered as similar to a vertical straight scale and read from the bottom upwards. There is no evidence to recommend either of these directions, though Naylor (1954) has shown that a left-hand sector scale is very significantly superior to a right-handed sector scale (presumably graduated from the top down), and Christensen (1952) has shown that circular scales numbered clockwise were superior to those numbered anticlockwise. In view of this, it is best to avoid a right-hand sector scale unless there is a very pressing reason for its use.

When a circular scale is being used solely for quantitative reading with no control movement involved, the position of zero appears to be relatively unimportant. A number of workers, particularly Kappauf and Smith (1950), have shown that the ease of reading a scale seems to be very little influenced by the portion of the scale which is being read. Grether (1948) suggested that there may be a slight preference for the 12 o'clock position because most people are influenced by the habit of reading a clock and this argument would particularly apply to a scale which covers the full 360°. On the other hand, when a 270° scale is used most people will be accustomed to reading a scale which is symmetrically arranged round 12 o'clock with the zero half-way between 7 and 8 o'clock.

It is only when circular dials are used with controls that the position of the scale and consequently of the zero becomes of any importance. With pressure gauges it is generally recommended that the normal working range should be at approximately two thirds the full scale reading. With the present arrangement of a scale this will produce a working arc down the right-hand side of the dial, an arc which Naylor (1954) has shown to be unsatisfactory. In other words, there will be a downward movement of the pointer which will be incompatible with an upward movement of a lever and may not be entirely compatible with the movement of a rotary control. There would seem, therefore, to be a case for moving the zero to 6 o'clock when the range of movement of the pointer would be across the top of the dial on either side of 12 o'clock, which would be compatible with the movement of a rotary control and with a left-to-right movement of a lever. If an up-and-down movement of a lever is involved, the zero would probably be best at 3 o'clock so that the pointer would move on either side of the 9 o'clock position. Placing the zero in either of these positions would also assist in the patterning of pointers.

A problem of special difficulty arises when pressure gauges are used with valves which turn anticlockwise to increase. With the present zero position the movement of the pointer will be in the main incompatible with the movement of the handwheel controlling the valve. This difficulty can be overcome to some extent by placing the zero at 12 o'clock so that the pointer will, in the normal working position, be moving from right to left across the bottom of the scale on either side of the 6 o'clock position. This movement could be compatible with an anticlockwise-to-increase movement of the handwheel of the valve. But however it is arranged, when a circular dial is used with a reciprocating control there must always be positions, in starting up from zero and moving to the normal working position, when the movement of the pointer is incompatible with the movement of the control. This is also true to some extent with rotary controls. From this it can be argued that when control movements are involved, the circular dial is less satisfactory than a linear dial. This has been confirmed for making settings by a number of workers, including, for instance, Graham (1952). How important these considerations are is very hard to assess. If there are other pressing reasons why the above suggestions should not be used little harm will probably occur except in an emergency when an incompatible movement between a pointer and a control may cause an operator to make a false movement and so lead to an accident. It is vitally important that all controls and their associated displays in a panel *must* have the same direction-of-movement relationships. This uniformity will often be most easily achieved by modifying the controls, but on other occasions it may be easier to modify the direction of reading of a scale or the position of zero. If this is not done it will be possible to predict with complete certainty that in an emergency the wrong action will be taken. Compatibility is discussed in detail in the next Chapter.

The Choice of Scale Maxima

It is at present normal in industry to see instruments with an immense variety of maxima, the choice of which seems to be determined largely by custom. This wide variety has been somewhat reduced in recent years through the issue of some British Standards specifying restricted ranges of maxima. However, old customs die hard so that the question of how many different scale maxima should be manufactured or used is one of the thorniest in instrument presentation. For instance, a 0 to 15 lb gauge with only four numeral groups including zero is probably one of the least satisfactory scales which conform to the general principles already laid down (see Fig. 72). It is certainly inferior to a 0 to 16 lb gauge which, because it can have five or nine numeral groups, will reduce substantially the chance of attributing the wrong value to the interval being interpolated. However, because atmospheric pressure of $14\frac{1}{2}$ lb/in.[2] has been associated, for many years, with 15 lb vacuum gauges, it seems to be assumed that the 15 lb gauge must always be best for this purpose. It is not unlikely, however, that the reading of vacuum on a 16 lb gauge will be superior to that of a 15 lb gauge since in practice the indication of commercial vacuum is likely to approximate closely to the 14 lb marking on the gauge.

It is clear that certain scale maxima must produce unsatisfactory scales and should be avoided if possible. Equally any reduction in the number of scale maxima will lead to quicker learning and greater accuracy. This is particularly true if choice is confined as far as possible to scales with similar scale divisions. In deciding how many maxima are required in a series of instruments, several previously discussed factors have to be taken into account. In an ideal series there should be:

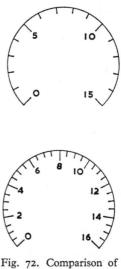

Fig. 72. Comparison of layout of 0-15 and 0-16 scales.

(1) a minimum number of ways in which the scale intervals are divided;

(2) the highest reading tolerance bearing in mind the probable interpolation;

(3) a minimum of different number systems.

The basic ways in which the three intervals 0-1, 0-2, 0-5, can be optimally sub-divided are shown in Fig. 73. In Table 14 these are related to the various m.s.vs which might be used (this is some of the information in Tables 12 and 13 presented in another way).

A series of scale maxima can be chosen in relation to several criteria. If a series is desired in which numbered divisions are sub-divided into halves the following is possible:

40, 50, 60, 80, 400, 500, 600, 800. 1 × 2 in fifths.

16, 20, 160, 200.   2 × 2 in fifths.

10, 100.   2 × 4 in fifths.

This will be called Series A.

A satisfactory feature of this series is that only two basic scale patterns are used (2 × 4 is a variant of 1 × 2 and need be used only if 1 × 2 gives too crowded a scale). Unsatisfactory features are that on the 0-40 and 0-50 unit scales the reading tolerance will probably be 2·15% and 2% respectively; that the reading tolerance on the 0-16 and 0-20 scales will be 0·2 units (this may or may not be a disadvantage according to the view taken) and that there is a gap in values between the 0-20 and 0-40 scales which do not overlap their normal working ranges.

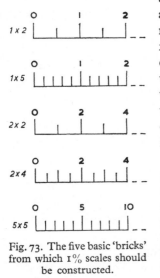

Fig. 73. The five basic 'bricks' from which 1% scales should be constructed.

It will be noted that in this series interpolation into fifths is required. If we accept this as our criterion we can add 0-25 and 0-30 scales from the 5 × 5 pattern. The 0-30 scale, with 30 scale divisions, will have to be half as large again as other members of this series and this would appear to be unacceptable. The 0-25 scale is also not entirely satisfactory; if it is interpolated into fifths it also should, theoretically, be larger. On the other hand the size increase of one fifth (since it is the same as the safety margin already allowed), might be ignored, unless the gauge were to be read at the limit.

In view of this, the 0-25 can be substituted in Series A, for the 0-20 scale to give a series consisting of 0-10, 16, 25, 40, 50, 60, 80, 100, etc. This will be called Series B, which will have three basic patterns.

An alternative series is possible based on subdivision of numbered divisions into five parts (i.e. 1 × 5 and 5 × 5), irrespective of required interpolation; this would give 0-15, 20, 25, 30, 40, 50, 60, which is rather unsatisfactory because 0-80 and 0-100 cannot be included. This third possibility will be called Series C.

Other possible criteria which might be adopted can include required interpolation into halves or quarters, or the use of one basic pattern such as 5 × 5, but none will be so satisfactory as those just given.

Table 14. Scale patterns with expected accuracy of interpolation

1 × 2 Range	%	1 × 5 Range	%	2 × 2 Range	%	2 × 4 Range	%	5 × 5 Range	%
0-10	1·0					0-10	1·0		
								0-15	3·3
								0-15	1·3
				0-16	1·25	0-16	0·6		
				0-16	3·1				
				0-20	1·0			0-20	1·0
				0-20	2·5				
								0-25	2·0
								0-30	1·7
0-40	2·5							0-40	0·5
		0-40	1·25					0-40	1·25
		0-50	1·0					0-50	1·0
0-50	2·0	0-50	2·0						
0-60	1·7	0-60	1·7						
		0-60	0·8						
0-80	1·25	0-80	1·25						
0-100	1·0					0-100	1·0		
								0-150	1·3
				0-160	1·25	0-160	0·6		
				0-200	1·0			0-200	1·0
								0-250	0·8
								0-300	0·71
0-400	2·5	0-400	1·2						
		0-400	1·0						
0-500	2·0	0-500	1·0						
		0-500	0·8						
0-600	1·7	0-600	0·8						
		0-600	0·7						
0-800	1·2								

Scale patterns which appear in Table 12 are shown in heavy type.

This takes care of objective (1) and (2) above. In order to achieve the third objective, the operating characteristics of instruments must be considered. For example, manufacturers of pressure gauges recommend that they should normally operate at not more than $\frac{2}{3}$ m.s.v. for fluctuating values, or $\frac{3}{4}$ m.s.v. for steady values (B.S. 1780:1960). Thus a working range from $\frac{1}{2}$ to $\frac{3}{4}$ m.s.v. can be regarded as reasonable. On this basis a series of m.s.vs can be constructed as shown in Fig. 74.

If we accept a slight lack of overlap, the minimum series of m.s.vs will be 0-10, 16, 25, 40, 60, 100, etc., five basic m.s.vs (Series D). If some overlap is considered to be essential, then we must choose 0-10, 15, 20, 30, 40, 60, 80, 100, etc., which comprises seven basic m.s.vs (Series E).

When all these considerations are combined, it is found that Series D

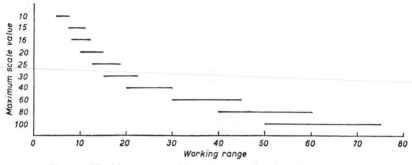

Fig. 74. Working ranges of instruments related to ½ and ¾ m.s.v.

can be drawn from Series B. All the numbered intervals in this series will subdivided into halves with the exception of the 0-25 scale which is sub-divided into fifths. This 0-25 scale is included in order to cover readings between 15 and 20 units; if, however, a 0-20 scale could be allowed to operate up to 17 units and the 0-40 scale down to 17 units then the 0-25 scale could be eliminated and a series drawn from Series A could be used. This would give greater uniformity with two basic scale patterns only and interpolation into fifths would be required.

Obviously there are other solutions besides these which will depend upon the characteristics of the instruments used and of the function being measured. The ideal arrangement is to work in arbitrary units, so that every instrument can be graduated 0-100 but when this cannot be done a range of instruments can be chosen by following the principles illustrated, which for the user could give more efficient dial reading and a smaller range of stocks and, for the manufacturer, a reduced variety.

Finally before ordering a 'special' a designer should be certain that it is really necessary. Just because a tank is full when the depth of liquid is 19 ft 7 $\frac{11}{16}$ in. does not mean that the m.s.v. of an instrument showing the depth should be at this value.

The Design of Scale Marks

The desirable length and thickness of scale marks have been the subject of a number of investigations, the results of which suggest that, within limits, some variation can be accepted without influencing the reading accuracy to any great extent. It seems clear that long marks are undesirable since they give a palisade effect which is confusing to the eye. Most recommendations agree that useful criteria are (i) that the minor marks should be equal in length to half the space between them and (ii) that the major marks should be twice the length of the minor marks. If intermediate marks are used they will fall conveniently half way between the two. The

thickness of the major marks should be between 5 and 10% of the scale spacing and minor marks should be about two-thirds the thickness of major marks. For a scale with a scale base length of 1·0 in. (for each 1·2 ft of reading distance) the minor marks should be about 0·02 in. long and 0·004 in. thick, while the major marks should be about 0·04 in. long and 0·005 in. thick.

No evidence has been published on the advisability of having boundary lines to the scale marks, the so-called 'tramlines', which simplify manufacture of hand marked dials. It must be a matter of opinion whether or not these lines should appear and recent practice appears to suggest that when the double line is no longer required from the manufacturing aspect, a single boundary line should be used. It is sometimes assumed that it is easier to interpolate in an open-ended space than in

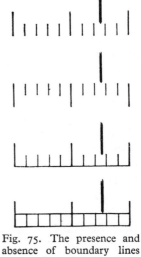

Fig. 75. The presence and absence of boundary lines ('tramlines').

a closed space, but no evidence of this appears to be available.

Whether scale marks and numerals should project inwards from the scale base, or outward in relation to the pointer, appears to be largely a matter for the designer to decide. Churchill and Allen (1955) have suggested that more accurate readings are made when the major marks project away from the pointer, since the distance between the pointer tip and the ends of the marks will then be uniform (Fig. 75). This finding does not, however, entirely agree with other findings that within limits the pointer clearance is not of major importance. An inward facing scale on a circular dial has the advantage that the maximum amount of scale can be wrapped round the dial blank; it has the disadvantage, however, that the numerals are more cramped together and can be obscured by the pointer. Conversely, if an outside scale is used on a circular instrument the dial blank will have to be larger for a given reading distance. This disadvantage may not apply to straight or eccentric scales, since the size of the instrument may not be influenced by the decision. On a straight scale the width of the visible portion of the scale will probably have to be greater if the minimum acceptable length of the pointer is to be used.

The Design of Numerals

The design of numerals should take into consideration two factors, their legibility and their aesthetic quality. While these two need not be mutually exclusive, it is quite likely that numerals which are designed solely to be legible may not always be aesthetically pleasing, so that some compromise may be necessary. A great deal of research has been undertaken on the legibility of both numerals and letters, but their aesthetic quality must clearly be left to individual judgement. The research on legibility covers a wide range of purposes from printed books to highways signs or direction boards on railways as well as for use as numerals on dials. The question to be decided is whether research on numeral shapes for uses other than on dials is relevant. It would seem that, providing the numerals are not intended to be looked at obliquely as were those investigated by Mackworth for use on raid blocks in aircraft control rooms, all the research may be considered to decide on the optimum design of numerals for use on dial faces.

Three characteristics of numbers and letters have been most often studied. These are:

(1) the ratio of the thickness of the stroke to the height of the numeral;
(2) the ratio of the width of the numeral to its height, and
(3) the shape of the numeral.

Stroke Width:Height Ratio. The stroke width of numerals depends on whether the numeral is black on a white background, white on black or transluminated. The results of investigations by a large number of workers

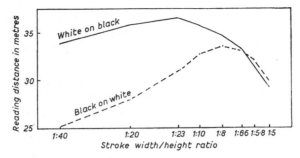

Fig. 76 The effect of stroke thickness on readability of lettering (after Berger, 1944).

are reasonably in accord and suggest that for black on white, the ratio stroke width:height should be between 1:6 (16·5%) and 1:8 (12·5%); (Berger (1944), 8-13%; Bartlett and Mackworth (1950), 14%; Kuntz and Sleight (1950), 20%). If numerals are white on black, a thinner numeral can be used with a ratio lying between 1:10 and 1:20. Berger's results are illustrated in Fig. 76, which shows the distance at which numerals of different thicknesses could be read. When numerals are transluminated, they should be even thinner and figures between 1:10 and 1:40 have been suggested. If the contrast between the numerals and the background is reduced by using a pastel shade or some other colour for the dial background (which neglecting aesthetic considerations is not to be recommended), it appears that the numerals should be rather thicker than they need to be when they are black on white and it has been suggested that the ratio should then be 1:5.

Height:Width Ratio.   The results of research on the ratio height-width of a numeral indicate that a ratio between 2:1 and 0·77:1 is optimal (e.g. Berger (1944), 1·9:1; Bartlett and Mackworth (1950), 1·4:1), most of the research workers favouring a ratio slightly more than 1:1, i.e. with the width greater than the height. There appears to be unanimity that tall thin numerals are less easily read than those whose proportions approximate to 1:1. Brown (1953) found the legibility improved as the width increased towards the 1:1 ratio. This ratio is normally considered to apply to the figure zero. When all the results are taken into consideration it seems that a ratio of about 1·25:1 is likely to be the most satisfactory and practicable.

Numeral Shape.   It should be the object of good numeral design to avoid confusion between one numeral and another. Unfortunately research results are conflicting though it appears to be agreed that Roman numerals whose thickness varies in different parts are more liable to cause confusion than numerals of constant stroke thickness. Whether *serifs* should be used or not is also a matter of opinion. It seems that they might with advantage be used where their presence could form the characteristic feature of a particular numeral. In general the numerals 3, 5, 6, 8 and 9 are likely to be confused when appearing in groups and that less confusion is likely to occur between 1, 7, 2, 0 and 4. Confusion between 3, 5, 6, 8, and 9 might be reduced by varying the ratio between the upper and lower parts of the numerals and in the case of **3** (when, as recommended in B.S. 3693/64, this has a straight top bar) and 5 by having different widths of the top bar. Berger (1944) suggested that the bar on the 5 should go across the whole width of the numeral, but that the bar of the **3** should go only half-way across, but squared numerals such as those proposed by Berger were shown by Atkinson *et al.* (1952) to be less legible than rounded numerals of similar design.

Confusion between numerals has also been studied by Soar (1955) whose results are shown in Table 15.

Table 15. Confusion between numerals

Numerals	Numeral read instead (percentage of occasions)		
	> 31%	30-16%	15-8%
0	—	6	9 and 4
1	—	4	—
2	—	8	6 and 7
3	—	—	—
4	—	—	6
5	6	8	9
6	—	4	—
7	—	—	—
8	—	6	9 and 3
9	—	—	7

In further work Soar (1958) started from these data and designed three new sets in which common elements were minimized and unique elements intensified on the hypothesis that readability may depend on the boldness of the stroke and the openness of the white space within the figure.

It will be realized that a very wide range of numerals could be designed with the idea of keeping confusion to a minimum and in fact quite a large number of different solutions to this problem have been proposed by different workers. Some of these are illustrated in Fig. 77. The British

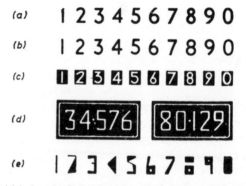

Fig. 77. British Standards Institution's digit designs (*a*) compared with (*b*) Granby Bold, (*c*) Mackworth for oblique viewing, (*d*) Berger for automobile licence plates, (*e*) Landsell for recognition under extreme conditions (from Spencer, Design of visual indicators, Engineering Materials & Design, Sept./Oct. 1961).

Standards Institution who have considered this matter in great detail have proposed the numerals shown in Fig. 78 specifically for instrument dials. These numerals are the copyright of the B.S.I., but the Institution will make copies freely available for reproduction.

All things considered it would seem that it is not of vital importance exactly what numerals are used provided that any proposed numerals are first tested to see if confusion between them is likely to arise.

The Height of Numerals. Reasonable numeral heights for viewing in normal levels of illumination vary between 0·020 in./ft of reading distance

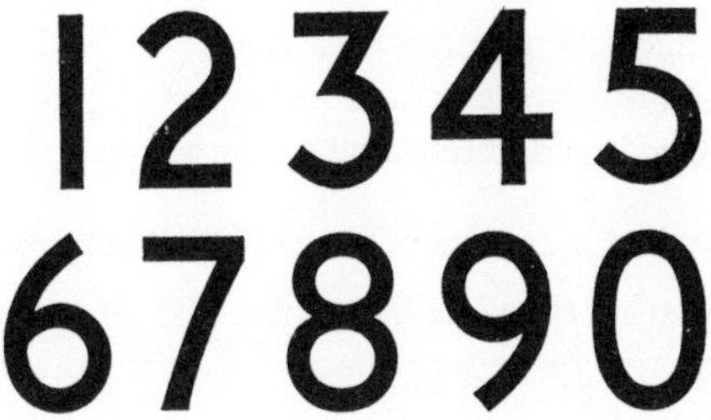

12345
67890

Fig. 78. Instrument digits recommended in B.S. 3693:1964 (by courtesy of the British Standards Institution).

(Smith, 1952), 0·032 in. suggested by Bartlett and Mackworth (1950), and 0·060 in. by Brown (1953). Smith's figure which agrees with that obtained by Forbes and Holmes (1939) was for outdoor signs with daylight illumination, and it is clear that this figure is too low for numerals viewed under artificial light. Murrell *et al.* (1958), using people with normal 6/6 eyesight obtained a figure of 0·031 in. Using individuals with 6/12 uncorrected eyesight, Murrell found that the factor of 1/5 had to be either added to the numeral height or subtracted from the reading distance and therefore it was suggested that a reasonable numeral height should be 0·035 in./ft of the reading distance. If the illumination level is very low, in the region of 0·001 ft/lambert, as might occur for white on black faced instruments associated with transport, the numerals would have to be in the region of 0·16 in./ft of viewing distance (Craik, 1941). What happens between these very low levels of illumination and the levels most normally found in machine shops, about 15 to 20 ft lamberts, is not known, but it seems that to allow for poor illumination or faded dial faces, the size of the numerals could be increased to 0·045 in. if there is room on the dial face.

Vernon (1946) and other workers have suggested that no numeral group should contain more than three numerals and this would appear to agree fairly well with the requirement that a numeral group should be read with one fixation of the eyes. In practice it would seem that the only exception to this rule need be the numeral group 1,000 when it is the maximum scale value. If instruments have a maximum scale value in excess of 1,000 it is better to use a multiplier than to attempt to figure the dial face from say, 0-10,000; in this instance the dial face might be numbered from 0-10 with the legend reading, perhaps, 'thousands of pounds'. Large numeral groups of this kind are likely to be wider than the gap between the beginning and end of a 270° circular scale and the dial face is likely to be so overcrowded that it will be difficult to read.

Naylor (1954) found that there was a significant difference between horizontal and tangential numbers on a curved scale. This agrees with the recommendation of others going back to Sewig in 1936. On the other hand Spencer (1962) found no difference on 95° sector scales.

General Formulae for Calculating 'Commercial' Parameters

To assist in calculating the scale parameters for circumstances other than those outlined in the preceding discussion, the following summary is given:

Let D = reading distance
L = scale base length
n = scale range
c = number of units to which the scale is to be read which must be 1, 2, or 5 (or decimal multiples)
i = number of parts into which a scale division is to be divided by eye
t = tolerance = $\dfrac{100c}{n}\%$
m = number of scale divisions
$c \times i$ = number of units in a scale division.

To find the number of scale divisions, calculate:

$$m = \frac{n}{c \times i}. \tag{10}$$

A satisfactory scale cannot be constructed unless m is a whole number divisible by 2 or 5 when its value is 32 or less, or by 5 when its value is 35 or 40.

Example:
Let $n = 135, c = 1, i = 5$

then
$$m = \frac{135}{1 \times 5} = 27$$

since this value of m is divisible by neither 2 nor 5 it is impossible to construct a satisfactory scale. A 0-140 scale with 28 scale divisions will, however, be satisfactory.

To determine whether a proposed tolerance will produce a satisfactory scale calculate:

$$c = \frac{nt}{100} \tag{11}$$

unless c is 1, 2 or 5 (or decimal multiples) a satisfactory scale cannot be constructed.

Example:

Let $n = 140$, $t = 1\%$

then
$$c = \frac{140 \times 1}{100} = 1 \cdot 4$$

with this value a satisfactory scale cannot be constructed. If the tolerance is 0·715% then $c = 1$; a scale with this tolerance can be constructed.

To find the scale base length for a reading distance D, calculate:

$$L = \frac{0 \cdot 06944 D}{t} \tag{12}$$

Example:

Let $D = 1,000$ cm and $t = 0 \cdot 715$ (for m.s.v. of 140)

$$L = \frac{69 \cdot 44}{0 \cdot 715} = 97 \cdot 12 \text{ cm}$$

To find the maximum reading distance for a given scale base length, calculate:

$$D = \frac{L \times t}{0 \cdot 06944} \tag{13}$$

Example:

A 'standard' 0-100 scale read to the nearest unit has an s.b.l. of 69·44 cm for reading at 1,000 cm. To find reading distance for a 0-140 scale of the same length, calculate:

$$D = \frac{69 \cdot 44 \times 0 \cdot 715}{0 \cdot 06944} = 715 \text{ cm}$$

Thus to read a 0-140 scale to the nearest unit the observer must be closer than when reading a 0-100 scale.

Dimensions of scales related to a reading distance of D units:

Minimum called spacing...............................	0·00069D
Major mark length.................................about	0·00278D
thickness .... ,,	0·00035D
Minor mark length................................ ,,	0·00139D
thickness .... ,,	0·00028D
Numeral height, from....................................	0·00208D
to	0·00313D

Pointer Design

Very little research on pointer shape appears to have been published; it has been summarized by Spencer (1963). Some work was under way in America about 1948 and a progress report suggested that it was advisable to use a thin pointer with a fine taper terminating in a relatively blunt tip whose sides made an angle of approximately 90°. The tip should be approximately the same width as the thickness of the minor scale marks. White *et al.* (1953) tried the effect of enlarging the tails of pointers when used for check reading. Their results were inconclusive. There was a suggestion, however, that fewer 180° reversal errors were made with a pointer of which the tail was suppressed. From this it would follow that if some form of counter-balance is required, it should be the same colour as the dial face, but the remainder of the pointer, from the collet to the tip should be in maximum contrast. Care should be taken to ensure that the pointer is not so thick that it unduly obscures the numerals. Pointers used on linear instruments must possess maximum attention-getting value in spite of very short length, therefore they should be fairly fat at the end away from the scale, thinning down, probably with a convex taper, to a fine thin point.

The length of the pointer should be such that its tip is close to the scale base. Vernon (1946) carried out an experiment to discover the optimum clearance between the scale marks and the pointer tip and came to the conclusion that the tip should be as close as possible to the scale and not more than half an inch away, but at what reading distance is not clear. With a reading distance of 28 in. Churchill (1956) found that reading time and interpolation errors decreased significantly as the pointer clearance was reduced from 2 to 0·125 in. (visual angle 4° 5' to 15'). At zero clearance no further improvement in performance was found. From these results it may perhaps be concluded that providing the pointer tip is not more than a distance equivalent to about 15 minutes of arc away from the scale its position is not critical.

Naylor (1954) found that on curved scales a pointer pivoted at the centre point of the scale was far superior (p < 0·001) to a pointer which is placed on the outside of a scale. This seems to be reasonable, since the pointer can have an attention-getting value in relation to its size and a pointer on the inside of a circle will be larger and longer than one moving round the outside of the scale. On the other hand, Spencer (1963) found no difference at all on 270° scales. Naylor also concluded that with vertical straight scales the pointer should be on the right-hand side of the scale with the numerals on the left, and with horizontal straight scales the pointer should be above the scale with the numerals below.

The Colour of the Scale Marks, the Dial Face and the Bezel

Maximum visual efficiency will be obtained when there is maximum contrast between the dial markings and pointer and the dial face. This usually means that one should be black and the other should be white. Quite a number of experiments have been carried out in order to determine whether black marks on a white background are superior to a reversed arrangement. Most experimenters have been interested primarily in air-craft instruments for night-flying, so that recommendations that dial faces should be black and the numerals and markings should be white are the rule. Kuntz and Sleight (1950) and Naylor (1954) both report that there is no difference at all in visibility with a well-designed scale whether it is black on white or white on black. On the other hand, Papaloïzos (1961) reports that light dial faces were read with significantly less errors (p < 0·01) than were dark faces. In view of this lack of unanimity we must make a decision on the best arrangement on other grounds. As will be pointed out when lighting is discussed, there is reason to believe that visual acuity is impaired when the surrounding to the task is lighter than the task itself (Lythgoe, 1932). From this it follows that for use in normal viewing conditions, the face of a dial should always be white. The suggestion has been made by Papaloïzos (1961) that major marks should be black and minor marks coloured. Using this arrangement his subjects made fewer errors than when the major and minor marks were the same colour (p < 0·05). When dark adaptation is involved, the use of instruments with black faces and white markings is indicated: to avoid sharp contrasts between the dial face and the panel background, instruments of this kind should always be mounted in black or nearly black panels.

The most important feature of a dial is its pointer and to avoid distracting the eye from the pointer, or 'killing' it, instrument cases or bezels should never be black when a white dial face is used, Fig. 52, p. 138. The bezel should be just sufficient to separate the dial from the panel and it should never be highly polished. Self-coloured anodized aluminium or sand-blasted brass are both satisfactory bezel materials. If coding is required the bezel can be anodized in colour to indicate either with which control a gauge is associated or in accordance with a familiar code such as that used for colouring pipes. Codes used in this way might well result in more rapid recognition of the functions of an instrument than would the use of labels.

Multipointer Instruments

The instruments which have been discussed so far have had, by definition, a reading accuracy not greater than 1% of the m.s.v. It has, therefore, been possible to give all the information required with one pointer. In comparatively rare instances it may be necessary to have instruments on which readings of substantially greater accuracy than 1% m.s.v. are required, but which, in addition, must be read 'at a glance'. Since they require both high speed and high accuracy, they fall between the industrial instrument requiring high speed and moderate accuracy and the test instrument which requires moderate speed and high accuracy. Since it is a requirement that readings are made with an accuracy greater than that represented by an interval of 1/100th of a revolution of a pointer, extended values, which are in effect a count of the number of revolutions made by the unit pointer, will normally be given by a second or a third hand. The commonest example of this type of instrument is the clock which is normally read to 1/60th of a complete circle, over a range of 720 minutes. Another common example of a different type is the electricity or gas meter, while the most notorious example is the three-pointer altimeter. Grether (1949) who has conducted one of the very few studies on this type of instrument, has suggested that it is not easy for a person to combine the information from two or more pointers into a single numerical value unless the design of the instrument helps him to do this in the proper order. With most long scale instruments with pointers of three different lengths, this would seem not to be the case. The attention-getting value of the large pointer might make the observer read this value first, whereas to obtain the values in their proper sequence the smallest pointer should be read first. Difficulties with this type of instrument may well stem from familiarity with the clock. In common usage it would seem that 'twenty-past-six' is more often told than 'six-twenty' or 'twenty-to-seven' than 'six-forty' but, be this as it may, there seems to be no doubt that this type of instrument is a source of difficulty when read rapidly, as Grether showed in his experiment. Using 97 experienced Air Force pilots and 79 Naval College students, he compared speed of reading and reading errors on a number of alternative designs for altimeters. With the commonly used three-pointer type he found readings were made in slightly over 7 sec with 11·7% errors of

1,000 ft or more by the experienced pilots and 17·4% by the students. Incorporating the smallest hand in a separate dial in the main face made little difference to the reading time and no difference to the errors for the pilots. The errors for the students were, however, reduced to 13%. Doing away with the third hand entirely, making a two-pointer instrument, the thousands of feet being indicated on an inner scale, the time fell to about 5 sec, the pilot error to 4·8% and the students error to 7·7%, but when the second hand also was done away with and the thousands of feet were given in numerals in a window on the face of the dial, the reading time for both groups of subjects fell to 1·7 sec and the errors to 0·7%. This shows very clearly the adverse effect of adding additional pointers to an instrument. In the same experiment he found, not surprisingly, that giving information by means of a counter led to almost zero errors and minimum reading times. But the altimeter is an instrument on which information may change very rapidly and it is used frequently as a qualitative instrument to judge the rate at which the altitude is changing. This is not a use for which a counter is suitable, so that the single-pointer instrument combined with a counter makes the best compromise for this use if the values are likely to change quickly and if qualitative information about the rate of change is going to be needed. In other instances, counters should, wherever possible, be given preference over multipointer instruments.

Where a pointer is used it would seem that all the foregoing design considerations which apply to 1% instruments would normally apply, and that the single-pointer should not be required to give information more accurately than 1% of a complete revolution.

Difficulties with the gas meter type of indication stem from the fact that pointers showing units, tens, hundreds, etc., revolve in opposite directions if simple gearing is used. This difficulty is so well recognized by the Utility companies that when consumers were asked to read their own meters they were given a card having a pictorial representation of the dial upon it, on which they were asked to mark the position of the pointers and to return the cards to the Utility company for interpretation. The only advantage that this type of presentation has over one having all the pointers on the same dial is that, at least, they can be read in the order in which the numeral is usually spoken, but this advantage should not lead to the use of this type of presentation if a counter can be used instead.

Test Instruments

Test instruments are those which have an intrinsic accuracy greater than 1% and which will normally be read by an observer who has plenty of time to be sure that his reading is correct. They are rarely, if ever, used for other than quantitative purposes. Virtually no research on this type of instrument seems to have been done, so any desirable design features will have to be deduced from first principles.

In the first place, since a snap reading is not required there seems to be no need to call upon the observer to interpolate into five parts, therefore, more markings will be permissible on test instruments than on industrial instruments. In the absence of any evidence to the contrary it may be assumed that adjacent scale marks should not be closer together than the minimum called spacing already determined for a particular reading distance. In practice, however, since scale marks as close together as this might well appear overcrowded, it seems not unreasonable to expect an observer to be able to interpolate a division equivalent to two called divisions with complete accuracy. Consequently, the minimum separation of two scale marks should be 4 minutes of arc. Whether this dimension is suitable will depend on the actual degree of reading accuracy which it is desired to achieve and the available scale lengths.

In order to reduce uncertainty, scale marks should be of three kinds: major, intermediate and minor; the major marks should be numbered and the intermediate marks may with advantage be numbered also. There should be a maximum of five intermediate divisions between each numbered major mark. Provided there is high contrast between the scale marks and the scale background, scale lines may be as fine as 0·005 to 0·003 in. for a reading distance of 18 in. It is important that the scale marks should not be overlong and it is suggested that the minor marks should not be greater in length than the spacing between adjacent pairs of minor marks, with the intermediate marks half as large again, and the major marks twice as long as the minor marks.

To see how these recommendations work out in practice let us assume that we must design a scale from 0 to 500 units to be read to the nearest unit. If interpolation into halves is acceptable, each minor scale interval

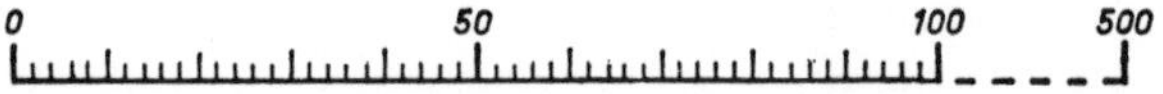

Fig. 79. Layout for a 0-500 test instrument read to $\pm$ 0·1% m.s.v.

will be equal to 2 units. It is convenient then to have intermediate marks every 10 units. Five intermediate marks between each major mark will enable us to number our major marks 0, 50, 100, up to 500. The layout of this scale and the relative proportions of the scale marks are illustrated in Fig. 79. A visual angle of 2'/scale unit gives a scale spacing for two units of 0·025 in. at a reading distance of 1½ ft. This is approximately equal to the distance between the scale marks on a 10-in. slide rule between 3·9 and 4·0. Examination of this interval will show that it is possible to interpolate it into halves with some difficulty at this reading distance, and it would seem that this dimension is an absolute minimum which should be used only when necessary. An interval of 0·040 in., almost 1 mm, can be interpolated with ease, so that the scale spacing should lie somewhere between these two limits at perhaps 0·035 in. In the scale under consideration there will be 250 intervals, so using the minimum dimension its length would be 6·25 in.; if the scale base length is less than this it cannot be read to the nearest 1/500th m.s.v. (0·2%). On the basis of the foregoing, the

Table 16.   Scale base length for different reading accuracies,
reading distance 18 in.

Reading Accuracy	Optimum Length (in inches)	Desirable Minimum (in inches)	Absolute Minimum (in inches)
1/150th or 0·66%	3·0	2·63	1·87
1/200th or 0·5%	4·0	3·5	2·5
1/250th or 0·4%	5·0	4·37	3·12
1/500th or 0·2%	10·0	8·75	6·25
1/1000th or 0·1%	20·0	17·5	12·5

minimum scale base lengths for different reading accuracies are given in Table 16 for a reading distance of 18 in. These dimensions are for reading with the naked eye; if powerful magnifiers are used, finer scales can be provided, but this is outside the scope of this discussion.

Dual Purpose Instruments

Certain instruments, such as pyrometers may be required to be read as 'commercial' instruments from a distance and as 'test' instruments from close to. When this requirement has to be met it would seem that the most satisfactory solution is to provide dual scales with the 'commercial' scale nearest to the 'body' of the pointer which can change its thickness sharply near the inner end of this scale to become a 'knife-edge'; alternatively, the two scales can be combined. By way of example, consider an instrument with a range of 0 to 1,200 which is required to be read to 2 units close to and to 10 units from a distance. Taking the fine scale first, the tolerance will be 0·167% and the optimum scale base length is obtained from the formula,

$$L = \frac{0·111 \times D}{t} \tag{14}$$

which at a reading distance of 18 in. gives a scale base length of 12 in. Shorter scales can be used within the limits shown in Table 16 if thought desirable. If interpolation is into halves, there will be 300 scale divisions, which is probably too many. Alternatively interpolation can be into fifths, which will require 120 scale divisions, a much more manageable number, each being 0·1 in. wide. With one intermediate mark between each five scale divisions and major marks numbered in 100s, a satisfactory scale can be laid out.

The 'commercial' part of the scale is to be read to the nearest 10 units to a tolerance 0·833%, thus from formula (13),

$$\text{Reading distance} = \frac{12 \times 0·833}{0·06944} = 144 \text{ in.}$$

or 12 ft approximately. If this reading distance is insufficient and the instrument must be read from about 35 ft, the scale base length will

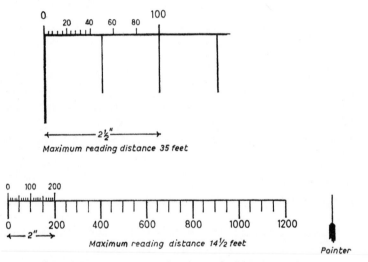

Fig. 80. Suggested layouts for 0-1200 dual purpose scales.

become 30 in. With this length, the fine scale could have the 300 scale divisions each 0·1 in. wide and interpolated into halves. The application of these data to the 0 to 1200 scale is shown in Fig. 80.

The Design of Instruments for Qualitative Reading

A qualitative instrument is required to show the direction or rate of change in a value, and in practice it seems to be extremely unlikely that this information will be given in the absence of quantitative information. There seems to be no doubt that counter-type instruments would be unsuitable for showing a rate of change, and open window type instruments would probably also be unsuitable, but where the full 360° of a scale are visible it seems unlikely that there is any difference between moving pointer instruments and moving scale instruments, in spite of several research recommendations to the contrary. These recommendations have all been based on comparing moving pointer instruments with moving scale instruments of the open window type. Provided the movement is not too fast, it would appear that rate or direction of change can be indicated on a straight horizontal or a vertical scale. In many ships, compass repeaters are of this type and it is not difficult to estimate the rate of change of heading from the movement of the scale. However, in this case the rate is comparatively slow, since a ship cannot make a turn very rapidly.

The direction of movement of the instrument may influence the comprehension of the meaning of the movement and, on this basis, it has been suggested that a vertical scale with a pointer moving upward may be the most satisfactory type. On the other hand, it has been suggested by other workers that when rate and direction of change are indicated by a pointer on a circular instrument, the estimation of the rate of change is made by observing the change in angle of the pointer. When all these points have been taken into consideration it appears that the actual shape of the instrument is comparatively unimportant compared with the need to understand the meaning of a change easily, and this is probably dependent more on the prominence of the index in relation to a well-designed scale.

The Design of Instruments for Comparison

The need to compare the values indicated on one instrument with those indicated on others may not occur very frequently in industry. It is not known whether, when comparisons are made quickly between the indications on two or more dials, the observer will read the exact values in order to compare them or whether he will attempt, on a circular dial for instance, to judge the agreement by observing the angular displacement of the pointers. If exact values are read, Vernon (1946) has shown that readings will be made less rapidly if the dials are graduated in different ways and this suggests that when comparisons have to be made, instruments should all have the same graduation system. It would seem to be unlikely that the need would ever arise to compare the readings of instruments with widely differing scale maxima.

Counters cannot be used when values are changing quickly but clearly there could be an advantage in putting together groups of counters which have to be compared, perhaps one above the other. If ordinary scaled instruments are used it would appear, there being no evidence to the contrary, that the considerations for the design of a quantitative instrument would apply.

The Design of Instruments for Check Reading

Instruments used for check reading will differ from those used for quantitative reading only if they are to be used exclusively for checking and are never likely to be used for setting or for quantitative reading. There is little research on any aspect of check reading instruments other than the shape, and on the arrangement of groups of other than quantitative types of instruments on panels.

On all check reading instruments the desired operating point or range of pointer positions, the 'safe' position, should be clearly marked whether or not it is a scaled instrument (Kurke, 1956). It is unreasonable to expect an operator to memorize the correct readings of a large number of instruments or to be able to notice if an unmarked instrument is deviating from its correct reading or range of readings. It would be easier to see the marking of the correct reading if there were no other markings on the dial, so that in deciding on the type of instrument to be used it is important to be sure that a graduated scale is really necessary. There must be many instances where the exact value of an indication is of no importance to the user; all he would need to know is that the indication is at its correct value, and this can be clearly shown on the dial face with a mark or a band of colour. Custom suggests that a band of green is probably the best with the use of red to indicate a danger zone. Care should be taken to ensure that any other colours which are used as well as the green do not swamp the 'safe' area and it would seem best to make the 'safe' area twice as wide as any other coloured area which might be used. The simple alternative of drawing limits in chinagraph pencil on the instrument glass often serves the purpose perfectly satisfactorily. This is quite a widespread practice in many control rooms which suggests that when the 'check' function is of importance, existing instrument design has been found to be inadequate.

As mentioned above, extensive research has been carried out into the best shape of instruments for check reading and there seems to be no doubt that an instrument with a moving pointer is greatly superior to an instrument with a moving scale, since pointer position is readily perceived. Recommendations which have been made as a result of this research depend to a large extent on whether the instrument has a control associated with it, and if it has, on the direction of the corrective movements of that control. It has been suggested that the shape of the instrument and position of the safe area may be influenced by the notion that an upward movement of the pointer tends to be interpreted as an increase in value. This, however, may be somewhat overrated. The results of experiments on the different sectors of a circular dial appear to depend on whether a control movement is required or not. Grether and Connell (1948) showed that when an up-and-down response was required the most satisfactory results were obtained with the safe area in the 9 o'clock position, that is when an upward movement of the pointer corresponded to an increase in value and was related to an upward movement of the response key. However, Morley and Suffield (1951) have shown that where a verbal response is required there is relatively little difference between the different sectors of a circular dial. From this we can perhaps conclude that all other things being equal, it might possibly be an advantage to choose a dial shape which will denote an increase of value by an upward or a left-to-right movement. This means that if a circular dial is used, it should be arranged so that the pointer when in the normal position is at about 9 o'clock or 12 o'clock. The choice will also be influenced by the type and location of any control which may be associated with the instrument. The two must be compatible and the importance of this probably over-rides any other consideration of shape when choosing a dial for check reading.

Virtually no work appears to have been done on the size of dials for check reading in relation to reading distance or on the size of the markings of these dials. White (1949) when working at a reading distance of 28 in. found that a scale of $1\frac{3}{4}$ in. diameter was the most satisfactory when compared with 1 in. diameter and $2\frac{1}{2}$ in. diameter. This gives 2·35 in. of scale base length to each foot of reading distance, which is more than two and a half times greater than the 0·833 in./ft recommended previously for quantitative instruments. It seems, moreover, that his subjects were reading closely graduated scales and were detecting deviations from fixed values which were not marked as a safe area, so that White's results are of little help and deductions must be made from other evidence. It can be argued that for an individual to be certain that the pointer has left the safe area it is necessary that it shall have deviated by the width equivalent to at least one called division, which for each foot of reading distance is 0·008 in. Thus if it is necessary to take action on a deviation of $\pm$ 1% of m.s.v. of the instrument, the scale base length should be the same as for a quantitative instrument, and it can be read from the same distance for both check and quantitative reading. If, however, the instrument is check read from a distance and read quantitatively from close up only, as when logging, or should the instrument be required for check reading alone, a smaller instrument might be used whose size would depend on the degree of tolerance allowed on the deviation from normal. For example, consider the case of a 200 lb pressure gauge with a 'safe' area extending from 140 to 160 lb from which the value may deviate by up to 5 lb above or below these values before action becomes imperative. If this instrument has to be read at a distance of 12 ft, the called spacing will be 0·10 in., and the scale base length will be $\frac{200}{5} \times$ 0·10 in. This gives a scale base length of 4 in. compared with the 10 in. which would be required for an instrument to be read quantitatively from 12 ft. It will thus be seen that it should be possible to use a 2 in. or 2½ in. nominal sized gauge instead of a 6 in. gauge.

The only indication whether this theoretical treatment is in accordance with actual performance comes from Murrell (1960a) who found that a pointer could quite clearly be seen to have deviated when it had left the 'safe' area by the equivalent of 0·015 in./ft reading distance which is nearly double the figure suggested above. This, however, was the result of observation, not experiment, so it would seem that pending further research it is probable that a deviation can be seen when the pointer has left the 'safe' area by 0·008 in./ft reading distance and that it is certain that it can be seen at 0·015 in.

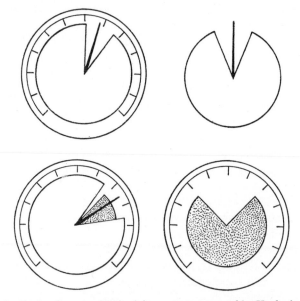

Fig. 81. Design for a special check instrument proposed by Kurke (1956), showing (right) the 'pointer' and dial face and (left) safe and danger conditions. This instrument can also be used for quantitative reading.

Since check reading depends on being able to notice the movement of a pointer, it is clear that pointer design is important and the pointer itself should not be overborne by dark bezels or masked by fussy inscriptions on the dial. It has been argued that the longer pointer on a circular dial will be more easily seen than the shorter pointer on a straight dial. However, this has not been borne out by experiment, since Murrell (1960a) has shown that deviations on vertical straight dials were noticed rather more quickly than those on circular dials, (p < 0·05). Although in this experiment the centres of all instruments were the same distance apart, in practice adjacent scales can be put closer together using vertical straight dials than is possible with circular dials and since scanning seems to take place mainly from left to right, the horizontal distance to be scanned will be smaller and this may well increase the advantage of the straight type of instrument (White, 1949).

An interesting design of a circular instrument for check reading which can also be used for quantitative reading has been suggested by Kurke (1956). He proposed that the pointer should be mounted on the centre line of a segment cut from a disc, the whole assembly rotating on the spindle of the instrument. The face behind the disc would be coloured in such a way that it would be exposed to give a high contrast wedge when a deviation had occurred (Fig. 81). This would, by its position, clearly show whether the reading was too high or too low.

Kurke tested his design against an instrument with no 'safe' indication and an instrument with a red 'safe' area; altogether 1,650 readings were made on each. His adjusted results, summarized in Table 17 clearly demonstrate the superiority of the proposed design on criteria of both accuracy and speed. All differences were significant at p < 0·01.

Table 17

Instrument	Errors	Time
No mark	39	27·8
Red area	18	20·5
Wedge	1	4·3

To sum up, it would seem that when check reading alone is required it is not of vital importance what shape the instrument should be, bearing in mind that if a control movement is also involved, as in check controlling, the movement of the instrument should be compatible with that of the control. There may be some small advantage to be gained by using the up-to-increase relationship or left-to-right across the top on a circular dial. It would seem that when deviations of the order of 1% must be noticed the size of the dial should approximate to that of a quantitative instrument, but if less accuracy is required, instruments could with advantage be smaller. The pointer is the most important part of an instrument for check reading and it must not be 'killed' by dark bezels or other dark objects in the immediate surround.

The Design of Instruments for Setting

It would appear that the design requirements of instruments for setting will follow those of quantitative instruments so long as the shape chosen gives a compatible relationship between index and control. No other special requirements appear to be necessary for this use.

The Design of Indicators and Warning Devices

An indicator gives non-quantitative information about the state of a system when it is under human control; two main types can be distinguished – shutters and pointer-type indicators. There appears to be no research on this type of visual display. In the case of shutters, the information, which may be given either in words such as 'on' or 'off' or by means of a colour such as red or black, should be of such a size as to give the information clearly and unambiguously to the operator. If a pointer-type instrument is used, the words or code which will indicate the state of the system should again be clear and unambiguous and the movement of the pointer in relation to any associated control should be compatible. It may well be that from time to time the function of indicating will be combined with the function of check reading; in this case the instrument would probably have to be larger than it would be for normal check reading, in that the inclusion of words indicating the state of the system in addition to a 'safe' area would require a larger dial blank than would be required for check reading alone. It seems that a pointer with limited travel such as would be found on a top sector scale would be more satisfactory for this purpose than a 270° scale.

Warning devices are required to call the attention of the operator to some action which he has to take in relation to the equipment. This can be done in several ways, the most familiar of which is by means of a shutter or similar device, or a signal light. A common example of the shutter type of display is the bull's eye on a manual telephone exchange but it is unlikely that this type of display has any great 'attention getting' value. Its main purpose is to warn the telephone operator which subscriber is calling at that particular moment. If the operator is not actually looking at the switchboard at the moment when the shutter falls the indication could be missed entirely since the change from dark to light is not very pronounced. This is not true when the contrast between dark and light is very great, such as occurs when a lamp lights, and if the light is flashing the maximum visual warning is obtained. Within limits, the brighter the light, the more superior is a flashing to a non-flashing light. The disadvantage of a light indication, however, is that the absence of light may mean either that all is well or that the electrical circuits have broken down, so that the operator can never be *quite* certain what is the true situation. This can be overcome by the signal light burning dimly all the time and changing to bright flashing when a warning has to be given. Alternatively two lights can be used, one green or amber and one red, the former showing that all is well and the latter flashing to give a warning. Lights which go out to convey a warning should never be used.

The meaning of a warning can be conveyed either by the shape of the light, by its position, by its colour or by transilluminated labels. When seen singly it appears that not more than five different colours can be identified with confidence, and then only if the colours are very carefully selected. Three or four colour systems could be used with confidence. Holmes (1941) has studied the chromaticity tolerances of these lights and concluded that blue and purple are not satisfactory since, unless the blue is very saturated, it can be confused with green and white. Red is a good colour and if no orange warning lights are included its wavelength should not be greater than 610 mμ. If an orange is used the red must be of longer wavelength. The green should have a wavelength of not less than 550 mμ, and yellow should be between 590 and 596 mμ, but it can be confused with white, especially if it is not saturated. If no white is present wavelengths down to 583 mμ can be used.

In 1959, the International Commission of Illumination (C.I.E.) published a report on the colours of light signals which gives recommendations for three, four and five colour signal systems. These are in the form of equations of the boundaries of the recommended colours as well as chromaticity diagrams in both the C.I.E. and R.U.C.S. systems. For a three-colour system, red, green and white/yellow are suggested and when a four-colour system is required the latter are used as distinguishable colours with a bluish-white, provided that they are shown either simultaneously or successively, or one or both are used as secondary signals. If they are to be distinguished separately the white and yellow signals must be large. For five-colour systems blue can be used but 'is to be seen at only relatively short range'. The use of purple is deprecated. Due to the difference in the sensitivity of the eye to different colours, blue and purple, in particular, will have to be from 15 to 20 times brighter than a white light in order to appear equally bright. Most of the earlier work on light signals has been reviewed by Stiles *et al.* (1937).

Bearing in mind that excessively bright lights cause discomfort and glare, a light which is normally on should not be more than three times the brightness of the background. Since warning lights will only flash very rarely, and attention must be gained immediately, glare is unimportant and a brightness of up to 50 times that of the background is quite acceptable. However if they are more than 100 times brighter than the background, they not only cause annoyance but can also impair the ability to see clearly. This effect is particularly noticeable at night with the stop and turning lights of a number of makes of motor cars.

If information is to be given by means of coloured lights, it should be remembered that a proportion of the population is colour blind. This has caused a certain amount of difficulty in America, where traffic lights are not always one above the other as they are in the United Kingdom, and experiments have been carried out with different shaped ribbing on the 'stop' and 'go' lights so that colour blind people can tell which light is on. This solution is unlikely to be possible in industry since most warning lights used on panels would be too small to be coded in this way. It follows, however, that if it is vitally important for the meaning of a code light to be immediately understood, the operator must be tested for colour blindness, particularly since red-green blindness is the most common.

Pictorial Displays

Pictorial displays will generally be used to give qualitative information about a situation; only rarely will they give quantitative information and then with a relatively low degree of accuracy.

In theory, pictorial displays might be taken to include a very wide range of methods of giving information, including not only instruments but tables, graphical presentations and even pictorial panels. With the exception of instruments, the information given by all these displays will be unvarying. In this chapter we are concerned only with presentations which give information about the state of a system and, since a change in a system must also change the presentation, only pictorial instrumental displays will be considered here. This does not mean that the design of other presentations is not important but presentations such as graphs or tables come more into the realm of typography than of ergonomics. Work has been done on many of these presentations, but in general it has been rather specific, and to date it does not seem to be possible to draw any very general conclusions.

A number of papers have been traced which deal with the design of pictorial instrumental displays. Almost without exception, however, they have dealt with instruments which are used in aircraft or which are used in connection with military gun laying or aircraft control. Instruments using this kind of presentation seem to be comparatively rare in industry, and as far as can be ascertained, no research has been done upon them. One type of pictorial instrument which has been studied is the artificial horizon whose vital requirement is that it should give information quickly about the attitude of an aircraft. This information is rather more qualitative than quantitative because the exact attitude within a few degrees is only infrequently required. To generalize from this example it appears to be a principle of pictorial displays that they should give the observer almost instant information about the functions involved and a clear indication of any action that must be taken. Because early designs of the artificial horizon violated these principles, they have been the subject of a number of investigations all of which have led to roughly the same conclusion. It would appear that when the idea of an artificial horizon was first conceived the instrument was thought of as being, as it were, a window to the outside world, therefore the aircraft was kept level and the horizon tilted, a sensation which would be received by the pilot if he were looking out and were not using his instruments. However, it became apparent that the information given by the instrument might cause confusion since the aircraft shown in Fig. 82 is banking to the left. When the aircraft has to be brought back on an even keel, the horizon should be turned anti-clockwise whereas the control movement involved should turn the aircraft clockwise. It was not unknown for pilots to turn their aircraft upside down! Investigations

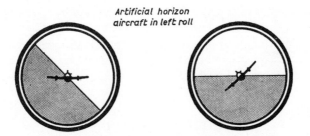

Artificial horizon
aircraft in left roll

Fig. 82. Comparison of incompatible (left) and compatible presentation of aircraft attitude (from Murrell, 1960; by courtesy of the British Productivity Council).

carried out by Browne (1954), Gardner (1954) and Creelman and Miller (1956), to mention just a few, have suggested that the alternative display in which the horizon is horizontal and the aircraft moves is more readily understood and involves a lesser chance of control movement errors than the conventional instrument. This example illustrates the difficulty of following the 'commonsense' approach to instrument design without considering all the factors which may be involved.

In industry a few displays which may be called pictorial are sometimes found. For instance, depth of liquid in a tank is often displayed by a moving column which gives a process operator very rapid information about the relative depths of the liquid in various vessels; these instruments are often graduated, it is true, but it seems doubtful whether the operator makes very much use of these graduations; he is far more likely to glance at the height of the column and say to himself that 'the tank is three-quarters full', rather than to say, 'there are now "x" thousand gallons in the tank'. This example illustrates very clearly the situation where a pictorial display may be used. If information such as 'three-quarters full' or 'increasing' or 'decreasing' is all that an operator really needs to know, then some form of pictorial display may be superior to a numerical display for speed and sureness of interpretation. To emphasize the point, if the quantity of liquid in a vessel is shown in gallons on a dial or a counter, the operator must still remember the total capacity of the vessel in order to arrive at an understanding of the actual depth of liquid in the vessel. Clearly, this is an unnecessary complication if all the operator needs to know is 'nearly full' or 'almost empty'.

To sum up, pictorial displays may be used when exact values of a quantity are not important for the operation of a piece of equipment; and they have the advantage that they may produce the information needed for the proper execution of a task more quickly and less ambiguously than a numerical form of display. The exact form a pictorial display must take depends on the function being displayed, but it is vitally important that any movement of the display must be compatible with any control movement which may have to be made as a result of observing the display. A pictorial display may also relieve the operator of the necessity to remember a series of complex, and for him possibly irrelevant, data.

AUDITORY DISPLAYS

The commonest method of transmitting auditory information from one person to another is speech, but apart from relatively few examples, the use of recorded speech to give warning and other types of information appears to have been little explored. More information might, in certain circumstances, be given in this way than by single tone warning buzzers. Although forms of auditory signals play an important part in everybody's lives, for example, the telephone bell, the motor car horn, the alarm clock, the door bell or the factory hooter, the full potentialities of sound transmission of information have been little developed in the industrial situation beyond conveying simple 'Yes/No' information. The importance of auditory signals is that they provide additional channels for the transmission of information which may be of the greatest use when other channels, particularly the visual channel, are already fully occupied. They also have a great advantage over the visual channel in that they are omni-directional rather than uni-directional. It is this particular feature which makes auditory warnings so valuable.

The simplest use of auditory signals is for warning, and warning signals essentially convey 'Yes/No' information. Thus the warning siren in the control room of a chemical plant tells the operator that one of the functions is out of limits, but he will have to depend upon visual signals to tell him which. These warning signals, therefore, are very simple devices which call an operator's attention to some other source of information.

The next step in the development from simple warning is to convey more information so that the signal now tells the operator not only that there has been a change in the system, but gives him information about either the part of the system which has changed or the nature of the change. Thus in the home, the ringing of a bell, if there is only one, could show that someone is at the front door, or on the telephone, or merely that it is time to get up. If bells with three different characteristics are used, the housewife will know immediately whether she should go to the front door or should answer the telephone.

Information may be conveyed by differences either of pitch or in the temporal distribution of sound. Examples of the latter are the Morse Code, while the telephone system which signals a caller 'dial', 'engaged', 'pay now', or 'number unobtainable' uses both.

The third type of auditory display is one which gives quantitative information in addition to the two previous types of information. A simple example of this is the chiming clock, which tells the hearer the time. Variations in the pitch of a sound can convey information about the extent of misalignment, as when making adjustments to electronic equipment or in Flybar (flying by auditory reference) which was used to give information to the pilot about his airspeed and his physical relationship to the horizon.

Finally, the most comprehensive form of auditory device could be the talking equipment already referred to.

Warning Devices

To be effective the sound intensity of a warning device should be at least 10 dB above the background noise. It has been suggested that intermittent sound is more attention-getting than continuous sound. For this reason bells have been found to be better than horns. Sounds with a frequency corresponding to the maximum sensitivity of the ear should preferably be used, and Houston and Walker (1949) recommended that the frequency should be about 3,400 c/s for maximum attention-getting value. Sounds of very low frequency appear to be more readily confused with noise. In this connection it must be remembered that maximum hearing impairment is usually found to be in the 2,400 to 4,800 octave, so if warning signals within this frequency range are used in situations where there is such a high noise level that hearing may have been impaired, it will be essential to have a louder warning signal in relation to the noise level, and preferably the signalled frequencies should in these circumstances be below 2,400 c/s. These general considerations probably apply also to signals giving alternative information.

Alternative Auditory Warnings

The effectiveness with which alternative items of information can be conveyed by auditory means will depend very largely on whether the alternatives are presented singly, simultaneously, or in sequence. When single pure tones are presented for absolute judgement, the average observer can identify about nine with certainty. If, however, multi-tonal signals are presented, 50 or more can be identified with a fair degree of certainty (Pollack, 1952). If pure tones are presented in successive pairs, the number which can be discriminated probably runs into thousands. In practice this type of discrimination is less likely to be used in industry than the identification of a sound which may be single or multi-tonal.

When sounds have different temporal distributions to provide a series of signals, the number of signals which can be identified will be still further increased. For instance, if the four sounds now in use in the telephone system were replaced by continuous sounds (as is the 'number unobtainable' sound) with the four messages being conveyed purely by pitch, the average telephone user would probably not know whether the bell of the called subscriber was ringing or whether the number was engaged, so that the present system which uses temporal and pitch differences gives more information than either used alone. The information capacity of temporally spaced sounds is well illustrated by the Morse Code.

There are so many combinations of pitch and temporal arrangement that no definite recommendations can be made and it would seem that in order to decide whether two or more proposed signals can be identified with certainty, an experiment would have to be carried out. It can, however, be said that, when two signals differ only in their temporal arrangement as in the Morse code, if identification is to be made with certainty, there should be a reasonably high signal to noise ratio of up to 30 dB or more.

Quantitative Information

The information given by a chiming clock or by the talking clock on the telephone is quite obvious, straightforward and unambiguous, and it is unlikely that the presentation would be much improved as a result of research. Its great advantage lies in the directness of the presentation of the information which has to be conveyed. The same principle would seem to apply when less obvious treatments are under consideration. Perhaps the most thorough study of this kind of auditory display is that of Forbes (1946). His problem was to find the best method of informing a pilot flying blind about the course, speed, and angle of bank of his aircraft. The research information was incorporated in the Flybar System. A number of different types of signal were tested and the one which was found to be most satisfactory combined all three indications in the same signal, which avoided the possibility of one particular signal holding the attention of the pilot to the exclusion of the others. For course indication, a steady signal in both ears told the pilot that he was on course; if a deviation from this course developed, the signal was made to move in the direction in which the aircraft was turning – sweeping, as it were, from left to right or from right to left. At the same time, bank was indicated by varying the pitch of the signal differentially to the two ears, thereby 'tilting' the signal to correspond with the degree of bank associated with a particular change of course. The airspeed was indicated by a frequency modulation super-imposed on the basic signal frequency to give a beat which was compared with a standard which was repeated at regular intervals, this standard corresponding to a predetermined airspeed. Variations in airspeed from this standard were indicated by increases or decreases in the rate at which the beat was given. It was suggested that this combination of signals 'sounded like the behaviour of the plane' and was very nearly self-explanatory. Even naïve subjects were able to control a link-trainer just as well as with visual instruments in the task of flying a straight course. This particular application is very specialized but it illustrates fairly clearly the general principles which have to be followed if this type of auditory display is to find use in manufacturing industry.

TACTUAL DISPLAYS

Finally, tactual displays should be mentioned. Giving an indication about the state of a system by means of touch is virtually non-existent in industry. It should be noted, however, that touch does play a part in the identification of controls.

CHAPTER TEN

Design Factors IV. Compatibility

THE DIRECTION OF MOVEMENT OF CONTROLS

Most people when switching on an unknown radio set with the controls arranged as in Fig. 83, will turn the knob in a clockwise direction, and if it is desired to move the pointer on the tuning scale from left to right will turn the tuning knob clockwise. We do not need to be told when we get into an unfamiliar motor car that we have to turn the wheel clockwise in order to make the car turn to the right, nor if we go into a strange room that we have to operate the electric light switch downwards in order to

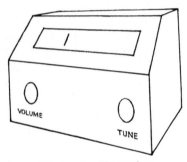

Fig. 83. A radio receiver.

switch the light on – if we are in the U.K. These are examples of what is known as the natural or expected direction of movement of a control. It will be realized that the switch knob on the radio and the electric light switch belong to one type of control, whereas the tuning knob and the steering wheel on a car belong to another. In the former, the movement of the control produces no corresponding movement in any part of the equipment. Switching on the radio starts the set working, while switching on the light causes a change from darkness to light, but there is no other effect visible. On the other hand, in the second group the turning of the tuning knob produces a physical movement of the marker on the tuning scale, and turning the wheel of the motor car produces a visible translation of the vehicle. There are thus two categories of control movement, one which produces an expected effect and the other which produces an expected movement in its associated display. Directions of movement which are expected by the majority of the population are often called *population stereotypes* and control movements which conform to these stereotypes are said to be *compatible*. Much research has been carried out both on population stereotypes and on compatibility, but the greater part of it has been on the relationship between controls and displays; in particular with rotary controls and circular dials. For this reason it is convenient to treat the two categories of stereotype separately, even though they may have some factors in common.

A question which is regularly raised in discussion on population stereotypes is whether these can be considered as being learned patterns of behaviour or whether some of them are inherent. In some instances the answer is quite obvious, for example, the direction of movement of an electric light toggle switch in the U.K. is down for 'on' and in the United States is up for 'on'. In other instances the stereotype might be inferred from the anatomy of the limb used for making the control movement. Sometimes it is difficult to conceive a different relationship being used. It would seem unthinkable for any person, however primitive, inexperienced or naïve, to expect the steering wheel of a car to turn anti-clockwise to turn the car to the right. The importance of determining whether a stereotype is natural or learned may perhaps be overlooked; there may be a conflict between a learned pattern and one which may be natural, and this can lead to difficulty or even disaster. It is well known that the movement of steam valves or of ordinary domestic taps is anti-clockwise to increase flow. If clockwise-to-increase is a learned stereotype, the anti-clockwise-to-increase movement can also be learned, and no harm is likely to come in an emergency whichever system is used, but if the former is natural and the latter learned, in emergency people may revert to the clockwise-to-increase stereotype. There is evidence that this will in fact occur.

Most experimental evidence on population stereotypes has been directed towards discovering what and how strong these stereotypes are. There appears to be relatively little evidence whether these stereotypes are learned or natural. One such study has, however, been reported by Cook and Shephard (1958), who tested child subjects with three different control-display relationships on the Toronto Complex Co-ordinator. On this apparatus the movement of a vertical joy stick controlled a circular marker which had to be moved to encircle a target: successfully holding the target for a short period of time was described as a match. A number of other scores were possible and one of these is the number of occasions on which the control was moved in the wrong direction when starting towards a fresh target. The compatible arrangement on this task was 'away' for 'up' and 'left-right' for 'left-right'. While the latter movement may have entered into the experience of young children, it seems highly unlikely that many children would have experienced the former movement. Alternative arrangements tried were with the movement of the joystick turned clockwise through 90° (3 o'clock position) and through 180° (6 o'clock position), the latter arrangement being directly opposite to the compatible arrangement. Three groups of children of both sexes were tested, of ages 5, 10 and 20 years. The experimenters found that the boys were on the whole better than the girls on the number of matches made in a given time, but this difference was less marked in the 5-year-olds with the compatible arrangement. Considering the 5-year-olds only, there were approximately four times more matches made on the compatible arrangement than on the six o'clock arrangement. There was a substantial improvement in performance with practice at this age, but the relative superiority of the compatible arrangement remained comparatively unchanged. When we consider the initial errors in the wrong direction, which can be a very good index of the strength of compatibility, the control was moved in the wrong direction in about $4\frac{1}{2}$ out of 10 matches with the compatible arrangement, compared with $6\frac{1}{2}$ out of 10 with the incompatible arrangement. With increasing age, errors of this type decreased substantially to only $1\frac{1}{2}$ in 10 at age 20, which may perhaps indicate that greater experience in the manipulation of controls of this type may make it easier to learn an incompatible arrangement.

The evidence of this one piece of research may perhaps be taken as indicating that even in young children of the age of 5 years who, it would seem, are unlikely to have experience in the particular control movement tested, there is a stronger tendency to operate the control in the compatible direction than in the incompatible direction. In the absence of other research of this kind we have to look elsewhere for evidence that inherent patterns exist and that learned patterns may be superimposed on them. Some evidence comes from laboratory studies, but probably the best evidence is from accidents; unfortunately these are rarely reported in sufficient detail to enable the contribution of compatibility to be assessed.

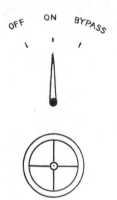

Fig. 84. Control for a lubricating oil cooler.

One accident of this kind came to the author's notice some years ago. A lubricating oil cooler in a ship was controlled by a valve which turned anti-clockwise to open, above which was an indicator with three positions (see Fig. 84). It will be noticed that as the control valve turns anti-clockwise, the indicator moves clockwise (from left to right) from the closed position to the open position, and that beyond the open position is a third position, 'bypass'. The valve was set in the second position whilst work was being carried out on the cooler.

When an emergency suddenly arose, someone yelled to a sailor who was standing near, to 'bypass the filter', and seeing that he had to move the pointer from the centre position to the right, he turned the valve clockwise, thereby closing it, with the result that the cooler was seriously damaged. This accident suggests that, in spite of the learned habit of turning the handwheel clockwise-to-close, in an emergency the sailor reverted to the stereotype of clockwise-to-open even although this movement was contrary to a pattern which he should have learned.

WHY COMPATIBILITY IS IMPORTANT

The accident reported in the preceding paragraph illustrates one reason why compatibility is important. But avoidance of accidents is by no means the whole story.

Let us consider the hypothetical case of two matched naïve men who are to be taught to drive cars, one with normal steering (N) and the other with reversed steering (R). Both will learn to drive, but R will take much longer to reach a standard of proficiency equal to that of N. Imagine now that both men have nagging wives and that both take these wives for a Sunday afternoon drive. The 'back-seat tactics' will have relatively little effect on N, but the driving performance of R will become rather bad. A child runs across the road: N takes the correct avoiding action but R reverts to the population stereotype, turns the wheel the wrong way and kills the child.

Although this example is hypothetical, the effects described have been found in a number of laboratory experiments (e.g. Taylor and Garvey, 1959) from which the following general conclusions can be drawn.

(a) Learning time for the operation of equipment on which controls are compatible will be much shorter than if the controls are incompatible.

(b) There will be a much greater risk of accident on a machine when incompatible controls are used, since in an emergency the operator will tend to revert to the expected direction of movement, however well the incompatible control movement has been learned.

(c) Performance on equipment with incompatible control movements will deteriorate when any form of stress, even one as simple as dealing with mental arithmetic during operations, is put upon the operator. It can be confidently expected, therefore, that the output of machines on which the controls are incompatible will be inferior to that of machines with compatible controls.

(d) All controls on a given piece of equipment should be consistent in their effect. If complete compatibility cannot be achieved, it is better to have all the controls incompatible than to have some compatible and some incompatible.

(e) All controls used for a given purpose, for instance for starting up or closing down a piece of equipment, should move in the same direction. This is especially important if the plant is liable to have to be closed down in an emergency.

(f) As far as possible the same compatible control movements should be used on all pieces of equipment, especially when an operator is likely to move from one piece of equipment to another.

(g) If in the initial design stages of a piece of equipment, it becomes doubtful whether a control movement does conform to a strong population stereotype, it is worth while to redesign the layout of the equipment in order to place the control in a position for which a strong stereotype has been established. This advantage will result from points set out, particularly in (a), (b) and (c) above.

(h) Where controls are associated with dials in a panel, the dials and controls should be laid out with the same spatial relationship, e.g., the left-hand dial of a row should be operated by the left hand control and so on.

(i) All the above effects, in so far as they influence performance, will become more marked as operators become older.

THE EXPECTED EFFECT OF A CONTROL MOVEMENT

When an expected relationship is likely to have been learned, it is necessary to consider the population which is expected to use the equipment being designed. Thus the expected movement of an electric light switch for an American is 'up' for on, while that for a Briton is the exact opposite, so that equipment going to America should have suitable directions of movement. On the other hand, clockwise for 'on' or 'increase' appears to be much more universal, and this might be deduced on anatomical grounds because it is easier to turn the right wrist clockwise from the neutral rest position, than it is to turn it anti-clockwise, moreover slightly more force can be exerted at the end of a clockwise movement. This could also explain the tendency to turn door knobs clockwise with the right hand and anti-clockwise with the left. An example of the effect of this stereotype came to the author's notice when he was carrying out a high speed photographic study of the operation of an aircraft crash tender. This vehicle had two doors at the back covering lockers which contained the hoses through which the foam would be played on the burning aircraft. The tests were carried out under realistic conditions with dummy pilots in actual burning aircraft. When the films were analysed, it was noticed that it took nearly 10 sec longer to open the left-hand locker than the right. Closer examination revealed that the man on the left was having difficulty with the latch type fastening which was on the right edge of the door and which turned anti-clockwise-to-open. This man knew perfectly well that the latch turned anti-clockwise, but in the excitement he reverted to the population stereotype and attempted to turn the handle clockwise, with a consequent loss of time which would have been very serious had a pilot's life really been at stake.

There appears to be a fairly strong stereotype of clockwise for 'on' with a rotary control, and 'up' or 'away' for increase with a lever type of control, see Fig. 85, but on the whole, relatively little research has been carried out on this type of stereotype. Some stereotypes which may be required for specific pieces of equipment may be determined as a matter of common observation, but in other instances it may be necessary to mount a simple

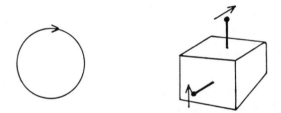

Fig. 85. Expected control movements for 'on' or 'increase'.

experiment in order to decide just what a particular stereotype may be, if in fact it exists at all. As an example of this type of investigation, a pilot experiment carried out by Kay (1958) during a course in Ergonomics conducted by the author may be cited.

The problem was to determine whether there was a population stereotype for vertical handles used for opening doors of cabinets containing electrical equipment. The hypothesis tested was that the expected direction of turning would be away from the edge of the door. Four arrangements were tested, comprising two doors hinged on the left with the handles pointing down and up (LD and LU) and two on the right with similar handle arrangements (RD and RU). Thirty subjects (two of whom were left-handed) were asked after briefing to open the doors, and the direction of movement and hand used were noted. The results are given in Fig. 86. All four arrangements confirmed the hypothesis when tested by χ^2; LD at $p > 0.01$ and the other three at $p > 0.001$. There was no significant difference between the least and the most favourable conditions, LD and RU, but in spite of this, inspection of the results does suggest that it might be better to use the RU arrangement. With LD more of the subjects used their left than their right hands, 18/12, and on LU the proportion is 14/16. Thus on neither of these arrangements is there unanimity on which hand to use, such as there is on RD and RU, when all subjects

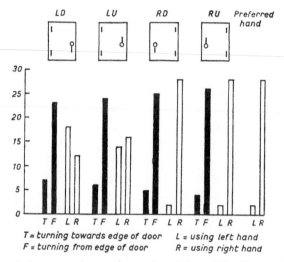

Fig. 86. Stereotype for opening cupboard doors (after Kay, 1958).

used their preferred hands. It is interesting to note that in turning inward on LU or RU the handle behaves as if it were a latch, but not on LD and RD. This point does not seem to have bothered the subjects, who appeared quite oblivious of the mechanical implication of the movements they were making. The importance of this kind of simple experiment is shown by the apparent inconsistency between the findings given above and the experience quoted on the aircraft crash tender. The two handles were, however, different: on the crash tender the handle consisted of a horizontal bar protruding on either side of the pivot and was normally gripped with the shaft between the second and third fingers. A moment's trial will show that with the palm horizontal it is not easy to rotate the wrist anti-clockwise.

CONTROL-DISPLAY RELATIONSHIP

We shall now consider the direction of movement of a control in relation to the effect which it is going to produce in a display, which may be the movement of parts of a machine, for instance the bed of a milling machine moving from left to right, the movement of a tracked vehicle when levers are operated to turn it, the turning of a knob to alter the setting on a dial, or one of many other control movements. It might, at first, be thought that many of these movements are so obvious that no research is necessary, but to judge by many pieces of equipment on which incompatible movements are found, it is clear that designers have given very little thought to the importance of having the control movements compatible with the effects which they produce. For instance, an engineering journal gave an example of a hydraulic press which was wrecked because the lever which operated it moved 'down' to raise the press and 'up' to lower it. In an emergency, the operator, wishing to raise the press, pulled the lever up, causing the platten to move down, and the press was wrecked. It is not impossible that this particular control movement was adopted in order to avoid the possibility of an accident in the event of some unauthorized individual inadvertently leaning on the operating lever and thereby bringing the press down, when the operator might have been working under the platten. This so-called safety device was in fact a trap, which caused an accident because of an imperfect understanding of what was likely to occur in an emergency.

It may sometimes happen that unsatisfactory control-display relationships have become established either for mechanical reasons or because of misconceptions in early stages of development. An example of the former is the movement of the rudder bar in an aircraft and of the latter, is the artificial horizon which was mentioned in the previous chapter. The presentation originally adopted was that in which the aircraft was kept level and the horizon was tilted. Thus if he is in a left roll, as illustrated in Fig. 82, p. 211, the pilot will, in effect, have to turn his aircraft clockwise whilst the instrument will be turning anti-clockwise. This possible incompatibility led Loucks (1945) to carry out an experiment on 555 subjects in a Link Trainer, using four alternatives to the standard type. The findings of this experiment were that the performance on the instrument in which the horizon remained horizontal and the aircraft tilted, was significantly superior to performance on other types. Loucks records also, that this particular type was consistently preferred by a large majority of the subjects. A finding of this kind can present a difficult administrative decision. On the one hand, there is the demonstrated superiority of a new type of instrument over the old, and on the other hand there will be a very large number of trained pilots accustomed to the old presentation. If the new presentation is adopted, will the learned pattern be too strong, or will the change to a display which is more compatible be rapidly learned, and will there be no tendency to revert, in an emergency, to the learned pattern? Unfortunately, there is very little experimental evidence on this point, and *a priori* it would seem that whether a change of this type should be made must depend on the strength of the population stereotype which it is proposed to adopt.

When it was first realized that the relationship between display and control might play an important part in the efficiency of operating equipment is not known, but interest in this subject certainly arose during World War II. A pioneering study was carried out by Warwick (1947a) who used five panels or boxes which are illustrated in Fig. 87. The display consisted in each case of a row of small lights and the control was a rotary knob. Turning the knob caused a light signal to 'move' by the lighting of the lamps in sequence. Fifty subjects were required to move the light to

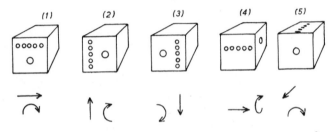

Fig. 87. Arrangements of control and display with strongest preferred relationship (after Warwick, 1947a).

the centre from various positions to right or left of centre using the control knob. This result would be achieved whichever way the knob was turned and the experimenter recorded whether the initial movement was made in a clockwise or counter-clockwise direction. With arrangements (1), (2) and (3), in which the control knob is in the same plane as the display, between 70 and 95% of the subjects moved the control clockwise to produce a left-right movement in arrangement (1), an up movement in arrangement (2) and a down movement in arrangement (3). However, when the control and the display were in different planes as in arrangements (4) and (5) the preferences shown by the subjects were nothing like so clear cut, and there appeared to be no general expectation of a particular direction of movement relationship. Warwick, therefore, conducted a second experiment using a further 42 subjects in which the relationship between the control movement and the movement of the light was fixed. The experiment was carried out under the conditions which would enable him to determine which arrangement would produce the greatest speed with the smallest number of errors. For arrangement (4), this condition was satisfied when a clock-wise movement of the knob produced a right to left movement of the light, and for arrangement (5), when the clockwise movement of the control would move the lights towards the subject. It was, however, pointed out that the differences between this arrangement and the alternative arrange-ments were so slight that they would have very little practical effect on efficiency. From this experiment, Warwick was able to lay down a general principle that the greatest degree of compatibility is achieved with a rotary control when the part of the control nearest to the index of a display moves in the same direction as does the index.

Since Warwick's time a great deal of research has been carried out by a number of investigators into the relationship between displays and con-trols. The extent of this work can be gauged by a bibliography produced as long ago as 1953 by Andreas in which he quotes over a hundred references to papers; a more recent review has been made by Loveless (1962). A large part of this work has been on studies of rotary controls associated with an index on either circular or straight scales. The use of joysticks and levers for different types of tracking has also been studied, but the use of levers for making settings on graduated scales has received relatively little attention.

All this research seems to confirm the original findings of Warwick, but there are some reservations. First, for there to be a 'natural expectation' it should appear to the operator that there is a continuity between the plane of movement of the control and the direction of movement of the display. Secondly, there may appear to be predictable expectations when the above condition is not satisfied, but these expectations may not be the same for all operators and in some instances the same operator may have different expectations on different occasions. Thirdly, there may be a conflict between the expected direction of movement of a control without a display and the same control movement when it is associated with a display. For instance, it may be that a lever placed to the side of the body may be expected to be pushed down to increase speed, whereas if the same lever is placed in front of the body and is associated with a vertical scale which moves upward to show an increase in speed, then the expected direction of movement of the lever may be upward. Fourthly, expectations which might be predicted on theoretical grounds may not be found in practice due to some peculiarities about the design of a particular machine or because of habits previously acquired by a particular population of in-dividuals. Fifthly, an expectation may depend on whether the movement produced is translatory (i.e., left to right) or rotary. A difficulty here will immediately be recognized; it is not easy to say at what degree of curvature the direction of movement of a pointer on a dial ceases to become rotary and becomes translatory. This point, which proves to be of some importance when dealing with valves and their associated gauges, was studied by Loveless (1956) to determine whether there was a tendency to regard the movement of a pointer as rotary or translatory and whether there was any inherent difference between reading the top or the bottom halves of a circular scale. The task used was compensatory tracking, with a display having a scale of 2·75 in. diameter and a control immediately below it of 2·5 in. diameter. He found in his first experiment that the subjects tended to respond to the rotary sense of the movement; that the clockwise control movement is expected to move the pointer to the right, and that there is a tendency for performance at the top of the scale to be inherently superior to that at the bottom. The second experiments were on the left- and right-hand quadrants of the circular dial, and he found that the rotary tendency was again dominant, the clockwise control movement being expected to move the pointer clockwise with a secondary tendency to respond in terms of up and down; i.e., a clockwise control movement was expected to move the pointer up. Summing up, he says that the general tendency is for performance to improve as the pointer is moved from the bottom to the top of the scale; that there is an expectation that a clockwise control movement will result in a clockwise pointer movement and an expectation that a clockwise control movement will move the pointer to the right or upwards according to the target used. A third experiment was conducted using a vertical straight scale and a horizontal straight scale. Clockwise-for-right shows the strongest tendency with clockwise-for-up as a second strong tendency, performance on the former being superior to that on the latter. The three best scales are said to be a horizontal straight scale with a clockwise control movement sending the pointer to the right, a circular scale, pointer at the top, with a clockwise-clockwise movement and a circular scale, pointer at the left with a clockwise-clockwise movement. With the pointer at the bottom of a circular scale, a clockwise-clockwise movement gave an error score of 168·9 and a clockwise-anti-clockwise movement a score of 175·3. The difference was not significant at 0·05% level.

The results of research on direction-of-motion stereotypes (for details see Loveless, 1962) indicate that there are a number of fairly well defined expectations which may be used in design and these are shown in Fig. 88.

It may be useful to consider some of the results of experiments in which the effects of expected and unexpected direction of movement on performance have been compared. Because compatibility is of such importance in the military situation, most of the useful comparisons have been done on tracking or on target acquisition. But even though these activities are not at first sight very relevant to industry, it is worth while considering them because the principles which they illustrate can be of general application. In pursuit tracking, the subject has to follow a moving index with an index under his control; this may take place in one or two dimensions. With target acquisition, the subject has to move a marker under his

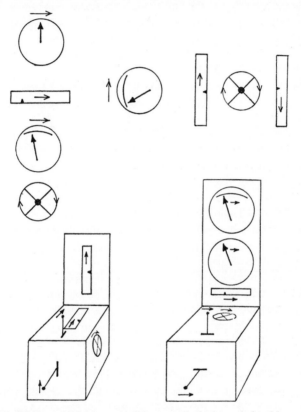

Fig. 88. Well established compatible control-display relationships.

control until it lies over a target. This usually takes place in two dimensions, the control used being a joystick. Two control systems with different dynamics have usually been compared; positional control and velocity control. In positional control the marker and the joystick always remain in the same spacial relationship to each other, that is to say, a left to right movement of the joystick will give a left to right movement of the marker and vice versa. In velocity control, the speed of movement of the marker depends on the extent to which the joystick is displaced from its central position, so that if a movement of the marker is being made from left to right, the initial movement of the joystick will be towards the right whilst the marker is accelerating, the joystick will be stationary whilst the marker is moving at a constant speed and it will be moved from right to left whilst the marker is decelerating. Thus in positional control the relationship between the joystick and the index will be compatible for the whole of the time but with velocity control the movement between the joystick and the index will be compatible only while the index is accelerating, when the index is decelerating the movement will be contrary to expectation. On this basis performance could be expected to be superior with the positional control and this has been confirmed by a number of experimenters. For instance, Andreas *et al.* (1955) have shown that for target acquisition with small tolerances, positional control took 4·0 sec compared with 5·3 sec for velocity control. When the tolerances allowed are somewhat larger, the relevant figures were 2·6 and 3·9 sec. This was after eight trials, during which the improvement in the positional control was quite small, while some improvement was shown in the velocity control.

This learning effect, particularly with velocity control, has been more fully studied by Taylor and Garvey (1959) in experiments which have already been quoted. It will be remembered that they trained subjects in tracking, using positional and velocity control. As expected, they found that the initial performance with velocity control was substantially inferior to that with positional control and that both improved as practice proceeded. After ten sessions, performance with the positional control levelled off, but it wasn't until after 17 sessions that performance on velocity control reached the same level. These learning curves are shown in Fig. 42, p. 122. The experimenters then introduced 'stress', additional tasks which had to be carried out at the same time as the tracking. These tasks included watching for and responding to a signal light, answering questions in mental arithmetic and so on. It was found that whilst performance deteriorated with both types of control, performance with velocity control deteriorated substantially more than did that with positional control.

Other confirmatory work was done by Simon (1951), who likewise found that a sufficiently long period of training would produce a satisfactory performance with a non-preferred relationship. But under conditions of stress an operator is liable to revert to a preferred relationship and Mitchell and Vince (1951) suggested that the effect of an expected direction of movement relationship became steadily less important as the subject became more practised or as the task became easier. Vince (1950), also found that subjects working with expected relationships performed better under 'stress' conditions such as forced pacing of performance, startling distractions and additional load on the operator. She also pointed out that individual differences in coping with these additional stresses appeared to be linked with the intelligence of the subject. The higher the intelligence, the better the performance under the stress conditions. This result was more marked when the display control relationships were unexpected.

THE PREFERRED HAND

Most of the experiments quoted appear to have been carried out using the right hand, though whether the right hand was the preferred hand of all the subjects tested is not always clear. There is relatively little evidence from research on whether the same relationships are expected when the left hand is used, whether this is the preferred hand or not. Such evidence as there is tends to suggest that, provided a strong stereotype is involved, the expected relationship between display and control will be the same whichever hand is used since the relationship involves the physical movement of two parts.

This view is supported by Simon *et al.* (1954) who used a task in which a target moved round the periphery of a circle and the following pointer was controlled by a handwheel; both direct and aided tracking were used. No significant differences in tracking accuracy were found between left- and right-handed individuals. Furthermore the accuracy of tracking with the non-preferred hand was not significantly different from the accuracy obtained with the preferred hand in either of the two groups of individuals.

Suddon and Link (1959) used the Toronto Complex Co-ordinator to test 15 right-handed and 15 left-handed subjects, using the non-preferred hand. The performance of the left-handed subjects was better than that of the right-handed subjects. They suggested that this may be because left-handed people have to live in a predominantly right-handed world and can more readily adapt themselves to working with the non-preferred hand.

If arguments on the anatomical reasons for the clockwise-to-increase stereotype are correct, a left-handed person using his left hand would naturally expect an anti-clockwise-to-increase movement. Experimental evidence seems to suggest that in fact left-handed people do not have this expectation. Moreover, the hand used will depend to some extent on the orientation of the control to the body and this cannot by any means always be predicted. Thus controls on the left of a person may well be operated with the left hand of a right-handed person and vice versa. It is clear that further experimentation is required on this subject, but in the absence of further evidence to the contrary, it seems to be a fairly good rule that designers should use the same expected relationships for both hands.

CONSISTENCY IN THE DIRECTION OF MOVEMENT OF CONTROLS

The controls on most electrical instruments turn clockwise to increase. On the other hand, valves turn anti-clockwise to open or increase. The instruments with which these controls will usually be associated will almost invariably move in a clockwise direction to increase if they are circular or eccentric. Thus if there are electrical and pressure controls mounted in the same panel, there will be a compatible clockwise-clockwise arrangement with one but not with the other. The effect of this kind of arrangement has also been studied by Warwick (1947b) who found that if a compatible and incompatible display-control arrangement were alternated the subject operated the controls far less efficiently than when both display-control arrangements are compatible, that is the most satisfactory results were obtained when the relationships were consistent. From this it follows that it can be more important to get a consistent relationship between a number of displays and their relevant controls, than to accept a compromise where some but not all of the relationships are expected. From this it follows that if for some reason it is quite impossible to get an expected relationship between one or more of the displays and their associated controls, it would be better to have all displays and controls incompatible rather than to have an inconsistent relationship between them.

To achieve a compatible arrangement when electrical and pressure controls are mounted in the same panel, those valves which produce an increase in pressure should be fitted with a left-hand thread. In fact it is vital that this should be done if accidents are to be avoided in an emergency. An objection to this has been that the fitting of left-handed threads to valves will conflict with the learned pattern of an engineer who has been brought up in the operation of steam or hydraulic equipment. Will such an engineer when he sees a hand wheel or knob protruding from the front of the panel still think of it as a valve or as a control which is associated with the pressure gauge above it? To test this point the author carried out an experiment (Murrell and McCarthy, 1951) in which a number of engineer officers and engine-room petty officers and ratings were asked to indicate which way they would turn pressure controls in the form of hand wheels protruding from a panel in order to cause a given pointer movement. The results of this test showed that a large majority had a 'follow the pointer' preference, although it was made quite clear to them in the test that they would be operating valves. From this it would appear to follow, that when a valve is mounted in a panel so that the pipe work cannot be seen the hand wheel controlling it becomes just another control with the normal display-control relationship. Left-hand threads in order to give a clockwise-to-increase relationship with valves would seem from this research to be usable with confidence.

What will occur if electrical and pressure instruments and controls are not arranged in a panel? In this situation the findings of Murrell and McCarthy will no longer apply, and the designers would seem to be faced with the choice of fitting a left-handed thread, which may tend to make the valve turn in an unexpected direction for those who are accustomed to operating valves, or taking other steps to produce a compatibility between display and control. One possible method of doing this, described in the previous Chapter, is to move the zero on a pressure gauge from its standard position to the 12 o'clock position, so that the pointer will normally function in the region of 6 o'clock and any increase in pressure will cause it to move from right to left, in the same direction as the nearest part of a valve handwheel when it is moved anti-clockwise to increase. This would appear to be in accordance with the principle laid down by Warwick which has been mentioned above. Whether this will be successful or not will depend on whether the movement of the pointer is considered as translatory or rotary. If the latter, the rotary movement of the pointer will be clockwise and may conflict with the rotary movement of the control anti-clockwise, as suggested by Loveless (1956). Unfortunately, Loveless used a rather small dial. When dials are larger, the segment over which the pointer moves begins to approach linearity and his findings for horizontal linear scales may apply. No experiments appear to have been carried out to elucidate this point and it would seem that in the absence of other evidence the use of a sector or straight scale, graduated from right to left, is likely to cause less ambiguity than a circular dial with a zero at 12 o'clock. An alternative solution proposed by Loveless (1962), based on the work of Bradley (1954) is that a moving scale instrument should be used with the numbering increasing clockwise so that the scale will move anticlockwise to increase. Such an instrument should probably have at least half the scale visible and being a moving scale type should not be used for check reading.

Another aspect of consistency is the arrangement of controls which are used to close down a plant in emergency. There seems to be very little doubt that all controls used for this purpose should turn in the same direction, whether they increase or decrease the function being controlled. It is highly unlikely that if only a few seconds are allowed before a plant blows up, the operator will be able to remember that the first three controls should be turned anti-clockwise and the fourth clockwise, even though under normal working conditions he may be able to remember these movements quite easily. Such an arrangement of controls may, of course, mean special treatment of the display in order to retain compatibility during normal working. An example of how this might be done in a particular panel is illustrated in Fig. 89.

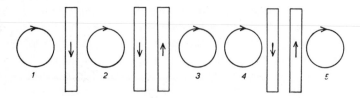

Fig. 89. In normal operation 1, 2 and 4 open to increase; 3 and 5 open to decrease. In emergency, all controls are operated clockwise.

THE SPATIAL RELATIONSHIP BETWEEN DISPLAYS AND CONTROLS

Many psychological experiments involve a subject responding to one of a number of stimuli, for instance by pressing one of a number of response keys, and it has been shown that performance is at its optimum when the spatial arrangement of the stimuli and of the response keys is similar. Thus if there are five lights in a row with five corresponding keys, the first light should be operated by the first key, the second light by the second key and so on. Performance will deteriorate substantially if some other arrangement is adopted such as a mirror image where the first light is operated by the fifth key, the second light by the fourth key and so on. Difficulty with this kind of spatial translation becomes steadily greater with age (Kay, 1954). Any arrangement of a number of displays in a piece of equipment will require some measure of translation before the correct control is selected, and if this is kept to a minimum, the operation of the equipment will be at its optimum. Learning will be facilitated and the chance of mistakes will be minimized.

CHAPTER ELEVEN

Design Factors V. Design Characteristics of Controls

In the absence of a computer, controls are the normal way in which an operator conveys his instructions to a machine. Research into the characteristics of controls has followed two main lines: in one instance, the shape, size, inertia and friction of controls of different types has been determined; in the other, the control dynamics have been considered. An example of the former is the crank; research has been carried out to determine the effects of radius on the rate of cranking, on the loads and on the torque produced; the effect of gear ratios on the accuracy with which a marker can be moved has also been studied and the viscous friction or inertia required. In the latter case, comparisons have been made on the use of joysticks for controlling markers, with rate-aided or velocity control, combined also with research to compare freemoving with pressure type joysticks.

A somewhat different approach to control design is to consider the anatomy of the limbs used in operation and from this to determine the best type of control for different uses. In some instances conclusions arrived at in this way have been confirmed by experiment, in others they are the best evidence available for the choice of a particular control characteristic for a particular purpose. In the discussions which follow this is the approach which will first be discussed.

CHOICE OF CONTROLS

A number of general principles of control design can be derived from the anatomy and physiology of the body. When a muscular movement is made, two or more muscle groups act in conjunction, one group acting as an antagonist to the other. In this way fine control of muscular action is obtained. The same principle can be applied to the operation of controls and when one limb acts as an antagonist to another, the most accurate control movement will result. The hands gripping the rim of a handwheel, such as a steering wheel, and the feet on a rudder bar or on a treadle such as is provided for the operation of a sewing machine, will produce the most accurate control adjustment available with these limbs. In the rotary movement of the forearm, involving as it does the crossing of the radius and ulna, the muscles of the arm have the advantage of moving as antagonists to each other, so that the operation of a knob by rotating the wrist is an accurate control movement. In contrast, the movement obtained by cranking, that is by revolving the hand or forearm, does not provide the same opportunity for muscular control and this movement is therefore less accurate than the turning of a knob. On many small machine tools a handwheel of perhaps 6 or 7 in. diameter is provided with a crank for quick slewing; this is thoroughly unsatisfactory, since if accurate adjustment is required, it is too large to operate as a knob and too small to operate as

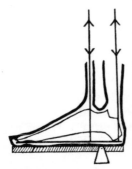

a handwheel. If the torque is not too high the best solution is to provide a large knob with a crank handle which will fold when not in use.

The characteristics of pedal design can also be derived from the same principles. The foot can be taken as pivoting around the small bones at the lower end of the tibia, so that muscles attached to the heel and in the vicinity of the instep can act as antagonists (Fig. 90). From this it follows that a pedal requiring accurate adjustment should have its fulcrum under the instep in line with the tibia. A certain amount of accuracy is lost if the fulcrum is under the heel but a substantial amount is lost if the fulcrum is under the toe.

Fig. 90. 'Antagonistic' movement of foot on a pedal pivoted in line with the tibia.

A second characteristic follows from the nervous control of the muscle fibres. The motor nerves in the arm and hand control a smaller number of muscle fibres than do those in the legs and the feet and it can be argued from this that a finer degree of control is obtained with hand control rather than foot control. At the same time the muscular strength of the leg enables a greater force to be applied with a greater speed, but where fine degrees of adjustment are required, the leg will be at a disadvantage.

The decision on which type of control is to be used in particular circumstances should also be influenced by its relation to the body. It is important that any movement which the operator is called upon to make should be in the direction following the free movement which a limb will make. Thus if controls are to be mounted on a horizontal surface, it is probably better to use a form of lever which can move in an arc in relation to the shoulder or the arm which is to operate it, than to use a rotary control in a horizontal plane. In the same way, if an operator has to stand it is better to avoid using pedals, especially if the pedal requires fine adjustment. The need adequately to support a limb which is operating a control has already been emphasized, and the feasibility of this may well influence the choice of the type of control which is to be used. A control which involves the continual exertion of a force should at all costs be avoided. For instance with a lever type control it should not be necessary to exert continual force to keep the control from the null position; rather the control mechanism should be arranged so that the control can be placed in a set position to obtain the required effect and will remain in this position without the exertion of further force. Fore-and-aft hand movements are slightly more responsive than movements across the body or rotary movements, other than knobs operated by one hand or handwheels operated with both hands. The weight of the arm can at times be used to exert continuous force, for instance on a dead man's handle: so can the weight of the foot, provided the operator is not standing. When he is standing, this would involve taking the greater part of the weight on one leg which means a great deal of postural work would have to be done in order to retain balance. With this general background, it is now possible to consider the different types of control in relation to their different uses.

Controls giving Continuous Adjustment

Controls used for continuous adjustment will be of two main types, rotary controls or reciprocating controls.

There are three principal types of rotary control: cranks, handwheels and knobs. Cranks may be used when rapid turning is required, but great accuracy cannot be achieved. The choice between a handwheel and a knob will depend largely on the torque which has to be produced. When the torque is low, a knob will be used, when it is high it is better to use a handwheel. As has already been mentioned, it is desirable to avoid rotary controls of between 5 and 9 in. diameter since these can neither operate effectively as handwheels or as knobs. Both knobs and handwheels are suitable for slow adjustment; they cannot be expected to give adjustment at high speed, except over very short distances – less than one turn for a knob or about a turn and a half for a handwheel.

Reciprocating controls for the hand are joysticks and levers. Levers give control in one dimension and joysticks in two dimensions and both types of control can give speed and accuracy. They can be used in a number of positions in relation to the body and their operational characteristics will depend to some extent upon the position chosen. It is probable that the best arrangement is a lever or joystick in the vertical position so that it will support the weight of the arm resting upon its end. This does not preclude the use of levers in other positions provided it is not expected that the operator will make adjustments with his arm unsupported. This suggests that horizontal levers facing the operator will be unsatisfactory while levers operated with the arm at the side may give satisfactory results with the pivot under the elbow. Reciprocating controls operated by the feet are pedals, treadles and rudder bars. Of these, treadles and rudder bars provide the greatest degree of accuracy. Treadles are quite common in many industries, and the control of sewing machines is very frequently achieved in this way, one foot being placed on either side of the pivot of the treadle. Where speed is important, a fore and aft movement of a foot on a pedal is probably the fastest but accuracy is likely to be low unless control is exercised by means of isometric force. The degree of accuracy of control achieved by rocking the foot depends on the position of the pedal in relation to the foot: in general the accuracy obtained in this way is not of a very high order.

Controls giving Discrete Movement

Controls required to give discrete movements will almost always be for some form of switching or to achieve a change from one mechanical state to another. The channel selection switch on a television set is an example of the first and the gear change lever on a motor car is an example of the second. As with controls giving continuous adjustment they may be either rotary or reciprocating.

Rotary controls used by the hand may be handwheels, knobs, handgrips, star wheels or rotary levers. As before the choice between these different types of control will to a large extent depend on the torque which is involved. When the torque is light, a knob will be used and there can be as many as 24 discrete positions selected by one knob provided that there are adequate detents. When a heavier load is involved, it will be possible to substitute a handwheel for a knob, but it is more likely that a form of rotary lever might be used, that is a handle or lever projecting from a central shaft. A special case of this type of control is a pistol-grip type handgrip but this is suitable only when the degree of turning required is less than 180°. A star-wheel will be used when it is desired to make extensive movements between stops, or between stops and a central position which is clearly and firmly located by means of a detent.

Reciprocating controls used by the hand can be levers, push-buttons or toggle switches. These can be used for multiple switching provided again that each position is clearly indicated by changes in the pressure required to operate or by a noise. In general, it will not be possible to operate this type of control quickly; since overshooting will be possible levers will in general be used for operating between two positions when movement in one direction is involved, or for four functions when the lever moves in two directions. This type of control can be very satisfactory for multiple switching operations and is probably quicker than using separate controls involving a movement of the hand from one control to another. A toggle switch can be thought of as a miniature lever, but its range of possible positions is confined to three. Push-buttons can normally operate in one sense only, to give two non-cancellable effects through successive operation.

Reciprocating controls used by the feet will be either pedals or push-buttons and the choice between them will depend on the load; and it is unlikely that they can operate more than two positions since, with the degree of control available, it is difficult to select positions other than right forward or right back. A foot pedal would be used on a machine when functions such as stop and go are required, the foot push-button might be used for electrical switching with a once-and-for-all effect.

In certain types of equipment it may be necessary to exercise continuous control in two dimensions. This does not appear to be very common in industry but because it forms an important part of many military tasks it has received a good deal of attention. The most commonly used type of control is a joystick and generally it seems not to matter very much whether the joystick is vertical and able to support the hand or horizontal. In the latter case it is possibly more compatible with a vertical display but some form of arm rest is required. Research has suggested that control by position rather than rate-aided or velocity control is a very much superior method of operation. When tolerances are large there is not much to choose between a joystick and the use of two knobs; the left-hand knob facing the operator and turning clockwise to move the marker from left to right and the right-hand knob sideways to the operator and turning clockwise to move the marker up. When tolerances are small the use of the two knobs appears to be superior to the use of a joystick. When a joystick is used it has been shown that control by means of pressure is superior to control by movement, but this is with velocity control and its superiority over positional control is not clearly demonstrated.

The main characteristics of different types of control are summarized in Table 18 in which controls are ranked in order of suitability from one to four. This is an aid to the selection of a particular control for a particular purpose. The details of the different controls in relation to human operation as they have been determined by research will follow. There are two methods of presenting data of this kind. The first method is to give details of the different experiments and then to summarize the conclusions which can be drawn from them. The other method is to summarize the results without giving references, a form which enables design recommendations to be used without the need for interpretation. The literature on the subject of controls is so vast (over 500 references have been consulted), that it has been decided to adopt this latter method of presentation in this instance.

Table 18.   Characteristics of controls
(after Murrell, 1960b)

Type of Control	Suitable for				Loads
	Speed	Accuracy	Force	Range of Movement	
Cranks – small	1	3	4	1	up to 40 in.lb
large	3	4	1	1	over 40 in.lb
Handwheels	3	1	2-3	2	up to 150 in.lb
Knobs	4	2	1	2	up to 13 in.lb
Levers – horizontal vertical	1	3	3	3	up to 30 lb
(to-from body)	1	2	short 3 long 1	3	up to 30 lb ★
vertical (across body)	2	2	2	4	one hand 20 lb two hands 30 lb
Joysticks	1	2	3	3	up to 5 lb
Pedals	1	3	1	4	30-200 lb†‡
Push-buttons	1	4	4	4	2 lb
Rotary selector switches	1	1	4	4	up to 10 in./lb
Joystick selector switch	1	1	3	4	up to 30 lb

Code
Nos. 1-4. Degrees of suitability.
★When operated by a standing operator depends on body weight.
†Depends on leg flexion.
‡Ankle only, 20 lb.

THE DESIGN OF CONTROLS

Some of the data on control characteristics are based on comparatively small numbers of subjects, while in other instances care has been taken to have an adequate sample. As in all aspects of human performance, there is a wide variability, so that it is not possible to state figures for diameters, loads and so on with absolute validity. Those which are suggested are the best estimates on the basis of the available data; some of the principles which have been included may not yet have been established beyond all reasonable doubt, nor has it been established that failure to implement the recommendations will necessarily lead to a loss of efficiency, serious or otherwise. However, at the best they give a reasonable estimate of the design parameters concerned, at the worst they are better than guesswork on the part of the designer, who if he follows them may rest assured that the equipment which he is producing is being designed in accordance with the best knowledge available at present.

In some of the data which follow, indications are given of the maximal applicable force. These forces should, however, be called for only under exceptional circumstances. It is possible to maintain maximal performance for only a very short period of time before some muscular fatigue sets in. As a result, the forces which can be exerted in normal repetitive operations, are likely to be rather less than half the maximum. The forces which can be applied by the non-preferred hand (the left hand of a right-handed operator) are about two-thirds of those of the preferred hand. Females can be expected to exert about two-thirds of the force which can be exerted by a male.

Cranks

Definition. A control in which the handle is offset from and parallel to the shaft.

Uses. Cranks are used when high turning speeds are required up to 200 r.p.m.; when rapid complete turns are required; and when heavy loads have to be moved manually.

Size. Under light loads up to about 5 in.lb and for cranking speeds of up to 150-200 r.p.m., the optimum diameter appears to be within the region of $2\frac{1}{2}$ to 3 in. At turning speeds of above 200 r.p.m., a smaller crank diameter seems to be indicated of between 2 and 1 in. Another study suggests that the optimum diameter is $1\frac{1}{2}$ in. The speed of turning is related to the diameter: in the smaller sizes the differences in rate of turning are comparatively small, but the velocity decreases fairly rapidly from a crank diameter of about 4 in. upwards. For example, the rates attained in one experiment are shown in Table 19.

Table 19.   Maximum cranking rates

Diameter (in.)	Max. Rate (r.p.m.)
1·2	260
1·6	270
2·4	275
4·0	255
5·6	235
9·6	185
19·2	140

These results were obtained with the preferred hand which appears to be faster than the non-preferred hand. For instance, at a diameter of 4 in., the rate of turning with the preferred hand in one experiment was found to be about 270 r.p.m. compared with 240 r.p.m. with the non-preferred hand. The direction of rotation does not appear to affect the rate. When accuracy rather than speed is involved, as in pursuit tracking, the cranks would need to be rather larger, between 4½ and 9 in. diameter. The optimum rate of cranking under these circumstances would appear to be about 50 r.p.m., at which point there appears to be little difference between the two diameters. If the speed is increased to 100 r.p.m., the smaller size appears to have an advantage, but if the speed is decreased to 20 r.p.m. or below, the larger diameter crank appears to be better.

The rate of cranking will decrease as the load increases. In one experiment, using a crank of 2 in. diameter, it was found that at a minimum rate close to 0 r.p.m. the forces which could be overcome ranged from about 14 to 32 lb. This force decreased, to range from 3½ to 17 lb, when the speed increased to 220 r.p.m. When quite slow speeds are used, with single turns, the minimum crank diameters for different torques have been determined for cranks in different positions and at different heights in relation to the standing operator. These are given in Table 20.

Table 20.   Minimum crank diameters

Torque (in. lb)	Vertical, facing, 36 in. from Floor	Vertical, facing, 48-58 in. from Floor	Horizontal, 36 in. from Floor	Side, 36 in. from Floor
0-20	3 in.	5 in.	3 in.	3 in.
20-40	5 in.	5 in.	9 in.	9 in.
40-60	9 in.	9 in.	15 in.	15 in.
60-90	9 in.	9 in.	15 in.	15 in.
90 upwards	9 in.	15 in.	15 in.	15 in.

Gear Ratios. Cranks will not normally be used for accurate positioning, a control movement for which they are rather unsatisfactory. For this reason gear ratios between cranks and pointers have not received very much attention. The choice of an adequate gear ratio must depend on the mechanical advantage required for the crank and should be a compromise between torque and speed of rotation.

It would appear from experiments on the use of cranks for tracking that it is necessary for winding speeds to be up to about 180 r.p.m. Speeds as high as this cannot be maintained for more than half a minute. Accuracy of tracking, and thus presumably accuracy of setting, will decrease rapidly as the winding speed becomes less than 70 r.p.m., so that for normal uses a speed of about 100 r.p.m. might be aimed at. Where the range of speed of turning is liable to be greater than three to one, it has been suggested that there should be a choice of ratios giving a high gearing for rapid slewing and a low gear ratio for making fine adjustments.

Friction and Inertia. The effects of friction and inertia on cranking are given in Table 21.

Table 21. Effect of friction on cranks

	Intermittent Winding	*Continuous Windings*
Friction (viscous)	Has an adverse effect	2-5 lb desirable
Inertia	No research known on this but may be assumed to be undesirable	Especially valuable in presence of friction. Maximum 10 lb force on crank to obtain maximum acceleration

Handle Shape. The handle of a crank should have parallel sides. It may have a slight indentation at the inner edge. Bulbous handles are not satisfactory. For light loads, the length may be $1\frac{1}{2}$ in. but for a heavier load it should be $3\frac{1}{2}$ in.

Type and Position of Crank. The preferred position for a crank is facing the operator, and located along a horizontal line at elbow height from the centre line of the body to the width of the shoulder of the operating hand and then in a vertical line to shoulder height. Thus, for the two hands, this will produce a zone approximating to a U of about the width of the shoulders of the operator. Except when heavy loads are involved, the evidence suggests that while this is an optimum position, it is not very important what the position of the crank is, provided that it is facing the operator and can be reached conveniently and comfortably. Cranks in the horizontal or side position can be operated reasonably well, but because a greater degree of arm and shoulder movement is involved, the speed with which they can be turned is substantially less than the speed with which a facing crank can be operated.

Cranks for Very Heavy Loads. Cranks are sometimes used sideways to the operator who has to move very heavy loads as, for instance, in hand operation of a crane. No work has been found on the optimum diameter of cranks nor on the torques which can be obtained. The force that a man can exert when cranking in this way appears to depend on body weight and from the forces which can be applied by the standing man in different

positions, it would seem to be about 80 to 100 lb. Clearly, it cannot exceed
the body weight, otherwise cranks could not be moved in a downwards
direction.

Handwheels

Definition. A handwheel is a circular control gripped by the rim, preferably
with both hands.

Use. Handwheels should be used when turning speeds are low (1 r.p.m.
or less); when very accurate partial turns are required and when torque
required is above 15 in.lb. Handwheels are best used when the amount of
turn required does not exceed 90° for fine positioning movement.

Size and Position. For a standing operator, the optimum position is with
the shaft facing the operator when loads are relatively light or with the
shaft vertical when loads are heavy. Another, but rather impractical position, is with the shaft at 40° below the horizontal. Normally wheels should
not be used sideways on to the operator, except when very high forces in
excess of about 120 in.lb must be exerted. In this instance, the operation
of the wheel becomes similar to that of a lever and the operator has to use
his body weight in order to turn it. Such wheels are sometimes found on
heavy control valves. For a standing operator the optimum height of the
centre of a facing wheel is at about elbow height, that is between 40 and
44 in. above the floor. For a horizontal handwheel where maximum forces
are to be applied, the optimum height of the rim of the wheel will be at or
rather below shoulder height, that is about 52 to 56 in.

For handwheels facing the operator which are to be used to make partial
turns, the minimum size for different torques at different heights above the
floor in relation to a standing operator are given in Table 22.

Table 22.   Optimum size of handwheels for partial turns

Torque (in. lb)	Wheel diameter at 38-48 in. from Floor	Wheel diameter below 38 in. or above 48 in.
20-40	6 in.	10 in.
40-60	10 in.	16 in.
60-90	10 in.	16 in.
90 upwards	16 in.	16 in.

For a seated operator using a vertical facing wheel gripped by both
hands, the best diameter for the wheel is between 12 and 14 in. with the
axis of the wheel 18 in. above the top of the seat and the wheel placed
14 in. from the seat back. For a horizontal wheel, the height should be

9 to 12 in. above the seat and the centre of the wheel about 22 to 24 in. from the seat back.

Maximum accuracy and force are applied to a handwheel when both hands are used on the rim of the wheel. The maximum tangential force which can be exerted with both hands in this position is about 80 lb. With one hand the force is about half this. A force by about 20 lb greater can be applied in an anti-clockwise than in a clockwise direction.

The diameter of the rim of a handwheel should be of such a thickness that it can be gripped comfortably. It is suggested that its diameter should be $\frac{3}{4}$ to 2 in.

Knobs

Definition. A knob is a rotary control that can be operated freely by gripping it on both sides with the fingers of one hand.

Use. Knobs may be used (i) for making fine adjustments when loads are light up to 22 in.lb, i.e. continuous function; (ii) as rotary selector switches for switching operations.

Continuous Function, Size and Load. While knobs can be used for loads above 30 in.lb, it is recommended that generally they should be used with lighter loads, and for this purpose a cylindrical knob is the best. In general, more force can be developed with a side knob than with a facing knob or a horizontal knob. A side knob should have a diameter of about 1 to 2 in. (optimum 1·6 in.), a facing knob may have a diameter of between 1 and 2·5 in. (optimum 2·0 in.), at which diameter the optimum load is about 2 in.lb. Where fine adjustment is paramount, loads up to 11 in.lb are possible. The maximum diameter of a cylindrical knob which can be gripped by the fingers is between 5 and $5\frac{1}{2}$ in.

For loads above 13 in.lb a knob shaped like a four-pointed star is superior to a cylindrical knob. Up to 17 in.lb the diameter should be 3 in. and up to 22 in.lb it should be about 4 in. These higher loads should, however, be considered exceptional.

Gear Ratios, Friction and Inertia. Knobs which are to be used for making a fine adjustment by turning to a null position, as for instance in tuning a radio receiver, should have a gear ratio such that they move approximately 80° between a just detectable mis-alignment in one direction and a just detectable mis-alignment in the other. The nature of the cues to the extent of the 'dead zone', whether tactual, visual or auditory, appears not to influence this requirement.

When making settings on a scale, the amount of pointer movement in relation to one turn of the knob will depend on the tolerance allowed. When tolerances are quite small, of the order of 1/100 in. or less, one turn of the knob should produce about 1½ in. of travel of the tip of the pointer, and if the travel is small this ratio will remain unchanged whether there is friction or inertia present. Settings to this degree of tolerance cannot be made by having a coarse ratio and increasing the size of the knob, the important factor being the ratio between the turning of the knob and the index movement, not the knob size. If too fine a ratio has been adopted, there is a temptation to use a crank because of the larger number of turns required for a very small amount of movement; in fact the use of a crank will not compensate for having a ratio which is too fine and in most cases the use of a crank will be less satisfactory than the use of a knob with the optimum ratio given above. When the tolerance allowed is about 1/50 in. the pointer travel may be increased to 4 in. for each revolution of the knob and when the tolerance is 1/10 in. the movement may be 6 in. When the distance of pointer travel is large in relation to the tolerance allowed, a knob with two gear ratios may be used. Under these circumstances, it may be an advantage to use a slightly larger knob than would be the case when short travel only was involved.

Friction applied to the shaft of the knob or in the mechanism operated will usually increase the travel time but will not affect the time taken to make the setting.

If heavy friction is present, inertia will tend to compensate for the friction and to aid performance. When friction is absent, inertia has very little effect, either on the travel time or on the adjustment time.

Shape. Cylindrical knobs should normally have serrations around the edge to assist gripping. If a knob used for fine adjustment requires that a reading should be made from an associated scale, it is best to have the scale engraved on the panel face rather than to add a skirt to the knob which moves in relation to a fixed index. It is important in the former case, however, that a clear index should be incorporated in the knob itself and in the latter case that the skirt should be wide so that the fingers do not obstruct the reading of the scale.

Step Function, Size and Load. For step functions, bar shaped knobs should always be used which should be not less than 1 in. long and not more than 1 in. wide. They should protrude about 1 in. from the panel. It is important that they should be nonsymmetrical so that the 'index end' can be readily detected by touch alone. Within these limitations, it is not very important what shape the knobs are. If coding by touch is required, a number of different shapes may be used which can be distinguished from each other by touch. The maximum force required to move the knobs from one position to another should be not more than 2 lb. It is undesirable to have four positions 90° away from each other since it is easy to confuse the two opposite positions by reading the wrong end of the knob. With a small number of positions, it is best that they should be arranged around a segment of a circle with stops at the beginning and end of the control position. When a larger number of positions is required, between 9 and 12, it is probably best to arrange the numbers in the position of the clock hours, in order to make use of the common familiarity with this arrangement. In general it is undesirable to have more than 12 positions on a selector switch, but up to 24 are acceptable, though not desirable. If more than 24 have to be used then the diameter of the knob must be such that there is at least $\frac{1}{4}$ in. between each position.

Friction and Inertia. Static friction and inertia should be kept to a minimum. Wherever possible, detents should be used to mark the control positions and there should be viscous friction between each detent position. This should increase at first as the control moves away from the detent and then decrease as the next detent is approached. This will ensure that the control will move positively from one detent to another.

If, in spite of their undesirability, round knobs are used as rotary selector switches, it is better to have a fixed index in one of the four cardinal positions with the switch positions marked on a skirt which will be numbered in a direction contrary to the movement of the knob. Alternatively, an open window type of display can be used. Only the very lightest detent should be used with this type of knob because the torques which can be applied are rather low for detent action.

Levers

Definition. A lever is a hand-operated rod-like control working in one dimension about a fulcrum.

Use. A lever should be used when the effective travel is small; when fast movement but no great accuracy is required; when there is a medium to heavy load; for step function for rapid switching with or without load.

Continuous Function: Horizontal Lever Pulled Up or Pushed Down

For a male adult, the maximum pull which can be exerted on a horizontal lever is at an average hand height, that is about 28 to 30 in. above the floor when standing or about 3 in. below seat height for a seated operator. At 30 in. an individual may be able to pull with a force of 105 lb with one hand. If the lever is lowered to 20 in. the pull drops by only 10 lb, but if it is raised to 40 in. the pull drops by 45 lb to about 60 lb. These forces are maxima and if a force of more than about 30 lb is to be exerted regularly, mechanical assistance should be given, especially if accurate movement is required.

The force which can be applied when a lever is pushed down will depend on the proportion of the body weight which the operator brings to bear, if the body weight is not used (and it is undesirable that a design should assume that body weight will be used), the maximum push at hand height of 30 in. is about 88 lb. The travel of a lever to be pushed should be not more than about 6 to 7 in. and the handle of the lever should be placed so that the arm is straight at the end of the travel. The greatest force is produced by the final straightening of the arm. Forces in excess of about 40 lb should not be demanded of either a sitting or standing operator if the lever is to be used frequently.

Vertical Levers Pushed Away From or Pulled Towards the Body

The optimum position for a lever to be pushed or pulled one-handed is opposite the shoulder of the arm which is to operate the lever. When a lever is operated with both hands, it should be placed centrally to the body. The force which can be applied to a lever falls off quite rapidly as it is moved to either side of the optimum position. For instance, if a lever has to be operated by the right hand opposite to the left shoulder, the force which can be applied is decreased by more than one third. The maximum pull can be exerted when the arm is fully extended and the maximum push when the arm is straightened from an elbow angle of about 160°. It should be noted that when vertical levers are operated using the body weight, accurate positioning will not be achieved and therefore levers used in this way should always be supplied with detents or operated between stops.

The seated operator should be provided with an adequate foot-rest. The optimum height of the handle of a lever is about 9 in. above the seat, though it may be as high as 15 in. In order to achieve the best arm position, the position of maximum pull of the average male operator is between 30 and 35 in. from the seat back and the position of maximum push is about 29 in. from the back. The maximum forces which can be applied by a seated operator to a vertical lever are given in Table 23 opposite.

It should be remembered that these are maximum forces and that they should rarely be called for. If a lever is placed in the optimum position and accurate movement is required, the force to be applied should not exceed 30 lb. It will be noted from the Table that the distance of travel over which a high degree of force can be developed is quite short, only about 3 in.

Table 23. Maximum forces which can be applied in a horizontal direction to a lever

Maximum pull by preferred hand   at 35 in. about 120 lb
Maximum pull by preferred hand   at 29 in. about 75 lb
Maximum pull by preferred hand   at 22 in. about 45 lb
Maximum pull by both hands   at 33 in. about 120 lb
Maximum push by preferred hand   at 29 in. about 130 lb
Maximum push by preferred hand   at 22 in. about 160 lb
Maximum push by both hands   at 29 in. about 240 lb

Vertical Levers Used Across Body

The optimum distance of a lever from the front surface of the body is said to be 17 to 18 in., and the force which can be applied falls quite rapidly as the distance is increased, so that 20 in. is the maximum distance a lever should be away from the body. It is easier, one-handed, to apply force inward across the body than outwards and the maximum force that can be applied inwards from a position opposite the operator's arm is about 45 lb and outwards about 30 lb. If both hands are used moving either to right or to left from the optimum central position, a force of about 70 lb can be applied, but this force decreases rapidly as the lever is moved away from the central position. For normal operation, the forces to be called for for one-handed operation should be not more than about 20 lb and for two-handed operation not more than 30 lb.

Gear Ratios. When a lever is used to control the movement of a pointer, the lever should be geared so that its tip will travel three or four angular units for each unit of movement of the pointer. With this ratio it has been claimed that settings can be made as accurately with levers as with knobs. The effect of friction does not appear to influence this ratio, though it does make the speed of the setting somewhat lower. The distance of travel of the tip of a lever should preferably be not more than half the length of the lever.

Levers Operating Through Pressure

Levers for tracking, which control displacement or velocity through the exertion of pressure against isometric forces have been shown to give more accurate control than free moving levers. This kind of control can be used only for rate-aided or velocity tracking, since positional control is not possible with a lever with no positional displacement. Levers with spring-loading may be used in the same way and they can give information about the control positions through kinaesthetic feedback. Data on the extent and efficiency of spring-loading do not appear to be available.

Levers Used for Switching

If levers are quite short they will, when used for switching, approximate to toggle switches. If they are larger they will have characteristics similar to those of joysticks but operating in one dimension only. Data on these two types of control given immediately below should be referred to when levers are used for switching.

Toggle Switches

Toggle switches are miniature levers used as selector switches. They may be up to 2 in. long and should be not less than $\frac{1}{8}$ in. or more than 1 in. diameter at the tip. The force which has to be applied to the tip of the switch in order to operate it should be between $\frac{1}{2}$ and 1 lb. Spring loading should be used which decreases to a minimum when the switch reaches its control position. Toggle switches should not be used for more than three switching positions, and there should be between 30° and 40° of movement on either side of the central position. Toggle switches should not be used for making discrete adjustments.

Joysticks

Definition. A joystick is a lever working in two dimensions.

Use. Joysticks should be used when continuous simultaneous control is required in two dimensions or for rapid multiple switching operations.

Short Sticks with Light Loads

It has been suggested that a vertical stick is superior to a horizontal stick because it supports the weight of the hand and the arm. If a horizontal stick is used the elbow of the operator must be supported. A vertical stick with the free end down held in the fingers like a pencil is superior to a stick with the free end up, provided that the arm is supported. A stick used in this position will appear to make use of the greater sensitivity of the fingers as compared with that of the shoulder and the elbow.

Movements towards the body can be made more accurately than movements away from the body. Push-pull is more accurate than left-right or up-down. For light loads, the maximum length of a joystick should be about 9 in. and the total travel of the free end about 6 in. Friction of about 2 lb may be used to minimize the effect of jolts. The optimum ratio for the movement of the tip of the control and a marker being controlled is similar to that for levers, that is, three to four angular units of joystick movement for each unit of movement of the marker. This ratio appears to be unaffected by the length of the joystick.

Joysticks may also be used as pressure controls in those systems where visual changes in the display do not clearly define the response required, as in complicated tracking with velocity control or rate-aided control. It seems likely that when adequate visual cues are available, the superiority of a pressure-control joystick over a free-moving joystick will be less marked.

Long Sticks with Heavier Loads

These joysticks will, as a rule, be floor-mounted and at least 26 in. long for a seated operator. The top of the joystick should be 8 to 10 in. above the seat height and be mounted about 19 in. from the seat back. Lateral movement should not be more than 7 in. each side of the neutral position, with 5 in. forward movement and 9 in. aft.

Force called for in the lateral direction should be less than 10 lb and that in the fore and aft directions should be not more than 20 lb. The maximum force which should be called for in either direction occasionally may be as high as 30 lb. If greater force is required, it will be necessary to give some form of mechanical assistance.

Joysticks Used for Multiple Switching

Joysticks should be short, between 6 and 8 in. in length, and the distance of travel of the tip of the joystick should be between 2 and 3 in. The optimum number of fixed positions is four, arranged in a star, the maximum number is eight. An H-shaped gate should not be used if rapid transit is required between a position on one arm of the H to a position on the other. This type of gate is suitable only when rapid switching from positions on the same arm of the H is required. Under these circumstances the H-shaped gate is superior to the X-shaped gate. For three positional switching a Y-shaped gate is superior to an E-shaped gate. The joystick should, in all cases, be spring-centred and the force required to overcome this spring-centering should be not more than 2 lb.

Joysticks Used for Discrete Positioning

Control characteristics of joysticks used for discrete positioning would seem to be similar to those used for multiple switching, except that there should be spring detents in each of the switching positions, and the joystick should not be spring loaded to the centre.

Rotary Levers and Starwheels

Definition. A rotary lever is a lever attached to a shaft so as to permit a complete turn (some authors call this a turn-style).

A starwheel is three or more levers attached to a shaft.

Use. A rotary lever is used when a fairly high force is required over a long range of movement with fairly accurate end control. A starwheel is used when rapid turns are required followed by the application of force, with or without accurate control.

Rotary Levers, Size and Load

Very little work on the size of rotary levers has been traced, but it seems that the principles which govern the design of ordinary levers may well apply. One study of the forces which can be applied with this kind of control shows that these will vary according to the position of the lever round the fixed turning circle. With a vertical lever, greater force can be applied than when the lever is approximately in the horizontal position, the difference being about 40 lb, though this difference is influenced by the height of the lever above the floor. It seems that the maximum force taking the different positions of the lever into account should not exceed 45 to 50 lb. Owing to the variability of position of operation which this type of operation requires, it would seem that it is not a type of control to be recommended.

Starwheels, Size and Load

No work on the size of starwheels has been traced. When moved with both hands, they will have similar characteristics to a handwheel and it is expected that they can be adjusted with a similar degree of accuracy. Owing to the nature of the grip which is more normal to the line of the arm, the forces which can be applied to a starwheel seem to be rather greater than those which can be applied to a handwheel of similar size. For instance, a horizontal starwheel operated by both hands can have an applied force of about 160 lb compared to 140 lb for a handwheel. When the starwheel is vertical and the hands grip two opposed horizontal spokes, the optimum height appears to be about standing hand height, that is 28 to 30 in. At 40 in. the operation is inferior; however, with a horizontal starwheel or a vertical starwheel gripped by vertical spokes there appears to be no difference between a height of 30 and 40 in.

When a starwheel is operated with one hand, it would appear to have many of the characteristics of a rotary lever, except that there is a choice of spokes with which to make any fine adjusting movements. For this reason starwheels would seem to be superior to rotary levers.

Handgrips

Definition. A handgrip is a rotary lever with a limited arc of movement which is gripped close to the shaft rather than by its end (also called a pistol grip).

Note: that part of a lever which is held by the hand is sometimes called the handgrip. Under these circumstances it is not a control in its own right.

Use. Handgrips are used for rotary multiple switching. They should not be used for continuous control.

Position and Load. No research on handgrips has been traced. The following suggestions are therefore made from personal experience and user comment.

When mounted on a horizontal or sloping panel, the handgrip should be away from the operator. This will avoid accidental operation by catching the sleeve on a handle when reaching for another beyond it. The movement also follows the rotary movement of the wrist. For this reason, the part gripped by the hand should be curved upwards and be shaped so that it can be comfortably held and turned by pivoting by the wrist. The arc of movement should not exceed 180° with an optimum of seven switching positions, giving 30° between each position. In vertical panels, the handle may point downwards if it is mounted at or about elbow height, but above this height the handle should point upwards. The design should therefore allow for operation in either position.

The switching positions should be clearly indicated by spring loaded detents, and the force required to move from one position to another should be about 2 lb but should not exceed 4 lb.

Pedals

Definition. A pedal is a reciprocating control operated by one foot acting independently.

Use. Pedals should be used when a very powerful force has to be applied; for continuous controlling when accuracy is not of a high order; for on/off or start/stop switching.

Pedals may be divided into two groups: those where the force is obtained by a movement of the leg, and those where the force is obtained by a movement of the ankle. A pedal should be operated by a movement of the leg when torques above 20 in.lb are required; when torques less than this are needed a pedal may be operated by the foot pivoted at the ankle.

Pedals Operated by the Leg – Seated Operator: (1) *Pedals in line with the lower leg*

The greatest force of which the human operator is capable can be applied by the leg of a seated operator with a good back-rest provided that the pedal is in a position in relation to the foot such that the knee angle is about 160°. In this position, forces between 700 and 800 lb can be applied by a man as an instantaneous thrust. The knee angle is, however, very critical; for instance with the knee angle at 115° the maximum instantaneous thrust drops to between 90 and 100 lb. The ability of the leg to thrust with different knee angles under different conditions is shown in Table 24, which shows that the distance over which a reasonable force can be applied by the leg to a pedal is very short; in practice forces of this order are generally isometric. It has been suggested that greater forces than those shown for the smaller knee angles in Table 24 can be applied if an operator brings the body weight to bear, particularly when the knee is bent in the region of 90°. This is probably true, but it involves a control action which is strongly to be deprecated, since under certain circumstances physical damage may be caused to the operator.

Table 24. The ability of the leg to thrust

Knee Angle	Back-rest Sustained Thrust Without Discomfort	Max. Instantaneous Thrust Without Strain	With Strain	No Back-rest Max. Instantaneous Thrust Without Strain	With Strain
135°	50 lb	80 lb	155 lb		
113°	45 lb	60 lb	90 lb		
90°	30 lb	50 lb	60 lb	30 lb	40 lb

The force should, as far as possible, be applied in line with the leg, thus the pedal should be positioned so that it can be operated by the heel or arch of the foot. Use of the toes should be avoided since it strains the foot extensors. To obtain a pedal in the optimum position for maximum force, the pedal should be between 4 in. above and 6 in. below the level of the seat top and its travel should be between 4 and 6 in. To cover a 90% range of men, the seat back should be about 41 in. from the fully depressed pedal with a range of adjustment of plus or minus 3 in. For women, the distance should be 38 in. with a similar range of adjustment.

Pedals Operated by the Leg – Seated Operator: (2) *Pedals at right angles to the lower leg*

A lever is sometimes operated by swinging the lower leg backwards or forwards. Such a pedal should be about as long as the lower leg, 17 in.

for men and 15 in. for women, and should ideally be adjustable for length. The pivot of the pedal should be at the level of the knee with the thigh horizontal. If these conditions are not fulfilled, the body must make complicated adjustments to allow for the displacement of the pedal from the ideal position. The maximum force which can be applied by extending the knee is about 90 lb for men and 65 lb for women. When the knee is flexed, the maximum force appears to be about 65 lb for men and 45 lb for women. It is the flexion therefore which is the limiting factor in this kind of control and the maximum force which should be called for under normal repetitive conditions should not exceed 20 lb for men and 15 lb for women.

Pedals Operated by the Leg – Standing Operator

The force which can be applied by a standing operator will be determined entirely by the operator's weight. If he is standing on both feet, the maximum force which can be developed will be rather less than half the body weight. If greater force than this is required, he will have to 'stand on one leg' on the pedal. The position of a pedal for the optimum application of force is when at the end of its travel it is at floor level and when the two feet are side by side. The further forward the pedal is from the non-operating foot the greater is the difficulty and muscular effort required to operate it. It is desirable that the pedal should have minimum movement and in any event it should be not more than 6 to 8 in. above floor level at the top of its travel. As the height of the pedal above the floor increases, it becomes increasingly difficult to place it in an optimum position, since the more the thigh is raised, the further the foot moves from the frontal plane of the body. As a result, a pedal which is pivoted in front of the operator will move in a different arc to the foot and if the pivot is high and the pedal is placed in a correct position relative to the foot when the leg is raised, it will be too far in front of the body when the pedal is depressed. Ideally a satisfactory pedal can be obtained only if it is pivoted behind the operator, with the arm of the pedal parallel to the thigh (cf. Fig. 49).

If the foot operating a pedal is to remain resting upon it when the pedal is not depressed, there should be sufficient force supplied by a return spring to support the weight of the leg. It is highly undesirable that an operator should actively have to keep his leg raised. It is a mistaken belief that more physical energy is required to depress a pedal against an adequate return pressure than is required to keep the leg raised off a pedal when the return pressure is inadequate. The return pressure should be somewhere between 18 and 20 lb, since the weight of the lower leg is about 6 lb and the thigh weighs about 13 lb.

Pedals operated by a standing operator should be used only if no other form of posture and pedal can be devised.

Pedals Operated by the Foot

Pedals which are operated by extending and flexing the ankle should not be used by a standing operator if this can be avoided. If such a pedal must be used in a standing position, it should be sunk into the floor so that both feet are level. It should be pivoted under the line of the lower leg and be set between horizontal and not more than 10° above horizontal. It should have a movement extending to not more than 20° below horizontal. Adequate stops should be provided so that the operator can rest his weight on his heels without operating the pedal. A ridge about the back of the pedal into which the heel can fit is desirable.

For a seated operator, the facility with which a pedal is operated will depend upon the position of the fulcrum. One study on speed of pedal operation suggested that the best position was with the fulcrum under the heel. Since however the purpose of a pedal of this type is not usually to produce a series of rapid successive movements, but to exercise gradual control, it would seem from an anatomical point of view that the proper place for a fulcrum is in line with the tibia. This position has been confirmed by later work using a Lauru force platform by an analysis of the forces involved in the operation of different types of pedal (Lauru, 1957). The maximum movement of the ankle is about 30° above and 45° below a position at right angles to the tibia. These, however, are maxima and approach the positions of strain so that foot pedals should be designed to have no movement greater than 20° up or 30° down if strain is to be avoided. Thus the maximum height of the free end of a pedal above the heel rest when the lower leg is vertical should not normally be more than 3 in. The nearer the foot is to a right angle the more comfortable is the operation of the pedal likely to be.

Pedals which are pivoted under the instep can be returned to a position of rest by depressing the heel, but this is usually unsatisfactory and pedals should be returned to a rest position by spring pressure. To support the weight of the foot when at rest a return spring pressure of 3 to 4 lb may be sufficient to support the weight of the foot if the pedal has its fulcrum under the tibia. If the pedal is hinged at the heel, a return spring pressure of 7 to 10 lb will be required. For accurate control when the operator is likely to be jerked, some friction should be provided.

Rudder Bars and Treadles

Definition. A rudder bar is a reciprocating control pivoted about its centre and operated by the foot in line with the lower leg. A treadle is a reciprocating control pivoted at or near its centre which is operated by the movement of the ankle.

Use. Rudder bars are used when accurate control is required with moderate force. Treadles are used when accurate control is very important but the amount of force required is small. (Treadles, together with foot-operated cranks, as on a bicycle, have been used in the past in order to obtain rotary movement. This use for these controls is unlikely to be found in industry at the present time, except under exceptional conditions.)

Accuracy. Operation by the legs of a rudder bar would appear to be rather less accurate than the operation of a treadle by ankle movement, but both are likely to be more accurate than the operation of a pedal by one foot alone. Rudder bars are operated with less accuracy than a joystick.

Position and Load. The optimum position for a rudder bar in the neutral position is at about 35 ± 3·75 in. from the back of the seat and 6 in. below the seat cushion. Rudders become more difficult to operate the lower they are relative to the seat, for instance at 12 in. the force which can be applied is only about two thirds of that which can be applied at 6 in., which will be between 400 and 500 lb. In practice the maximum which should be called for should not be more than 150 to 180 lb.

No recommendations on treadle position seem to be available. From first principles it can be argued that the surface of the treadle should form a cord to an arc about the knee of a seated operator when the thigh is horizontal. This arc should have a radius equivalent to the lower leg length of the operator when wearing shoes.

No research on treadles as such has been found but since two feet are acting as antagonists, the control should be reasonably accurate. This view seems to be supported in practice, since in the garment and shoe industry the control of sewing machines exercised in this way has been shown to be quite precise provided that variable speed gear is fitted.

Push-buttons

Definition. A push-button is a reciprocating control, small in size which has a positive action in one direction only.

Use. Push-buttons should be used for switching in one or two senses only.

Push-buttons Operated by the Foot

A push-button will normally be operated by the toe of the foot and as such will have most of the characteristics of a pedal hinged at the heel. The diameter would seem to be relatively unimportant, but if the push-button is too small, too great a pressure will be applied to a small area of the foot and the minimum diameter should be $\frac{1}{2}$ to $\frac{3}{4}$ of an inch. The movement of the push-button should be not less than $\frac{1}{2}$ in. since a smaller distance than this would be difficult to detect; to limit ankle movement it should be not more than 1 to 2 in. Force required to operate the push-button when the foot is not resting on it should be 3 to 4 lb and this resistance should start low, build up rapidly and then decrease rapidly so that it indicates that the control has been operated. If the foot has to rest on the control all the time, a force of around 9 to 10 lb should be required to operate it, as for a pedal. Push-buttons can be operated by the side of the foot, but this is probably undesirable and the load would have to be fairly light with a very positive action.

Push-buttons Operated by the Hand

The surface of a push-button should be larger than the surface of the finger which is to operate it; it should therefore be not less than $\frac{1}{2}$ in. in diameter and should preferably have a slightly concave surface. Push-buttons which are to be pushed by the palm of the hand, such as emergency stop buttons, may be as wide as the hand, but should never be less than 1 in. in diameter. The minimum movement should be about $\frac{1}{4}$ in. and the maximum movement about 1 in. The movement should have positive action and the operator should be able to know when the movement has been completed by a change in load or by a click. The maximum pressure which is required to operate a push-button should not exceed 30 oz and in order to have adequate positive control it should probably be not less than 7 to 10 oz. If there is a danger of a button being pressed accidentally, it should be recessed into a panel or guarded by a raised rim. This should be large enough to permit the entry of the finger or thumb without obstruction.

General Note

There are a number of specialized controls which will not be dealt with here. For instance, the combined lever and segmental handwheel used in aircraft or the handlebars of a bicycle. It will generally be possible to reduce the movement of such controls to one of the basic types which have already been discussed: for instance, the operation of the handlebars of a bicycle may be likened to that of a horizontal starwheel in which opposite spokes are gripped.

CHAPTER TWELVE

Environmental Factors I.
Environmental Temperature and Humidity

Temperature and humidity affect workers in industry in a number of ways. In ordinary jobs, the temperature of the work room may influence the efficiency and/or safety of the workpeople. Studies of performance have been carried out, both in the laboratory and in industry, the former predominating; industry has also yielded some accident statistics.

Of the laboratory studies, an extensive series on the effect of both heat and cold have been carried out by members of the Applied Psychology Research Unit at Cambridge on artificially acclimatized men (Mackworth,

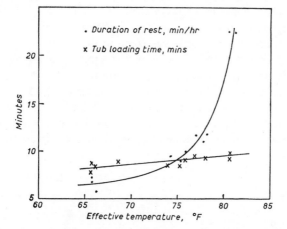

Fig. 91. The effect of temperature on work and rest when loading coal (after Vernon and Bedford, 1927a).

1950) and in heat on naturally acclimatized men by members of the R.N. Tropical Research Unit in Singapore (Pepler, 1958). This work suggests that skilled performance would deteriorate sharply if the effective temperature passed beyond the region 27° to 30°C (81° to 86°F). This same region of temperature was found both in Cambridge with men living in a temperate climate, and with men acclimatized to working in heat in Singapore. Most experimental work related to activity in cold, has involved the testing of special clothing at extreme temperatures (Burton and Edholm, 1955). Manual dexterity at temperatures which can be withstood in normal clothing, are said by McFarland *et al.* (1954) to fall off markedly below 10°C (50°F). Tracking performance is impaired both by long (13 hours) and short (½ hour) exposure to an ambient temperature of 11·5°C (55°F) (Teichner and Wehrkamp, 1954; Teichner and Kobrich, 1955).

Industrial studies of performance, in relation to temperature, are rather more scarce. In an investigation into coal mining, Vernon and Bedford (1927a) showed that there was a slow but steady increase in the time taken to load coal tubs as the effective temperature increased from 19°C (66°F) to 28°C (82°F). Time taken 'resting' also increased, but more markedly at temperatures above 24°C (75°F) (Fig. 91). The working efficiency at the higher temperature was 41% less than that at the lower. In a weaving shed, the average hourly output by pick-count on 44 looms fell from 7,163 with r.h. 77·5 to 77·9% to 6,832 when r.h. rose to 82·5 to 84·9% (dry bulb 24° to 27°C, 75° to 80°F). When the r.h. was between 75 to 80%, the count fell from 7,335 at 21·2° to 22°C (72·5° to 74·9°F) to 6,995 at 27·5° to 29°C (82·5° to 84·9°F) (Wyatt *et al.*, 1926). On the industrial effects of cold, little has come to light. Bedford (1940) gives details, but no reference, of an industrial experiment on bicycle chain assembly in which reduction of temperature from 17·5° to 10°C (62° to 50°F) caused the time to complete a task to increase by 12%. Some data on accidents in relation to temperature are available (Osborne and Vernon, 1922) which show an increase in accidents both with decrease and increase of temperature from an optimum of 19° to 20·5°C (65° to 69°F) (Fig. 92),

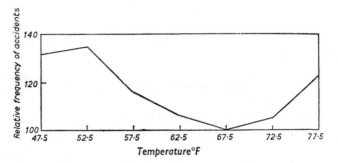

Fig. 92. Accidents and temperature (after Osborne and Vernon, 1922).

the increase with cold being somewhat the greater. A similar effect is shown by Vernon and Bedford (1931) in relation to age of coal miners; below 21 °C (70°F) there is very little difference between the age groups, but the effect of increasing temperature is more pronounced over the age of 30 years particularly on the over-50s (Fig. 93). If accidents reflect in any

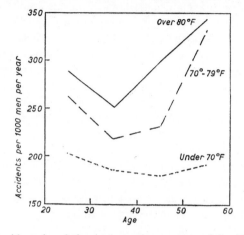

Fig. 93. Accidents in relation to age and temperature (after Vernon and Bedford, 1931).

way the effect of temperature on efficiency, then it might be taken that older men are adversely affected by higher temperatures. The age at which heat will begin to have an adverse effect is not known. Men aged 40 to 45 years, who could be engaged on mine rescue work, had survival times which differed little from those of men 20 years old (Weiner and Lind, 1955), but, as the authors say, the men were 'subject to a continuous process of elimination by regular tests until they retire from rescue work at the age of 45' so the result is not very surprising. They go on to suggest that, taking into account individual variability, an increasingly larger number of men will be less tolerant to heat by age 50. When these effects of age are taken into account, these results clearly demand that findings on 'fit acclimatized young men' should be treated with the utmost caution until they have been adequately validated in relation to an industrial population of *all* ages.

So far we have been discussing relatively small temperature deviations from the comfort zone, but more extreme conditions of heat and cold may be found which will give rise to special problems. Heat may be troublesome under two conditions: warm-moist and hot-dry. Other combinations of these variables could be warm-dry which will be no problem and hot-moist in which it is likely to be impossible to survive; these two will not therefore be discussed.

In a warm-moist environment, the heat is mainly convected and work becomes difficult because of the high humidity which reduces the evaporation of sweat, which in the absence of convection is the main way in which the body can lose its heat. The amelioration of this condition is primarily a problem for the ventilating and air conditioning engineer; it will be for the ergonomist to define the limiting conditions which should be met.

In the second condition, the heat, being in the form of radiation, cannot be dealt with by air conditioning. However, the length of time a man can work when subjected to radiation, will again depend upon the amount of sweat he can evaporate. Amelioration of this condition will depend on preventing the heat from falling on the body or, where this is impossible, removing the workman from the environment at appropriate times to allow him to cool off.

Work in extreme cold will usually be found in Britain in the cold storage industry. In winter people working out of doors will sometimes have to withstand cold for which they are frequently ill prepared, but since this is largely fortuitous, it will not be dealt with here.

INDUSTRIAL HEAT

In many industrial processes, men are subjected to radiant heat from furnaces or similar sources; this can rarely occur without in some way affecting the efficiency and output of the operatives. In some instances the effect may be quite obvious since men have to take rest away from the heat in order to cool off. In other instances the effect may not be so obvious since the men's cooling off period may be disguised by breaks in the process. For instance, in a forging shop with one or perhaps two furnaces per machine, the men and the machine may be idle for quite long periods while the forgings are being heated, and the men usually spend this time away from the source of radiant heat. If, however, the speed of heating were increased, and assuming that the forging press could be sufficiently cooled, the men would probably be unable to tolerate the increased heat load unless measures were taken to give them protection.

There are other situations in industry in which individuals may be subjected to heat stress; and these will arise not because the air temperatures are extreme or because radiant heat is excessive, but where the nature of the process causes individuals to work in high humidity. Warm-moist processes of this kind are to be found, for instance, in textile mills, dye works and laundries. Here the problem arises not so much because of the rapid rate of heat gain into the body, as occurs when men are working in radiant heat, but from the difficulty of dissipating from the body even the heat generated by metabolism during work.

We have already seen that the body attempts to maintain heat balance and a constant body temperature, and that this can be done so long as the regulating mechanism is working within its capacity, and heat gains or losses are not excessive. Heat is gained by the body through convection, radiation, the production of body heat through muscular activity and basal metabolism; it is lost through evaporation, convection and radiation. This heat balance may be expressed thus:

$$M \pm R \pm C - E = O \qquad (15)$$

Where M = heat gain from metabolism; R = heat exchange by radiation; C = heat exchange by convection; and E = heat loss by evaporation. Strictly speaking, a term for conduction should also be included in this formula, but the amount of heat actually gained or lost by this means is so small that it can be ignored. In the kinds of environment which we are considering, where temperatures and/or humidity are in excess of the normal comfort range, there is generally heat gain from either radiation or convection or both, so that the principal source of heat loss becomes evaporation. The formula may therefore be re-written as:

$$E = M \pm R \pm C \qquad (16)$$

The rate of evaporation depends upon the sweating mechanism of the body and upon the relative humidity or water vapour pressure of the environment in which the individual is working. In the hot-dry environment in which the main source of heat is radiant, the problem is primarily one of heat gain, while in the warm-moist it is primarily one of heat loss. In a hot-dry environment, the treatment of the situation will depend in the first instance on the reduction of the heat load; in a warm moist environment it is primarily a problem of ventilation and dehydration of the ambient air. An individual who is responsible for the ergonomic aspects of a particular factory should be in a position to diagnose conditions under which people are working unfavourably, so that the appropriate action can be taken. The methods of measurement which will be described will be applicable to both types of problem.

Should the body be gaining heat either through muscular work, radiation or convection at a rate greater than it can lose heat by evaporation, then the body temperature will rise and it is generally accepted that the maximum rise which can be permitted is about 1 °C or 2 °F. Should the body temperature be allowed to rise too much severe heat stress can lead to heat stroke. But before this occurs, there may be a number of indications of heat strain; these will include an increase of blood flow to the skin surface with an increase of heart rate and an increase in skin temperature, thirst, lassitude and faintness. Of these indications, the most readily recognized is the increase in heart rate and this means older men are less well able to withstand working in hot environments, due to a progressive impairment of the cardio-vascular function. For the same reason young people who are suffering from any form of heart trouble should not be called upon to work in hot environments. Man's tolerance to heat increases with time so that conditions which would be quite intolerable to unacclimatized individuals can be reasonably tolerated by those who are acclimatized.

In all the foregoing it has been assumed that the individuals involved are doing some form of physical work. The effects of hot environments on individuals doing very light or clerical work can also be of importance, although the occasions when this kind of work will be undertaken in excessive heat or humidity should be relatively infrequent in the U.K., although they are a very real problem in the tropics.

The Estimation of the Heat Load

Measurements may be made either upon the man or upon the environment and these will be dealt with separately. These measurements can be used in order to identify those jobs on which some action needs to be taken, to confirm the effectiveness of any action taken, or to calculate the length of time an individual should work in a particular environment which cannot be modified.

Estimation by Measurements on the Man

Physiological measurements on the man are of two kinds. First, pulse rate and oxygen consumption and secondly, temperature. Considering the use of pulse rate first, the calories expended in doing a particular physical job will remain almost unchanged in the presence of radiant heat. Therefore, measures of oxygen consumption alone will not be an indicator of heat load (Christensen, 1953). The heart rate will, however, increase and the normal relationship between heart rate and oxygen consumption in the absence of heat will no longer apply. The amount by which the heart rate exceeds that which is associated with a given task in the absence of heat is a measure of the heat load on the individual, and as a working rule the pulse rate should not increase by more than 15 beats/min on the average above that for the same work in the absence of heat, or an average of 45 beats/min above the resting pulse rate. A calculation of the working and rest periods can be made using the heart rate in the same way as it is used for estimating working and resting periods in the absence of heat (see Chap. 17). The pulse recovery sum can also be used as an indication of heat stress (Müller, 1959-60) and the pulse recovery time as an indication of the length of recovery time required (Brouha, 1960).

The pulse rate can with advantage be used also to indicate situations which require some amelioration; the more rapidly the pulse rate increases, the more drastic are likely to be the remedies which will have to be taken. Although pulse rate is a rather rough and ready method, it is probably sufficiently accurate for many industrial purposes. But when really accurate measurements are needed, the second method of measurement by recording deep body temperature should be used. This would have to be done under physiological guidance by an expert; the most common method is to use a recording rectal thermometer, for this reason it is not a method

Fig. 94. Temperature sensitive radio pill – external appearance (Wolff, 1961).

which is likely to be used for routine measurements in industry. 37·1°C (98·7°F) is the mean rectal temperature: work should stop when this has risen to 38·6°C (101·5°F). In studies of mine rescue personnel, rectal temperatures were allowed to rise to 38·8°C (101·8°F), a temperature above which the subjects might be expected to collapse (Lind *et al.*, 1957; Lind, 1960).

Recently a new method of measuring deep body temperature has been described by Wolff (1961), which involves the subject swallowing a small radio pill which will transmit the body temperature to a remote receiver (Fig. 94). The pill, which is rather less than 1 cm in diameter by 2½ cm long, consists of a temperature sensitive element, and a transistor oscillator powered by a sub-miniature mercury battery (Fig. 95). The pill takes about 48 hours 'in transit', the battery life is sufficient for the pill to function during this time. Once the pill has been swallowed its presence is unnoticeable and the subject can go about his daily work without any encumbrances such as a small transmitter or leads from a rectal thermometer. This pill is now in production and has been tested extensively in military experiments (Edholm, 1962). Clearly it can open up interesting possibilities for the measurement of the effects of industrial heat (or cold) upon workpeople.

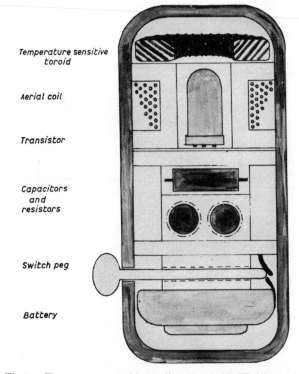

Temperature sensitive
toroid

Aerial coil

Transistor

Capacitors
and
resistors

Switch peg

Battery

Fig. 95. Temperature sensitive radio pill – detail (Wolff, 1961).

Estimation by Measurement of the Environment

An object of carrying out measurements on the environment is to attempt to express in terms of a single index the variables which may contribute either to comfort or to heat stress. Such an index could be used to compare different environments, or else to lay down limits for work under extreme conditions. The variables which must be taken into consideration are the dry bulb temperature, the humidity, the air velocity, and radiation from surrounding solid bodies. The instruments used in making these measurements will be described at the end of this section, but it should be pointed out that humidity is obtained from the wet and dry bulb readings, while a measure of the radiant heat is obtained from the globe temperature.

Many people have proposed scales which purport to measure two or more of the thermal factors involved, but the only one which seems to have come into general use is that originated by Houghten and Yagloglou (1923) which is known as the scale of *effective temperature*. This was devised by collecting a large number of subjective judgements by individuals stripped to the waist under varying environmental conditions, and by comparing the sensation of warmth experienced with that in still saturated air. Thus any combination of environmental conditions which feels as warm as still saturated air at a temperature of 21 °C will have an effective temperature of 21 °C. This scale is now known as the 'basic scale', subsequently a second series of determinations was made on individuals lightly clothed which is now known as the 'normal scale' (Yagloglou and Miller, 1925). In preparing these scales only temperature, humidity and air speed were taken into consideration, and no allowance was made for radiant heat. Effective temperature, therefore, is of value only for measurements for heating and ventilation purposes under normal factory conditions, it cannot be used to define conditions in the presence of furnaces. For this reason, in 1940 at the request of the Royal Naval Personnel Committee, a revised scale known as the corrected effective temperature scale was produced by Bedford (1940) in which an adjustment was made to allow for radiant heat (Macpherson, 1960).

Both the effective and the corrected effective temperature scales have been very widely used, since they cover conditions which are to be found in the home or in light industrial work. However, they make very little allowance for differences in clothing worn or in the amount of physical activity being undertaken, and for this reason they may, under certain circumstances, give a totally false picture of the needs of individuals working under conditions which differ very much from those under which the scales were determined, that is on lightly-clothed people carrying out light work. Thus if these scales are to be used for heavy work, it would be necessary to establish scales for varying degrees of energy expenditure. Scales of this kind have been produced (Smith, 1955) which cover a range of energy expenditure from 54 to 175 Kcal/m²/hr, to give a scale of adjusted effective temperature. Other deficiencies of the effective temperature scale are that the harder men work the nearer the effective temperature seems to lie to the wet bulb temperature, that the effective temperature scale does not make sufficient allowance for low airspeed in hot and very humid atmospheres, and that, except under conditions similar to those under which the scale was determined, undue weight is given to the dry bulb temperature.

A scale of warmth known as the *equivalent temperature* is used in the United Kingdom by some heating and ventilating engineers. Although it takes into account dry bulb temperature, air movement and radiation, it ignores humidity and therefore has very little value except over a very restricted comfort range and certainly not above 21 °C (70 °F). It is, therefore, of no value for our present purpose and will not be discussed further.

Two major attempts have been made to overcome the deficiencies of the corrected effective temperature scale. The first of these was proposed in 1945 by McArdle and his co-workers and was based on the sweat rate during four hours (P4SR). Their results were originally published in a Naval Report, but two reports including later work have now appeared (Smith, 1955; Macpherson, 1960). The other index is known as the *heat stress index* and was proposed by Belding and Hatch (1955). The background of these two proposals was somewhat different; the work of McArdle and his colleagues was done during the war and was aimed at a measure of the environmental load in H.M. ships. Subsequent development of their work has been mainly under Naval patronage and is aimed primarily at defining limiting conditions. It does not, so far, appear to have been applied to a situation where it is desired to estimate the length of working time and of rest time in a given environment. On the other hand, the Belding and Hatch index of heat stress was worked out against an industrial background and is intended to be used for defining both the environment for a given amount of work and the amount of work in a given environment. Both measures differ from the effective temperature scale in that they take into account the metabolic heat produced in the individuals. The P4SR can take into account differences in clothing, but the Belding and Hatch index is based on a 'standard nude man' extrapolated to cover the wearing of a light covering of clothing. Both measures are based on the concept that the rate at which a man sweats is an adequate index of the heat stress which he has to withstand.

McArdle and his co-workers carried out a great number of experiments on men in different environments, measuring their rate of sweating. From their results they have prepared a nomogram from which, knowing the globe or dry bulb temperature, the wet bulb temperature and the speed of air movement, it is possible to obtain figures for the *basic* four hour sweat rate. The *predicted* four hour sweat rate (P4SR) is equal to the basic four hour sweat rate for a man sitting and dressed in shorts. But when radiant heat is present or when the subject is working or when he is wearing more clothing than shorts, an adjustment has to be made. This is done by making corrections to the wet bulb temperature by adding a given amount for differences between globe temperature and dry bulb temperature, additions varying between 1°F and 7·5°F for activity which produces from 55 to 250 Kcal/m²/hr and 1°F for every 300 grams of clothing above the standard clothing weight of 600 grams. Using the modified wet bulb temperature, the basic four hour sweat rate appropriate to the clothing and work being done is obtained. Further additions to this basic sweat rate are then made to allow for the increased clothing or increased rates of work to give the appropriate predicted four hour sweat rate, so that two adjustments are made for each factor.

McArdle and his co-workers found that the P4SR could be used to predict the increase in rectal temperature in degrees Fahrenheit at the end of a four hour period in any given environment, by multiplying the P4SR in litres by 0·4. It has already been mentioned that it is generally accepted that heat exposure should cease when the rectal temperature has increased by 1°C (1·8F) and this is thus directly related to the maximum probable limit of endurance for fit acclimatized men at a P4SR of 4·5 litres. Although it might appear that, by using this relationship, it would be possible to predict the time of exposure under given conditions, this is not necessarily so, since it has not been established that it is possible to extrapolate from the four hour rate to other periods. The rate of sweating over a particular period appears to depend on the length of time that sweating has continued. For instance, for brief periods men can sweat at well over 2 litres per hour, but if exposure continues for about 24 hours, the rate of sweating is likely to be below ½ litre per hour. This means that it cannot be assumed that if a P4SR of nine is found, exposure should continue for 2 hours only. On the other hand, Lind and Hellon (1957) found a linear relationship between pulse rate and P4SR between values of 1·75 and 6·5 P4SR. The pulse rate value corresponding to 4·5 P4SR is about 103 beats/min, which is about the value accepted as maximal in the absence of heat. The subjects used were resting, nude men, who are not usually to be found in industry. It seems, therefore, that sufficient research has not yet been done on this index to enable it to be readily used as a means of estimating the length of working period and length of rest. For this, the authors can hardly be criticized because it was not intended for this purpose, but as a means of defining thermal stress. But in industry the problem can be somewhat different; conditions may need to be specified, but more often when everything possible has been done, the question which has to be answered is how long should work continue and how long is required for recovery?

This has been largely side-stepped by work physiologists preoccupied with military problems. An early attempt to determine safe exposure time was made by the American Society of Heating and Ventilating Engineers and the U.S. Bureau of Mines in 1922-26 (Hatch, 1958). Tolerance time was defined as the duration of exposure required to raise the rectal temperature to 101 °F or the pulse rate to 125 beats/min. Curves are given relating exposure time to environmental temperature at three levels of relative humidity and at two levels of metabolism (rest and 240 Kcal/m²/hr). The use of this relationship in the presence of radiation is not likely to prove very satisfactory, based as it is on data similar to that from which the effective temperature scale was constructed. Since this A.S.H.V.E. tolerance table was published, the only new approach to industrial heat assessment has been that of Belding and Hatch (1955), who related the research on the sweat rate to an index which could be used for industrial purposes.

Belding and Hatch's approach is based on the notion that heat stress will exist whenever the heat produced by metabolism or absorbed from the environment, exceeds the loss from the body. Under normal conditions of comfort, when no heat stress is present, radiation and convection will be able to dissipate all the heat which is produced. A certain amount of heat is also lost through insensible perspiration, by which means some 12 to 24 Kcal/hr are evaporated by means of the moisture in the breath and on the skin surface but since this insensible evaporation is relatively quite small it is ignored in the computation. When radiation and convection are unable to maintain heat balance, the only other means of losing heat from the body is by the evaporation of sweat. From this it follows that it should be possible to estimate the evaporation which is required in order to maintain heat balance. This has been called $E_{req.}$, where

$$E_{req.} = M + R + C \qquad (17)$$

The greatest amount of heat which can be evaporated under any particular combination of environmental conditions will depend on the air velocity and the vapour pressure of the ambient air; this has been called $E_{max.}$ and in theory so long as the $E_{max.}$ exceeds the $E_{req.}$ an individual should be able to get rid of all his excess heat by sweating. He should not start to store heat until he can no longer lose heat into the environment, that is, until $E_{max.}$ and $E_{req.}$ are approximately equal. Unfortunately this does not work out entirely in practice because as $E_{req.}$ approaches $E_{max.}$ the cooling efficiency of the sweat will decrease.

In order to obtained the value of $E_{req.}$ it is necessary to know the metabolism and the values of R and C. The metabolism may be obtained either by direct measurement in a cool environment of the work normally undertaken in heat or may be estimated from synthetic data. To assist in obtaining values of R and C, Haines and Hatch (1952) have produced formulae which have been incorporated in nomograms* using globe temperature and air speed. (The figures are given in Btu per hour as they are intended for engineers – for comparison 1 Kcal can be taken as being equivalent to 4 Btu, or more exactly 3·95 Btu.) These figures are for an average nude man with skin temperature of 95°F. This nude man is an arbitrary standard man who is considered to weigh 154 lb and have a surface area of 20 sq ft. He will be of above average physique and be acclimatized to heat.

The $E_{max.}$ is calculated from the formula:

$$E_{max.} = 10·3 \ V^{0·4} \ (42 - VPa) \tag{18}$$

where V is the air velocity in feet per minute and VPa is the vapour pressure of ambient air in millimetres of mercury. Again the authors have done the hard work and have produced nomograms* from which the $E_{max.}$ can be derived knowing the wet and dry bulb temperatures and the air velocity.

In order to tie these figures to the known tolerance of men to different environments, the authors have examined the measured sweat rates found by other workers and from these they have concluded that, taken over the normal working day of 8 hours, the sweat rate of a fit acclimatized nude young man would be approximately 1 litre of sweat per hour. They have therefore tied their index of heat stress value of 100 to this sweat rate by saying that the evaporation of a litre of sweat requires 2,400 Btu. For stresses below 100 they have proposed that the heat stress index should be:

$$h.s.i. = \frac{E_{req.} \times 100}{E_{max.}} \tag{19}$$

or should the $E_{max.}$ be greater than 2,400,

$$h.s.i. = \frac{E_{req.} \times 100}{2,400} \tag{20}$$

Comparisons of this index with results from other workers, which are given in the Belding and Hatch paper, show that there is a fairly close fit when relative humidity is low but that the heat stress index produces lower results than those found experimentally when the humidity is high. For this reason, Belding and Hatch proposed that the heat stress index could

*The nomograms will also be found in Hatch (1958).

be used for individuals wearing light clothing, although the calculations have been made on the basis of a standard nude man.

In theory, so long as $E_{max.}$ exceeds $E_{req.}$, that is when the heat stress index is less than 100, an individual should be able to get rid of all the heat generated or received from the environment and should be able to maintain heat balance without a rise in body temperature. When, however, the required evaporation exceeds the maximum which is possible, heat will be stored in the body and it will cause the temperature to rise. The permissible rise in body temperature should be $1°C$, which will permit the standard man to store approximately 250 Btu, at which point exposure to heat should cease and he should be allowed to cool off. On this basis the maximum spell of work (TW) in hours has been suggested as being:

$$TW = \frac{\Delta H}{E_{req.} - x} \tag{21}$$

where x can equal $E_{max.}$ or 2,400, whichever is the smaller. The time which it will take a man to dispose of this excess heat (TR) will depend on the amount stored, the clothing worn, and the difference between the maximum amount of evaporation which is possible in the recovery environment and the evaporation required in that environment, so that:

$$TR = \frac{\Delta H}{E_{max.} - E_{req.}} \tag{22}$$

where $E_{max.}$ does not exceed 2,400 Btu.

A difficulty which arises in making practical use of this formula is that it seems to be applicable only to fit acclimatized young men, capable of withstanding a $h.s.i._d$ above 100. Now Belding and Hatch have suggested that there may be justification for averaging the heat stresses when exposure is intermittent and from this it may be argued that where stress and recovery are intermittent it might be possible to use, in arriving at an average, the h.s.i. of the recovery environment which is likely to be negative. Based on this argument Murrell (1957) has suggested that if an overall heat stress index for the day is decided upon, the following relationship could hold:

$$h.s.i._d = \frac{h.s.i._w \times t_w + h.s.i._r \times t_r}{t_w + t_r} \tag{23}$$

where $h.s.i._w$ and $h.s.i._r$ = heat stress index while working and resting. t_w and t_r = total time of working and resting in hours.

This may, for a working day of x hours, be reduced to:

$$t_w = \frac{x \, (\text{h.s.i.}_d - \text{h.s.i.}_r)}{\text{h.s.i.}_w - \text{h.s.i.}_r} \qquad (24)$$

To show how these formulae may be used consider an environment in which a fit acclimatized young man is working in heat, when $E_{req.}$ is 1,250 and $E_{max.}$ is 1,000 to give $\text{h.s.i.}_w = 125$. When not working, this man rests in an environment which has a dry bulb temperature of 65°F, a wet bulb temperature of 60°F and an air speed of 50 ft/sec to give an $E_{req.}$ of -600 and an $E_{max.}$ of 1,300 with a h.s.i._r of -46. Applying these data to formulae 21 and 22:

Maximum spell of work (TW)

$$= \frac{\Delta H}{E_{req.} - E_{max.}} = \frac{250}{1{,}250 - 1{,}000} = 1 \text{ hour.}$$

Time required for recovery (TR)

$$= \frac{\Delta H}{E_{max.} - E_{req.}} = \frac{250}{1{,}300 + 600} = 0 \cdot 132 \text{ hour.}$$

Where ΔH is heat stored for a temperature rise of 1°C = 250 Btu for an average man.

$$\text{Since } t_w = \frac{x}{TW + TR} \cdot TW$$

$$\text{Working time in 8 hours} = \frac{8}{1 \cdot 132} \times 1 = 7 \cdot 067 \text{ hours.}$$

Using these same values in formula 23 to average the heat stresses, an h.s.i._d of 105 is obtained. If this value is substituted in formula 24:

$$t_w = \frac{8(105 + 46)}{125 + 46} = 7 \cdot 064.$$

It seems that the use of formulae 21 and 22 will give h.s.i._d of above 100 whenever h.s.i._w itself exceeds 100 by an amount which requires some rest. If it is accepted that h.s.i._d should not exceed 100, this value should be used instead of 105 in formula 24 to give:

$$t_w = \frac{8(100 + 46)}{125 + 46} = 6 \cdot 85 \text{ hours.}$$

Lower values of h.s.i._d can be taken for individuals who are considered not to be able to withstand an exposure of $\text{h.s.i.} = 100$ throughout the working day. For instance, at levels above 30, Belding and Hatch suggest that there may be some risk to health if men are not reasonably fit and have not had time to become accustomed to working in heat. On this basis and using the same conditions as above:

$$t_w = \frac{8\,(30 + 46)}{125 + 46} = 3{\cdot}55 \text{ hours.}$$

This calculation does not, however, give the distribution of working and resting periods for an individual who should have an h.s.i.$_d$ of 30. Nor does it imply that he can withstand nearly four hours at 125 provided sufficient rest is given. Such an assumption might be highly dangerous if applied to older people or others with poor heat tolerance.

Unfortunately, as Belding and Hatch point out, there are relatively few physiological studies of the tolerance and recovery times of a normally clothed industrial population under conditions found in industry against which to test the validity of these procedures. While, in theory, a method which depends on estimating the time required to store a given amount of heat and then to lose it again should produce a practically useful result, too many assumptions may have to be made for reliability. It is on these grounds that the method has been criticized by others (e.g. Lind, 1960) but so far, no equally practical alternative has been proposed which does not involve making measurements on the workmen and which could be used as a routine method to decide the length of time work should continue in a particular environment.

Some suggestions have been made for improving the Belding and Hatch index. Dutch workers at the Netherlands Institute of Preventive Medicine, who have been working with a modified version, propose (Murrell, 1958, p. 159) that the heat stored should be obtained from the relationship:

$$\text{Heat stored in Kcal} = W \times 0{\cdot}83 \times X \qquad (25)$$

where $\begin{aligned} &W = \text{body weight in kg} \\ &X = \text{permitted temperature rise in °C.} \end{aligned}$

As the result of later work, Belding et al. (1960) have come to the conclusion that the h.s.i. overestimates the heat stress. This they attribute to the difference between the actual insulation of work clothing, in particular woollen clothing, and that which was allowed for when the calculations leading up to the development of the h.s.i. were made from data on nude men. In consequence, they now suggest that exposure times obtained by the Belding and Hatch method may safely be increased by 50%. A few investigations have been made in industry to assess the value of the Belding and Hatch index. Turner (1958) studied heat stress in non-ferrous foundries and compared the rate of heat loss by evaporation which could be observed with that which could be predicted on the basis of the Belding and Hatch index. Under practical industrial conditions, it is not very easy to organize the weighing of workers in the nude. Several checks were, however, made and it was thought that weighing subjects in their working clothes would produce figures that are sufficiently close for the loss of moisture through sweating to be estimated by weight. Each subject was therefore weighed clothed before starting work and immediately on finishing work: a careful record was kept of all the fluid consumed and excreted. The necessary environmental measurements were made in order to estimate E_{req}.. The metabolic rate of the work was not measured, but was estimated to be about 1,200 Btu/hr/20 ft² (165 Kcal/m²/hr). The work

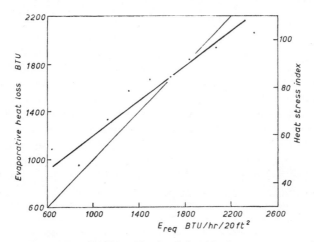

Fig. 96. Experimentally determined relationship between evaporative heat loss and E_{req} (after Turner, 1958).

being done varied from time to time, but Turner claims that a practised observer can make subjective appraisals of the metabolic cost of heavy work (Turner, 1955). Ten workmen in all took part in the study and 98 observations were made upon them. The results are shown graphically in Fig. 96, in which values for the observed evaporative heat losses are plotted against the E_{req}.. The mean heat stress indices, according to the Belding-Hatch formula are also given by Turner and these varied from 27 to 105, the three highest values are at about 100 and Turner comments that when these values were observed the men were complaining constantly about the heat in the foundry. The weather on these occasions was hot and fine with mean external temperatures of 25°C (77·5°F) and 28°C (82·5°F). It will be noted in Fig. 96 that the slope of the curve fitted to Turner's data differs from that which would have been obtained if there had been perfect agreement between the estimated and the observed figures. Below 1,700 Btu there is a progressively greater tendency for the $E_{req.}$ to underestimate the heat load, whereas above this level the tendency is reversed. It is suggested that this may be due to variations in the metabolic cost of the work in the cooler parts of the foundry on the cooler days. Under these circumstances men may work harder and Turner points out that an increase in the rate of working of only 5 Kcal/m²/hr would have increased the $E_{req.}$ by between 5 and 10% in the areas in which the heat stress was low. Within the limits of this study it would seem that the agreement between the measured and theoretical values for the evaporative heat loss is not unreasonable. In other work, agreement has not been as close as this. Three very hot environments with heat stress indices of over 300 have been reported by Lewis et al. (1960). Physical measurements were made in order to estimate the heat stress index by the Belding-Hatch method, and subjects were exposed to these environments for varying periods of time, after which the pulse rate and oral temperature was taken. The results are summarized in Table 25. The subjects were unacclimatized healthy males between 30 and 40 years of age, with weights between 72 and 89 Kg. They wore socks and shoes, short-sleeved undershirt, cotton undershorts and light trousers. The work carried out was light standing work with a metabolic rate of 600 Btu/hr. Using a 2°F increase in temperature (or an increase in pulse rate of 40 to 45 beats/min) as an end point, the Belding-Hatch method consistently underestimates the exposure time, a conclu-

Table 25. Comparisons of observed and predicted exposure times
(after Lewis et al., 1960)

h.s.i.	$E_{req.}$	$E_{max.}$	Exposure Time Predicted B. & H.	A.S.H.V.E.	Actual	P.R. Increase	Oral Temperature Increase
524	2,445	467	7·6	20	15	26	0·9
					20	30	0·8
					15	22	0·8
397	2,440	615	8·4	20	15	40	1·6
					30	62	3·7
					22	48	1·7
395	3,808	965	5·3	15	15	34	2·0
					17	28	1·7
					20	28	2·1

sion also reached by Belding *et al.* (1960), although the amount of under-estimation is substantially greater than the one third suggested by Belding. On the other hand, the A.S.H.V.E. predicted times were reasonably close to the times which were found. In addition to the effect of the insulation of clothing, underestimation of exposure time by the Belding and Hatch method may arise because their calculations are based on a skin tempera-ture of a nude man of 95°F. It is likely that the skin temperature will rise above this figure in exposures similar to those used in these experiments, which means that although, theoretically, loss of heat by radiation and conduction should cease at ambient temperatures above 95°F, it actually continues due to the elevated skin temperatures.

The authors point out that the heat stress index is not in itself a measure of the permissible exposure time. The former depends on the *ratio* of $E_{req.}$ and $E_{max.}$ whereas the latter depends on the *difference* between $E_{req.}$ and $E_{max.}$. Thus two separate environments can have a similar h.s.i., but the permissible exposure can be several times greater in one than in the other. Thus the ratio of 1,500 : 750 is the same as that between 3,000 : 1,500, i.e. the heat stress index will be 200 under both circumstances, but the difference in the first case is 750 Btu/hr, whereas in the second case it is 1,500 Btu/hr. Exposure time in the second case would therefore be half that in the first.

Control of Heat Stress

Although it has been convenient to discuss first the limitations of the length of working time or the assessment of environmental limits, strictly speaking the first step which should be taken after it has been determined that heat stress may be present is to attempt to reduce this stress. The action taken may remove the heat stress almost entirely, so that a deter-mination of the length of exposure will no longer be necessary.

In a warm-moist environment, amelioration will usually consist of removing the excess water vapour from the air. This is a problem in air conditioning and should be handled by an expert in this subject. When even quite moderate radiation is present, very little can be done by air conditioning. For instance, in a factory investigated by Grieve (1960), the amount of radiation through the glass roof was quite small as shown by the difference in one shop of some 6°F between the dry bulb temperature of 80°F and the globe temperature of 86°F. With an air speed of 24 ft/min she found that the corrected effective temperature was 76·5°, i.e. 5·5°F above the level of 71°F proposed by Bedford (1936) as the limit of comfort for light work in summer. If no steps were taken to deal with the radiation, it would be necessary to increase the air speed to over 1,000 ft/min in order to lower the corrected effective temperature to 71°. This, however, might not have had the desired effect because the difference between the dry bulb temperature at the working level of 3 ft and at 12 ft was more than 3°, so that if the increased air speed was obtained simply by increasing the turbulence in the factory itself there might have been an increase in the heat gain from convection. This solution was therefore not likely to prove very successful, nor was the alternative solution of dehydrating the air to a point where the wet bulb fell to 51°F. This would have meant reducing the relative humidity of the air from about 45% to less than 10%, which would have been impracticable. In the second shop investigated by Grieve, which had a globe temperature of 89·5° and a corrected effective temperature of 79°, no increase in the air velocity could have reduced the corrected effective temperature to 71°F, while the wet bulb temperature would have had to be reduced to 45°F, which would have been a physical impossibility. The only solution was to deal with the primary cause of the trouble, the radiation through the glass roof.

Since air conditioning can barely cope with very small amounts of radiation, when moderate or substantial radiation is present other means will have to be introduced to reduce the thermal load; these will consist either of screening the radiant source or of protecting the worker. Under extreme conditions, both methods will be used. The most effective and economic material to use for screening is aluminium foil, which has an emissivity of about 0·03 compared with 0·07 for sheet aluminium or polished steel, 0·05 for polished brass and 0·22 for stainless steel. A form of screening used by the present author, which was found to be the most effective, was to mount sheets of asbestos cement in a light angle frame and to attach the loose sheets of 0·05 mm soft-tempered foil to either side by means of paper fasteners. This gives a small air gap between the two sheets of foil and the asbestos sheet and is more than 25% more effective than when the aluminium is stuck directly on to the asbestos. A pair of billet furnaces were screened in this way, as it had been found that there was a mean radiant temperature of 296°F at the working position between them. Under these conditions the men drawing billets could work for only 7½ minutes at a spell. When the screens were installed, the working time was increased to half an hour and instead of sitting out near the door to cool off the men were doing other work. In fact, the last time the job was seen (in mid-winter), the men were working in pullovers and complaining that they were too cold! In addition to primary sources of radiation, such as furnaces, substantial quantities of radiation may come from secondary sources such as cooling forgings. An ingot at 1,200°C will be radiating at the rate of 77,000 Btu/ft^2/hr and thus stacks of cooling work may often radiate more heat than the furnaces themselves. Heat load from these sources can be reduced by the use of portable screens or curtains made from foil-faced fabric.

When radiation is from one direction only, personal protection can be provided by aprons or bibs made of aluminium-faced fabric, or by a species of aluminium chain mail which has been developed in Germany. It is important that garments used for protection in this way should cover only those parts of the body upon which radiation is likely to fall, leaving the remainder of the body uncovered for the purpose of ventilation. Bibs, which will usually have attached sleeves, should have a gusset in front projecting beyond the chin so that the air warmed by the reflected heat will rise clear of the face. Sleeves should be attached so that, whatever the position of the arms, no clothing beneath is exposed. Provided that the radiation is not so severe that the clothing will catch fire, the protection afforded by ordinary textiles is quite substantial. The insulating property of clothing depends on the air trapped in the fabric and to some extent on the amount of moisture in the fabric which can be evaporated, so that it is possible that woollen garments will be more effective than cotton, although this is not necessarily the case and depends on the nature of the weave. The greatest amounts of heat are lost from the body by the limbs, which have the greatest surface area in relation to the volume. Therefore the ideal clothing for work in environments where there is moderate radiant heat is brief shorts with the trunk and arms covered by a woollen garment, if these are the parts of the body on which the radiation will fall. The alternative method of dressing, wearing trousers and stripped to the waist, means that the parts of the body not receiving radiation, from which there should be the maximum of evaporation, are covered, whereas the parts which should be protected are left bare.

When radiation comes from all round the worker, protective clothing of this kind is not very efficient. If aluminium fabric is used it may be very effective at keeping the heat out but it is equally very effective at keeping the heat in; the worker will therefore be living in a micro-climate of his own which may rapidly reach a level at which he could have heat stroke. Completely enveloping garments of aluminium-faced material are not a practical proposition for such jobs as repairing furnaces, however, the Shirley Institute, Manchester, has developed a permeable aluminium fabric, licensed for manufacture under the name Flectavent. This is protected by British Patent No. 856,341. An alternative approach has been the development of air-ventilated clothing as at the Dupont Laboratories in America (Brouha, 1960) and at the Climate and Working Efficiency Unit at Oxford, where both impermeable (Crockford et al.,

1961) and permeable suits have been developed (Crockford and Hellon, 1962). Impermeable ventilated clothing is not entirely acceptable to some workers because of the discomfort which can be produced when air, warmed by the body, escapes at the wrists and neck. This objection is overcome in the permeable suits, since the air escapes from the whole surface of the suit and this in itself acts as additional protection. The suit described by Crockford and Hellon (1962) is designed for use in mean radiant temperatures up to 200°C. They give details of the conditions pertaining when an open-hearth furnace is being cooled from the working temperature of about 1,600°C. Without special protective clothing, men can first get into a furnace to start repairs when it has cooled down to about 150°C, at about five to six hours after the heat has been turned off. But at this temperature, work can continue for only a very few minutes at a time, and anything up to 14 to 16 hours may have to elapse before the temperature has reached a point where work can be continuous. Using the permeable suit, work can start after about four hours cooling and can continue without pauses for rest because of the heat stress. Heat exchange experiments conducted with impermeable and permeable suits suggest that with the permeable suits the temperature rise of the air will be about four times that which can be achieved with the impermeable suit. An additional advantage of the air-ventilated permeable suit compared with the reflective suit, is that it can be equally effective in very high ambient temperatures when very little radiation is present.

INDUSTRIAL COLD

Cold in industry may have to be taken into account in three main situations. Firstly, the temperature may fall in a factory, because the heating equipment is unable to maintain an adequate working temperature when the weather becomes cold. It will be characteristic of this situation that the workpeople will normally be clothed for warmer temperatures and that they will not be able to protect their hands with gloves since this will prevent them doing their jobs. In the introduction to this Chapter, details are given of the findings of research into the effects of cold under these circumstances, and it seems that when the temperature drops to between 10° and 13°C (50° and 55°F) there may be a loss of manual dexterity. A useful summary of the effects of cold has been written by Provins and Clarke (1960). It is to be hoped that circumstances which will permit the temperature to fall below 13°C will occur very rarely, but it must be recognized that performance is going to suffer when this happens and that the chances of accidents will be increased. The remedy obviously is to provide an adequate heating system. Secondly, low-ambient temperatures may influence working conditions in the transport industry. Here again, the principal effect is likely to be loss of manual dexterity. It does not appear, however, that the cold likely to be experienced under these circumstances will affect the speed of reaction, unless the driver has been exposed to extreme cold for so long that his body temperature has started to drop (Provins, 1958). As in the factory, difficulties in transport will occur because conditions of extreme cold are comparatively rare and either the vehicle or the driver are not adequately equipped to meet them.

The third industrial situation in which the effects of cold may be of importance will be in the cold storage industry. With the growth in the sale of frozen food, the number of individuals who have to work in temperatures at or below zero is steadily increasing, but since these conditions can be predicted, adequate steps can be taken to provide the workers with the proper clothing. Due to changes in the technique of cooling, there has been a tendency to increase the velocity of the air circulating around a store instead of distributing refrigerant through pipes. Low temperatures are maintained by refrigerating the air outside and pumping it through ducts to all parts of the store. Where the air system has been installed, it is found that clothing which was quite adequate under the previous conditions now becomes inadequate. This is due to the increased chilling which occurs when wind speeds are increased by even quite small amounts and the effect is pronounced on unprotected parts of the body, such as the face. Thus, however well the clothing is adapted, it may be necessary for the workers to leave the cold stores at regular intervals, in order to have relief from this chill. In order to determine the amount of insulation required in clothing, the term 'clo' has been introduced by Gagge, Bazett and Burton (1941). The average insulation provided by clothing worn by people sitting comfortably at rest, in an environment of 21 °C (70°F), is said to be equal to one clo.

$$\text{The clo unit has been defined as} = 0.18 \frac{C^\circ}{\text{Kcal}/\text{m}^2/\text{hr}} \qquad (26)$$

The amount of insulation provided by one clo permits a heat transfer of 50 Kcal/m²/hr with a temperature gradient of 1 °C. Thus if a metabolism is known or can be estimated and the ambient temperature is known the number of clo units of clothing required can be determined. Insulation against cold depends on the air layer trapped in the clothing. If the garments are in any way permeable to air and the insulating layer of air becomes disturbed by air movement, the heat loss becomes very substantial (Burton and Edholm, 1955). An empirical wind chill factor has been developed by Siple and Passel (1945) which enables the cooling effect of air movement to be expressed. The effect of wind can be minimized by making the material impermeable to air, but this in turn may cause difficulty if the clothing is also impermeable to water vapour, since the sweat evaporated from the body will condense in the clothing and this in turn will make the fabric conduct heat very much more readily than when it was dry. Thus the problem of providing adequate working conditions for people who work in cold storage is not just a matter of providing layers of thick woollen clothing. When chilling effects are reduced by making the fabric of the clothing as impermeable as possible, arrangements may be necessary for the clothing to be dried out at regular intervals while the workers are resting in a suitable environment. If this is not done, much of the value of the clothing is likely to be lost.

One large firm which has introduced the air blast system of cooling has taken great care to ensure that workers are not unduly stressed (Davy, 1961). The average temperature in a store is −26°C with an air speed of 5 m.p.h. and under these conditions men work for 40 to 45 minutes followed by 20 minutes rest in the warm. Clothing worn consists of a string vest, a garment of plastic foam between cotton, a knitted pullover and a windproof outer garment of nylon and cotton. Special boots, mitts and helmets complete the outfit. Jobs which have to be undertaken include driving stacking trucks, a job which is relatively inactive. Men on this job have to be watched carefully because it has been found that if work continues too long, efficiency falls rapidly with consequent danger of accident. The men are carefully screened by the medical department which keeps an active watch on them while at work. This is an example of the application of low temperature physiology in an important industry.

Methods of Measuring Environmental Conditions

In order to measure environmental conditions, four factors have to be obtained: the air temperature, the relative humidity of the air, the air velocity, and the radiant heat. The methods of making the necessary measurements are fully described by Bedford (1940). Charts and nomograms for converting the instrumental readings into the form required and for the compilation of effective and corrected effective temperature are published as a supplement to this monograph.

Air temperature is normally measured with a mercury-in-glass thermometer. If a stationary thermometer is used, it can be influenced by any radiation which is present and therefore the bulb should be screened. If, however, the air temperature is taken from the dry bulb of a whirling hygrometer the chance of error is substantially reduced. This practice is very common, and so the air temperature is usually called the 'dry bulb' temperature. The relative humidity is also measured with the whirling hygrometer, the wet bulb temperature being obtained and the humidity determined by reference to a psychometric chart.

There are a number of instruments available for the measurement of air velocity, and these range from a rotating vane anemometer, which can be used when high air velocities are uni-directional to a hot-wire anemometer which will measure very small air currents. The simplest means of measuring air velocity is by using the kata thermometer, which is an alcohol filled thermometer, with a large bulb about $1\frac{1}{2}$ in. long and $\frac{3}{4}$ in. diameter, coated with polished silver. Kata thermometers are normally manufactured in two ranges, one having two graduations on the stem indicating 125° and 130°F respectively, while the other has graduations at 145° and 150°F. An earlier type of kata thermometer, which had an unsilvered polished glass bulb and a cooling range of 95° to 100°F, was used in order to measure the warmth of the environment. It was thought that the 'cooling power' of the environment was measured by the rate of cooling of the dry kata. This term will be frequently found in some reports of the Industrial Fatigue Research Board, but further work has shown that cooling power is not a measurement of heat loss from the human body, so the kata is now used only in its silvered form for the measurement of air velocity. The method of using the silvered kata thermometer is to heat the bulb in hot water (it is usually convenient to keep this in a thermos flask) until the small bulb at the top of the stem is about half full. The bulb is then wiped dry and the thermometer is best suspended from a retort or similar stand, in the airstream which it is desired to measure. As the thermometer cools, the meniscus at the top of the column will move slowly down the stem, and the time taken for it to pass from the top mark to the lower mark is measured with a stop watch. This should be repeated between three and five times, depending on how closely the readings agree. On the stem of each instrument will be found a number preceded by the letter F, which is the kata factor. When this factor is divided by the average cooling time in seconds the cooling power in $mcal/cm^2/sec$ is obtained. The air velocity is then calculated from this cooling power and the air temperature. This procedure is simplified by the use of one of the nomograms published with Bedford's monograph, which will give air velocities up to 1,200 ft/min.

Radiant heat is expressed in terms of the mean radiant temperature. This implies an average value of all radiation being received from every direction simultaneously. It is as if the individual, subjected to a particular mean radiant temperature, was at the centre of a hollow sphere, the whole of whose surface was radiating at that temperature. For this reason, any uni-directional method of measuring radiation, such as with a thermopile and a sensitive galvanometer, would become extremely tedious, unless it was desired to know the radiation contours from a particular direction. A simpler instrument for obtaining the mean radiant temperature is the globe thermometer. This consists of a hollow metal sphere, about 6 in. diameter, coated matt black on the outside, with a mercury in glass thermometer fitted to the neck so that the bulb is at the centre of the sphere. The globe thermometer will normally be mounted in a retort stand and it will take about 20 minutes for the temperature inside the globe to rise above air temperature as a result of radiation. The calculation of mean radiant temperature from the reading of the globe thermometer involves knowing the air velocity and the dry bulb temperature and again the task has been made easy by the publication with Bedford's monograph of a suitable nomogram. The globe thermometer was originally described by Vernon (1932); its use for the measurement of radiation from surroundings follows the work of Bedford and Warner (1934). A disadvantage of the globe thermometer is the length of time which it takes to come to equilibrium: to overcome this Hellon and Crockford have (1959) suggested improvements. First they reduced the thickness of the wall of the globe from about 0·4 to 0·2 mm. This caused no significant improvement in the time taken to reach equilibrium. The next step was to use a 30 s.w.g. thermocouple, in place of the mercury-in-glass thermometer; with this the globe reached equilibrium in 10 minutes, compared with 18 minutes for the glass thermometer. Finally, in order to provide full convection in the interior of the globe, a lightweight globe was fitted with a small fan driven at 3,000 r.p.m. When this stirring was introduced, the globe came to equilibrium in about one third the time required by the standard globe, i.e. between 6 and 7 minutes.

These four measurements made on three basic instruments are all that are necessary to define the limiting conditions of a particular environment, to calculate the heat stress imposed by an environment, or to calculate the exposure time, using a method similar to that of Belding and Hatch. The equipment is simple to use and the measurements can be carried out by an individual with a small amount of training. Using measurements of this kind, Grieve (1960), investigated a single-storey glass-roofed factory, as the result of complaints from the workpeople when the summer external temperature rose above 70°F. As a result of the measurements and the determination of the corrected effective temperature she found that during most of the day the conditions were outside the comfort zone for light work. Measurements of energy expenditure showed that some tasks were in the 'moderate' category. She concluded that the workpeople were subject to a significant thermal load due to heat radiation from the glass roof and the low air movement.

CHAPTER THIRTEEN

Environmental Factors II. Noise

Noise can best be described as unwanted sound. This definition can cover a wide variety of situations from a high ambient noise level which can cause actual damage to hearing, to the radio in the garden next door, of which it can be said that one man's music is another man's noise! The nature of sound transmission has already beeen described in Chapter 1, and the structure of the ear in Chapter 5, but for convenience the salient details will be repeated here. Sound consists of pressure waves originating from a vibrating object which pass through an elastic medium as successive waves of compression and expansion. These waves travel through air at about 1,100 ft/sec (760 m.p.h.) and the sound energy is transferred from molecule to molecule in a sphere of steadily increasing radius. Pure tones are sine waves which can be defined in terms of amplitude and frequency; complex tones are made up of superimposed sine waves.

The intensity of a sound depends on the amplitude of its constituent waves, the greater the amplitude the greater will be the sound pressure (or the sound energy) transmitted. Sound pressure is expressed in dynes/cm² and is usually measured on the decibel scale which is referred to 0·0002 dynes/cm² (1 microbar) and the level of sound expressed in this way is known as the sound pressure level (s.p.l.).

The tone of a sound is determined by its frequency, measured in cycles per second (c/s), and the frequencies which are audible to the normal ear will range from about 20 to 19,000 c/s. The octaves used in sound measurement are not the same as those used in music; they start from 37·5 c/s and range to 19,200 c/s.

The ear consists of three parts, which convert the sound pressure waves in the air into hydrostatic waves in the fluid medium of the cochlea, thereby causing the basilar membrane to vibrate at different positions along its length in sympathy with the frequency of the tones being received. Its sensitivity is least at the low frequencies and increases to reach its greatest sensitivity at above 1,200 c/s. The risk of damage to the ear is therefore not uniform over the whole frequency spectrum, but appears to be greatest for sounds between 2,400 and 4,800 c/s. In addition to any damage caused by noise, hearing loss, known as presbycusis, will also occur naturally with age. This loss seems to be greatest at the high, and least at the low frequencies. For instance, at the age of 50 years, a hearing loss of 40 dB might be expected above 12,000 c/s, of 12 dB at 12,000 c/s and 5 dB at 125 c/s compared with the hearing of an individual of the age of 20 (Hinchcliffe, 1958).

When a sound is made up of frequencies covering a major part of the sound spectrum, it is usually called 'white' or broad-band noise. In many instances sound of one wavelength or a group of wavelengths may predominate, in which case, to use the same analogy, the noise might be said to be slightly coloured! For this reason it is probably better to use the term broad-band to describe this kind of noise, although the term white is also frequently used in the literature. Unless it is specifically stated to the contrary, the two terms can be taken to be synonymous. Broad-band noise may be steady or intermittent, and upon this will depend to some extent the effects which it will produce on the hearer; these have received a great deal of attention from research workers, which has led to the definition of various effects of noise, both physical and mental, in terms of s.p.l. and continuity.

Another type of unwanted sound which has received very little attention, but which cannot be overlooked, is meaningful noise, which will have an effect upon the hearer which is related to the information content of the noise rather than to its s.p.l. As a general rule, the effects of meaningful noise will be specific to a particular situation which probably accounts for the relatively small amount of research which has been done upon it. Meaningful noise will usually be intermittent rather than steady.

In industry broad-band noise may cause deafness, may reduce working efficiency or interfere with communications. Meaningful noise may interfere with efficiency and may under certain circumstances make communication difficult.

THE DAMAGING EFFECTS OF NOISE

Aural damage can be caused by broad-band noise to which the hearer has been exposed for a number of years, the extent of the damage depending upon the susceptibility of the individual, the amount by which the noise exceeds the damage risk level, the length of exposure, and whether the noise is steady or intermittent. Steady sound may be found in such industries as weaving, where the noise from the looms will remain substantially level throughout the working day. In other industries such as heavy engineering, the noise may be mainly intermittent, caused by successive blows of pneumatic hammers or swageing machines. Another form of intermittent noise which is unlikely to be found very frequently in industry is a high intensity noise produced at irregular intervals, as, for instance, by gun fire. Unless the noise is so intense that one burst is sufficient to cause aural damage, the effects of this type of intermittent noise are more likely to result in interference with efficiency rather than to cause damage.

Continuous Broad-Band Noise

It has been long recognized that individuals who have been exposed to high noise levels over a long period of time will suffer a hearing loss. What has not been so well known is how much noise for how long will cause how much damage. This has been due not so much to lack of research effort, but rather to the many difficulties which are involved. In the first place, hearing loss must take a number of years to develop, during which time some presbycusis will also have set in. In this, as in most other human functions, there are wide individual differences, so that for any particular individual it is not always clear how much of an observed hearing loss is due to age and how much is due to the environment in which the individual has worked. In the second place, individuals for whom hearing loss has been observed have not always worked under exactly the same conditions for the whole period to which this loss may be related. They may have changed their jobs, or the nature of the process may have altered either the s.p.l. or the sound spectrum of the noise. Uncertainty on these matters

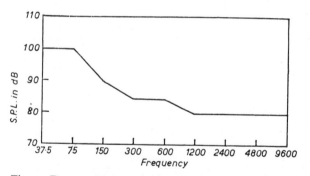

Fig. 97. Damage risk levels (after Burns and Littler, 1960).

makes the retrospective use of audiograms rather difficult and it would seem that of available data, only 10 to 15% of audiograms are of any use even under the most favourable conditions (Burns and Littler, 1960). Longitudinal studies on selected people in constant noise would be more valuable than the retrospective studies at present available, but the practical problems of obtaining these are very great. Workpeople cannot be deliberately deafened in the interests of science! In spite of these difficulties, several damage-risk levels have been proposed which have been getting progressively lower as more data become available. In the late 1940s it seems to have been generally accepted that the damage-risk level was about 110 dB in the 37·5 to 75 c/s waveband, falling to about 95 dB in the 300 to 600 c/s waveband, and remaining substantially level thereafter. The American Academy of Ophthalmology and Otolaryngology (1957) lowered this level to 85 dB between 300 and 1,200 c/s. Based on an extensive summary of available data on hearing loss, collected by the American Standards Association (1954), Littler (1958) lowered the level even further to 70 dB for frequencies between 1,200 and 9,600 c/s. This level is rather low except for people who are unusually susceptible and subsequently

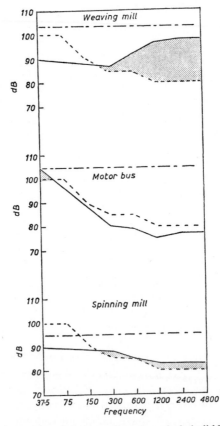

Fig. 98. Overall s.p.l. (dot-dash), and octave analysis (solid line) compared with the Burns-Littler damage risk curve (dotted line) (adapted from Bonjer and Van Zuilen's data given in Murrell, 1958).

Burns and Littler (1960) have suggested a modification which is shown in Fig. 97. This damage-risk curve relates to exposure for eight hours a day, five days per week, for a working lifetime. There are insufficient data to make recommendations above 9,600 c/s.

A broad-band noise will very rarely be made up of equal intensities in all the frequency bands, that is the noise will rarely be truly 'white'. It is not sufficient therefore to take only the overall s.p.l.; before the damage-risk can be judged the noise must be analysed into its constituent frequency bands. Should the s.p.l. in any of these bands exceed the damage-risk level, it will be necessary to take some remedial action. The effect of the distribution of the frequencies on the damage risk in three different jobs with similar s.p.ls is illustrated in Fig. 98 which is based on data given by Bonjer and Van Zuilen (1958). The overall s.p.l. in the first two instances was about 105 dB, but whereas in the weaving shed the s.p.l. exceeded the damage-risk level at frequencies above 1,200 c/s, in the motor-bus the greater part of the sound is concentrated below 300 c/s, so that the damage-risk is slight. In the spinning mill shown in the third curve in Fig. 98, the overall s.p.l. is 10 dB lower than in the motor-bus, but the frequencies are so distributed that there is a damage-risk in the 2,400 to 4,800 c/s band. The extent to which the risk of damage increases as the s.p.l. of any specific frequency band exceeds the damage-risk level is not entirely clear. It can be taken, however, that the greater the excess the shorter will be the time which will elapse before hearing is impaired. In practice this should only be of theoretical consideration, since if in any situation the s.p.l. in any of the frequency wavebands exceeds the damage-risk level, steps should be taken either to reduce the level or to provide a personal protection.

It appears that the frequency at which hearing is lost is not associated with excessive s.p.ls at this frequency, but with frequencies in the octave below; thus with excess s.p.l. in frequencies above 1,200 c/s, a temporary hearing loss is likely to show itself first in the region of 4,000 c/s. When an individual is exposed to noise which exceeds the damage-risk level, the first effects are likely to show as temporary hearing losses most of which will occur in the early hours of the work period and from which there will be complete recovery overnight. This temporary loss may be as much as 20 to 30 dB and may be accompanied by tinnitus. If exposure continues for a substantial time, irreversible damage can occur to the hair cells of the organ of Corti and the hearing loss will become permanent. For individuals continuously exposed to excessive noise, hearing loss appears to be greatest in the early years of exposure. For instance, in a study of weavers, Bonjer and Van Zuilen (1958) found that in the first four years in a group of 35 women there was an average hearing loss of 42 dB in the 4,000 c/s frequency, the 80% range being from 25 to 65 dB. For women who had worked from 4 to 25 years in noise the median increased by only 15 to 57 dB, the 80% range being from 20 to 90 dB. With women who had worked for more than 25 years the median hearing loss was 73 dB with an 80% range from 62 to 90 dB.

A similar result was found by Keating and Laner (1958) who examined the audiograms of workers who had been exposed to noise above the damage-risk level for a period of between one and three years. These workers were all under the age of 40. Their results suggested that the greatest hearing loss occurred during the initial period of exposure of up to three years. After three years exposure, the rate of hearing loss appeared to slow down, or possibly even to cease altogether. This suggests that individual protection should be worn from the start of work in the noisy environment.

Since early hearing loss is incurred in frequencies which are only peripherally involved in speech, the social effects of progressive deafness may not be apparent to the victim until it has proceeded to a point where the lower frequencies are substantially affected. Because of this, it is difficult to bring home to workpeople the need for conservation of hearing when other steps cannot be taken to reduce the risk to hearing.

Intermittent Broad-Band Noise

The noises produced by many industrial processes are intermittent, and since they are usually the result of an impact, their wave form will be characterized by high peak intensity coupled with a sharp wave front. It is thought that, due to the impulsive nature of the sound, noise with these characteristics is more damaging than is continuous noise. Satisfactory data on the effects of intermittent noise are not available since no adequate method of its measurement has been devised. The normal meter, which functions on the root-mean-square sound pressure will not normally indicate the levels reached by the sudden peaks of energy. Therefore an ordinary sound-level meter will probably substantially underestimate the peak levels reached by these noises and may therefore give a more optimistic picture of the situation than actually pertains. The most that can be said in the present state of knowledge is that any intermittent noise which even approaches the damage-risk level for steady noise should be considered a potential hazard.

Hearing Conservation

Hearing may be conserved by two methods of approach. Firstly, if individuals must work in irreduceable noise levels which exceed the damage-risk level they should wear some form of personal hearing protection. This may take the form of ear plugs which will adequately attenuate noises in all frequencies up to an s.p.l. of 110 dB, or ear muffs, which will attenuate noises up to 125 dB above 600 c/s and up to 115 dB below this frequency (Burns and Littler, 1960). Since permanent hearing loss is to an extent related to the length of time of exposure, some reduction in the risk may be achieved by alternating jobs and so reducing the length of time that any one individual is exposed to high noise levels. It seems that the effects of excessive noise on the ear will depend on the total energy which has been received during the day. Burns and Littler (1960) suggest that halving the time of exposure could permit a doubling of the energy which will produce an increase of 3 dB in the s.p.l. This means that if a noise is found just to exceed the damage-risk level in any of the octave bands the risk would probably be eliminated if the exposure time is halved.

It is recognized that it is very difficult to persuade individuals to take steps to protect themselves, but this may be made easier by the second approach of making regular audiometric examinations of all individuals who are exposed to potentially damaging noise. These regular examinations should show the development of hearing loss which at the outset may be only temporary, and act as a warning both to the individual himself and to the management that all is not well. Audiometric tests should also reveal individuals who are particularly susceptible to noise. In one study in a trade school where operatives were trained before going into a textile mill a noise level of 90 dB was measured (Bonjer and Van Zuilen, 1958). Three individuals were found who, six weeks after moving into the mill, had already suffered hearing loss in the 4,000 c/s frequency band, which was not found in audiograms taken on leaving the school. This suggested that the individuals involved were particularly susceptible to noise, and that they should not continue to work in such a noisy environment as a weaving shed. The authors suggest that it is rather disturbing that this should have been discovered only after the individuals had been trained.

In a number of countries deafness due to noise has become recognized as a compensatible industrial disability (Murrell, 1958, appendix IIb) so that it is often the practice, when there is likely to be a risk of damage to hearing, to take audiograms of all employees on engagement. In this way people who had already suffered hearing loss could be detected, and if it appeared that they were particularly susceptible to noise their place of employment might be selected so as to avoid risk of further damage. Should there be any claim for compensation these audiograms could show the extent of the hearing loss which had occurred before employment in the particular firm. It is interesting to note that this is probably the only industrial disability for which compensation can be paid which will not necessarily affect earning capacity. It may even improve the efficiency of the victim since hearing loss in the higher frequencies above those involved in speech may improve an individual's ability to hear speech in a noisy environment. Ear defenders will also act in the same way by selective attenuation so that speech can be heard better when wearing ear defenders than without them. The development of hearing loss compensation in America has been discussed by Nelson (1957).

So far we have been dealing with broad-band noise in which the frequencies, while there may be some peaks, are reasonably distributed. It may sometimes occur that within a specific frequency band there may be intense, narrower bands of noise which will have a greater effect upon the cochlea than would be suggested by the s.p.l. for that particular frequency band. Burns and Littler (1960) suggest that if this is the case, the noise is likely to be potentially more harmful than would be indicated by the levels which can be measured. This is particularly true in the lower frequencies, especially between 300 and 1,200 c/s. Under these circumstances the damage-risk criterion should probably be reduced between 5 and 10 dB to allow for the effect of these very narrow bands. It is probably advisable also, if this situation is suspected, that more frequent audiometric tests should be given to ensure that the effect of this noise is not under-estimated.

THE EFFECTS OF NOISE ON EFFICIENCY

All the three kinds of noise which have been specified may, under certain circumstances, have an effect on efficiency. Continuous noise, if it is broad-band, may have an effect which is related to its intensity. If it is narrow-band and the frequencies are near the top of the spectrum, it may cause irritation (and, indirectly, inefficiency) which may not be directly related to its intensity. Intermittent noise, if it is regular, may have effects which differ little from those of continuous noise, but if it is irregular and unexpected it may cause a 'startle reaction' which can be most disturbing. The influence of meaningful noise will be related to a large extent to particular circumstances and is likely to depend on the nature of the noise and what it means to the hearer.

Continuous Noise

The quite extensive literature on performance in noise and quiet has been summarized by Kryter (1950) and Broadbent (1957). No useful purpose will be served by repeating the details of the many experiments which are covered in these papers since the results show that it has not been possible to demonstrate clearly that noise will influence performance either on tasks involving such activities as mental arithmetic or clerical checking, or on sensory motor tasks involving the use of controls. One area in which noise has been demonstrated to have an effect is in studies of vigilance, to which Broadbent himself has made a substantial contribution, in which noise has been shown to have some influence on performance at levels above 90 dB. In an attempt to validate this finding the process of film perforation was investigated by Broadbent and Little (1960) by acoustically treating some factory bays and comparing performance with that in untreated bays before and after the change had been made. The s.p.l. of about 99 dB before treatment was reduced to 89 dB. Over the period studied there was an improvement in performance in both treated and untreated bays; there was no significant difference in output between the two bays, but on one measure of quality over which the operators had some control, the improvement in the treated bays was significantly better ($p < 0.5$) and on another measure there was a difference but the probability was $p < 0.10$. While too much weight should not be placed upon the magnitude of the changes, the improvement in the quality which was found confirms the prediction from the laboratory experiments that noise above 90 dB can induce the commission of errors even in workers who are accustomed to noise. It will be realized that in order to conserve their hearing operators should not be working in a noise at or above 90 dB, so that the practical significance of these results should not be very great, but unfortunately, men are very often to be found working in s.p.ls as high as this and once they have become accustomed to the noise, it is unlikely that they will think that it will have any effect upon their performance. Acclimatization to noise may take place quite rapidly, in some experiments probably in less than five minutes, and in general there are very few complaints from workers about continuous noise. That it can produce a higher incidence of errors and accidents is an added reason for taking steps to reduce the s.p.l. and/or to provide personal protection.

Intermittent Noise – Unexpected

Sudden loud noises which are entirely unexpected can produce a startling effect which is probably a relic of the more primitive emergency reaction which prompts an organism to 'get the hell out of it' at very high speed. This emergency state is produced by an increased secretion of adrenalin, which in turn causes an increase in blood pressure, in sweating, in heart rate, in respiration rate, and in muscular contraction (Landis and Hunt, 1939). Sudden noises will also produce a decrease in bowel activity and in the flow of saliva and the digestive juices; thus repeated exposure to sudden and unexpected noises may affect an individual's digestion (Smith and Laird, 1930). Because of the physiological effects associated with the startle reaction, this type of noise is likely to have a detrimental effect on both the health and the efficiency of workers who are subjected to it.

Woodhead (1958) studied a visual matching task which was carried on for four minutes, during which four bursts of noise of peak intensity of 100 dB and lasting four seconds were given at irregular intervals. The duration of the task seems to be rather short and it is questionable whether the noise given under these circumstances could be said to be entirely unexpected. Nevertheless, these bursts of noise caused immediate deterioration in the performance of most of the subjects, followed by complete recovery during the next half minute of quiet. This effect was found even though the subjects knew that the noise was likely to occur some time, and in one condition were given a warning when the noise would occur by means of a light. However, the startling influence was perhaps not quite so extreme as it would have been if the noise were completely unexpected.

The present author had experience of the effects of unexpected noise when he was asked to advise on an office which was overlooking a steel stockyard. Occasionally a crane would drop a bundle of steel rods into a cradle and the effect on the office staff was unfortunate. Complaints about this noise were frequent and included statements that health was being affected, particularly the digestion, which confirms the suggestion of Smith and Laird (1930). Due to impending reconstruction it was not practical to move either the stockyard or the office at that time, so it was suggested that the crane driver should be provided with a hooter sufficiently loud to be heard in the office, but not so loud as to cause startle and he was instructed to sound a warning before dropping a load. While members of the staff never completely reconciled themselves to the noise, this procedure put an end to the complaints.

Intermittent Noise – Repetitive

Intermittent noise of a repetitive nature will frequently be found in industry arising from processes involving mechanical impact. We have already noted when discussing the damaging effects that this kind of noise has the characteristic of high peak intensity which makes it possible to underestimate its effects. Unfortunately, there does not appear to be any evidence on the effects of impact noise on efficiency. In most experiments the intermittent noise has either been of irregular length or of slow intermittency and has been given on horns or hooters. Experiments using actual impact noise do not appear to have been carried out. On the face of it, it would seem that the effects of this kind of noise would vary between the condition in which the intermittency rate is high, when the effects might be analogous to those of a continuous sound with due allowance made for the peak intensity, to a condition in which the intermittency rate is low when the 'startle effect' may come into play. It might be tentatively concluded that as the repetition rate gets lower a point will be reached at which it will begin to have an effect on performance.

The results of laboratory studies on the effects of intermittent noise have been summarized by Plutchik (1959). These showed first, that performance impairment is more likely to occur during exposure to high intensity intermittent sound than when the sound is of lower intensity or when it is steady. Secondly, that the effects of noise on performance are likely to be determined by the relative difficulty of the task which is being undertaken. The more difficult the task, the more likely that it will be disrupted by noise. Thirdly, that individual differences in the reaction to noise are of great importance when evaluating the results of research. Therefore the use of the mean performance of a group may be misleading if there are wide individual differences between the subjects being used. In most of the reported experiments on intermittent noise, as in those on continuous noise, the time under test is usually quite short and bears no relationship to the length of time an individual would be exposed when working in a factory.

Annoyance

Under some circumstances some individuals may become irritated by certain types of noise and this may, in turn, have an effect on their work. High-pitched noises having a frequency above 1,500 c/s appear to be more annoying than lower-pitched noises of equal intensity (Pollack, 1949). Low-pitched noises below 256 c/s were also said to cause annoyance (Laird and Coye, 1929) and they set the upper limit at 1,024 c/s. Other workers set the upper limit as high as 2,000 c/s. From these results it seems that annoyance is associated with the higher or lower frequencies, and that between 200 and 1,200 c/s, covering two octaves, annoyance is at its minimum.

Noise which is modulated either in intensity or in frequency seems to be more annoying than noise in which these two parameters are constant. Individuals do not seem to be able to become as accustomed to this kind of noise as they do to noise which is steady. It seems probable that modulation in intensity causes greater annoyance than does modulation in frequency (Miller, 1947). Other factors which appear to be associated with annoyance are intermittency, unexpectedness, inappropriateness and reverberation (Laird and Coye, 1929). Several workers suggest that the annoying effect is enhanced when the sound pressure level is above 90 dB, and that very little annoyance is caused below this level. If this is so, then, ideally, factory workers should not be annoyed by noise, but unfortunately people often have to work in conditions which are far from ideal, so that the possibility of annoyance cannot be disregarded.

Relating a subjective feeling such as annoyance to an objective measure of performance would seem to be just as difficult as trying to relate, for instance, boredom to performance. Therefore it is difficult to say whether or not annoyance does cause loss of output, since it will be necessary to rely on the reports of individuals that they found certain conditions annoying, always assuming that they know what that term means. This approach was attempted by Pollack and Bartlett (1932), but their results are, not surprisingly, rather inconclusive; they suggest that under certain circumstances performance in noise may be either better or worse than normal. Whether annoyance caused by noise does or does not affect performance may become irrelevant if steps are taken to avoid conditions under which annoyance is said to occur.

Meaningful Noise

Meaningful noise is any sound which conveys an unwanted meaning to the hearer; it may range from your wife talking when you wish to listen to the radio, to a door banging in the night. Since meaningful noise transmits unwanted information to the hearer, its effect may have no relationship to its loudness. This makes a scientific study of meaningful noise rather difficult because the effects of any particular noise are likely to be specific. It is important, however, that it should be recognized that this type of noise exists and that its effect on performance may be more disrupting than that of broad-band noise. For instance, in a large office, which has received the full acoustic treatment, footsteps in an adjacent, uncarpeted corridor may convey a meaning to the occupants of the office if the footsteps are recognized as being those of a person attractive to the male population. On the other hand, if the footsteps are heavy, they may be recognized as those of 'the old man' and the work activity may be intensified. The private conversation of an individual with a 'quarter-deck' voice talking to Stockholm on the telephone can bring the activity of an office almost completely to a standstill.

It will be recognized that this situation will arise primarily when the working environment is so quiet that sounds which would normally pass unnoticed become obtrusive. From this it follows that care should be taken to establish an s.p.l. in offices which is sufficient to isolate individuals from their neighbours but which is not of a sufficiently high level to cause difficulty in other ways. Under certain circumstances it may be necessary to introduce sound artificially into quiet working places in order to act as a barrier between one worker and another. This sound may be 'white' or it may take the form of specially selected quiet music. The author has been told, in a private communication, of the introduction of music into the offices of an industrial concern. The noise level is not specified, but in one office the experiment could not continue because telephone communication was found too difficult, particularly if the music started suddenly in the middle of a telephone call, which would suggest that the music was too loud. In other offices, however, the use of music was continued at the wish of the occupants though there was no evidence of an increase of productivity as a result of its introduction. Any attempts, therefore, to introduce sound as an isolater between individuals and their surroundings must be very carefully balanced between too much sound and too little sound. Any sound introduced must be just sufficient to mask the unwanted sounds without destroying the illusion of peace and quiet which, to some people, can be very beneficial.

'Music While You Work'

Since music played in factories is usually assumed to be for the benefit of the workers, it may seem a little out of place to discuss this subject in a chapter on noise. However, there is evidence that not everybody enjoys music, so that for those who do not it is a noise. The possibility of using music as a sound barrier has just been mentioned; music in factories is not intended to act as a barrier but to increase output and to improve the workers' feeling of satisfaction in the job. The evidence which has been published on music in factories has been very ably summarized by Uhrbrock (1961), from which it would appear that, in spite of popular belief, there is a lack of evidence to show that music (including piped music) does increase output. Because of the recent development of the use of canned music over public address systems, it is often thought that the use of music in factories is a development of the last 20 years, but in America at any rate it was quite a common custom for live musicians to be employed in factories during the years before World War II. For instance, Clark (1929) reported that 267 bands, 182 orchestras and 176 choruses were employed in a sample of 679 plants studied.

The research reviewed by Uhrbrock consists mainly of popular articles which seem to accept that music is a good idea without requiring any proof, of surveys of the opinions of the workers, of studies carried out for commercial purposes with the publication of selective and unproven results and of investigations by scientists whose methodology is open to question, either because the groups of individuals whose output was compared were not strictly comparable, or because they were inadequately trained. One piece of research which cannot be criticized on these grounds is that of Wyatt and Langdon (1937) who conducted a 24-week study of the effects of gramophone music as part of a larger study of boredom. The job being studied was the assembly of paper decorations which was a repetitive job demanding very little mental effort. Difficulties associated with experiments of this kind were minimized by obtaining production data for 30 days before any experiments started and then alternating periods of music and no music for varying lengths of time and at different periods of the same day. In their initial studies of boredom, Wyatt and Langdon formed the opinion that the greatest incidence of boredom was in the middle of the morning, and they therefore played music, in one experiment, from 10.0 a.m. to 11.15 a.m. and obtained increases in output of about 6%. When they tried the effect of playing music twice during the morning the output was not as high as that achieved with the single period, but was still about $2\frac{1}{2}$% higher than it was in the reference period. Finally they tried the effect of playing music in alternate half hours throughout the work period and under these conditions they obtained an increase in output of about 4·5%. Although this is less than the increase obtained with the single longer period of music, this arrangement was the one which was preferred by the workers.

A more complex industrial task involving rug manufacture was investigated by McGhee and Gardner (1949). They studied 142 women who had already been fully trained. The experiment lasted five weeks, on four days in each week music was played and there was one day without music. The investigators were unable to find that the music had any effect, good or adverse, on the production of these women. The same result was obtained by Smith (1961) in an experiment on key-punch operating, a task involving complex mental activity. Industrial music had no significant effect on output, errors or absentee rates, but then, there is no reason to expect that it would since the music was played only during the lunch and break periods.

As it cannot be demonstrated with any degree of certainty that music has any effect on output, is there any good reason for it being used, or equally, are there reasons for it not being used ? Opinion surveys have been conducted which suggest that varying numbers of people, up to 10% of those questioned, thoroughly dislike music. Even more people may dislike particular kinds of music and this may account for some of the findings that quality of work may suffer when music is played and in some instances output may decrease. Nevertheless, it does appear that the majority of people prefer to have some music played at some time and this may be a good reason for adopting the practice, in order to produce a happy atmosphere in the workshop. It is quite clear, however, that very little in the way of increased output can be anticipated with any degree of confidence.

One point which is not mentioned in the literature is the effect of superimposing music on an already high noise level in a workshop. If this noise level is bordering on the damage-risk level the addition of music, which must be above the noise level by perhaps 5 to 10 dB to be heard, may render the general noise level dangerous. On the other hand, exposure will be only for a limited period during the day. Whether employees like music which may make the noise level so high that speech becomes impossible is a point which is not touched upon in the literature, and no attempts appear to have been made to relate experimental results in this field to the overall sound pressure level in the workshop when music is being played. Clearly this is an area which requires further investigation.

THE EFFECT OF NOISE ON COMMUNICATION

Noise may interfere with speech communication from person to person, by loudspeaker or by telephone. The first two will be influenced by the ambient noise level in which the listener is situated, while the third will, in addition, be influenced by noise in the communication channel. In all cases the extent to which the noise will interfere with the communication will depend on the complexity of the message and the extent to which the content of the message can be predicted. Thus, digits can be understood with 70% intelligibility when the signal is 10 dB less than the noise, but for 100% intelligibility the signal must not be less than about 3 dB than the noise. Seventy per cent intelligibility is usually sufficient to understand most messages and this can be achieved when the signal-to-noise ratio is parity, but should the nature of the message be difficult to comprehend or if proper names are used, as when they are called over the public address system, the level of the message may approach that required for nonsense syllables, about 20 dB above the noise (Miller *et al.*, 1951). Since the main components of speech are concentrated in the lower frequencies, particularly for male voices, interference by noise will depend upon its spectrum: the nearer the spectrum approaches that of speech the more likely it is that there will be interference by noise. In general it may be taken that the extent of speech interference may be determined by taking the arithmetic average of the decibel levels of the noise in the three octave bands, 600 to 1,200, 1,200 to 2,400 and 2,400 to 4,800 c/s. Hawley and Kryter (1957) have suggested noise levels which will just permit reliable conversation at different distances between the speaker and the listener. These are shown in Fig. 99. If noise levels above these are found, then some difficulty in communication may be expected. Shouting will permit communication at higher noise levels, but shouted speech is not always very intelligible. When a noise has a large component above 1,500 c/s the wearing of ear plugs can assist voice communication.

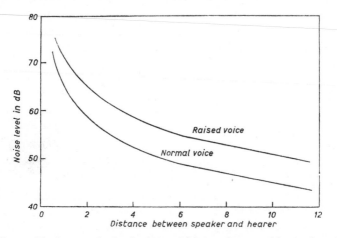

Fig. 99. Maximum noise levels above which interference with speech can occur (after Hawley and Kryter, 1957).

Noise can also interfere with communication between man and his equipment by masking the sound of warning signals or bells; where this is likely to happen, auxiliary visual signals should be considered. An operator may also have to use auditory cues from his equipment in order to ensure either that it is working correctly, or as signals for action. Care should be taken to ensure that, where it is known that these cues are likely to be used, the noise level is not such that they are rendered inoperative, which could easily occur at levels well below those at which aural damage might be expected.

NOISE REDUCTION

There are three main ways in which the level of noise at the ear can be reduced. The first, and probably the most important, is the reduction of the noise at the source. The second is the isolation of the equipment from the surrounding structure or its total enclosure to prevent the noise spreading, and the third is the absorption of noise to prevent either direct transmission or reflection from surrounding objects. Reduction of noise at source will usually have to take place at the design stage, although there are a number of noise-producing activities which may be under the control of the production engineer. What is often not realized is that, quite apart from any influence which noise may have on workpeople, noise reduction can often be a good economic proposition· A good deal of noise which emanates from factories is impact noise; this implies that wear on tools and moving parts must be greater when they are very noisy, and further, that under certain circumstances, greater power is required than when, by adequate design, noise is reduced. For instance, punches are often ground so that they have a level or almost level surface (it is much easier to grind a punch in this way and the initial cost is lower), but if the punch is ground with a sheer, not only will the noise level be substantially reduced, but the life of the tool may increase between three and five times; additionally it is often possible to use a lighter press requiring less power to drive it. Another common impact noise is in barrelling; the noise level in this process can be as high as 110 dB. If, however, the barrel is lined with a suitable material such as neoprene, the noise level will be substantially reduced; up to 30 dB has been claimed. But again, the benefits are not all in noise reduction; the life of the barrel will be increased so that the cost of the lining is paid for more than two-fold. The design of gearing and other parts of machinery to reduce noise is primarily an engineering problem and cannot be dealt with here; the subject is very adequately covered in Harris (1957).

Noise is often produced by resonance and this may occur, for instance, when parts which are being ground are vibrated or in office machinery mounted in sheet steel cabinets. In the first instance, suitable damping devices can often be attached to parts which would resonate and, in the second instance, damping material can be applied to the inner surface of the sheet metal. Complaints about the noise of office machinery can often be eliminated by treatment in this way.

While noise may emanate directly from a machine, a further substantial proportion emanates from the structure to which the machine is attached. Useful reductions in ambient noise can be achieved by isolating the source from the remainder of the structure, thereby preventing it from acting as a sounding board. There are a great many ways in which this can be done. One method often practised is to fix a machine to the floor with a layer of felt between the machine and the floor. A common form of noise transmission, with which everyone will be familiar, is the noise produced by domestic plumbing, this can be substantially reduced by isolating ball valves and so forth from the pipe work by means of plastic connectors and by isolating the pipes themselves from the structure by means of insulating material placed under the retaining clips. Even when insulation has been carried out to the greatest possible extent, it may still not be entirely effective if the building itself has not been designed with the non-transmission of noise in mind, so clearly noise prevention must start when a building is first erected.

If, after everything has been done to prevent the production and distribution of noise from the source, the noise level is still too high, some form of acoustic absorption will be necessary. This can be done in a wide variety of ways and is a task for an expert. The object of this treatment is to impose absorbing barriers between the noise source and the hearer and/or to prevent the reflection of sound from smooth surfaces.

THE MEASUREMENT OF NOISE

Measurements of noise in industry will be made with a noise-level meter. Firms in the U.K. which supply such equipment include Dawe Instruments Ltd., and Standard Telephones and Cables Ltd. Their instruments are illustrated in Fig. 5; more complicated equipment will be used in laboratory experiments. The simpler instruments for use in industry will, in their basic form give an overall measurement of the sound level in dB and will usually include some weighting for differences in the sensitivity of the ear to different frequencies at different s.p.ls. The range of these instruments will normally lie between 30 and 130 dB. If a true assessment of the damage-risk of a particular noise is to be made, a measurement of the overall s.p.l. is not adequate, and for this purpose a supplementary octave wave-band analyser is required which will function with the basic instrument (Anon, 1963).

A fundamental difficulty with instruments of this kind is that, because of the characteristics of their damping, they give the root-mean-square s.p.l. and not the peaks of intensity which may arise from impact noise. As has already been mentioned, the effects of impact noise on the hearer may, therefore, be greater than would be indicated by readings taken on instruments which give root-mean-square s.p.l. Allowance should be made for this condition in the manner which has been suggested above.

CHAPTER FOURTEEN

Environmental Factors III. The Visual Environment

LIGHTING

The majority of industrial tasks will depend for their efficiency on adequate vision, and therefore lighting may play an important part in determining the efficiency with which tasks are carried out. Other factors which can also be of importance are the contrasts between the surroundings and the task, which may be influenced by colour, and the presence or absence of glare. To a large extent the recommended levels of illumination which have been current at different periods have depended upon the amount of light which could be obtained economically. The standards of lighting considered adequate in industrial plant 30 years ago would be very low judged by the standards current at the present time; and even as recently as 1958 the standards suggested by the American Illuminating Engineering Society have increased as much as ten-fold some of the standards which were set in 1949 in the U.K. But even now, the levels which can be economically obtained with artificial light do not approach levels which may be obtained by natural daylight under ideal conditions. It is comparatively rare for lighting in a factory in this country to exceed 200 lumens per square foot (lm/ft²) whereas natural north daylight may well exceed 1,000 lm/ft². On the other hand unless the building is very modern with very large window areas, the daylight factor, that is the amount of daylight falling on the work, is unlikely to exceed 20%, so that under these conditions the levels of artificial illumination are now approaching the levels which are to be found in natural daylight and exceed the levels to be found in the winter or in old-fashioned buildings with very small windows.

Despite all the research which has been carried out in recent years, it is still undecided exactly how much light is required for a particular job. At normal levels of illumination, the ability to see increases as the logarithm of the illumination, so that from the practical and economic aspect a point is reached at which large increases in illumination will produce relatively small increases in efficiency or output. This effect was shown by Weston

(1949), with a test object of one minute of arc an increase of the illumination from 5 to 10 lm/ft² produced an increase of 10% in visual performance, a further increase of 10 to 20 lm/ft² produced an additional increase of 10%, but an increase from 20 to 50 lm/ft² was required to produce a further increase of 12%. Kuntz and Sleight (1949) found that individuals with normal visual acuity had no significant increase in acuity above 31·6 ftL, although the experiment continued to 1,000 ftL. The actual increase in performance resulting from the increase in luminance from 31·6 to 1,000 ftL was about 7%. On the other hand, a subject with sub-normal vision reached the same level as the normal subject at about 100 ftL. Similar results have been obtained by Lythgoe (1932) (Fig. 100) who found that the relationship between visual acuity and the logarithm of the luminance began to depart from linearity at 10 ftL. McCormick and Niven (1952), working with a task requiring speed and precision, found a

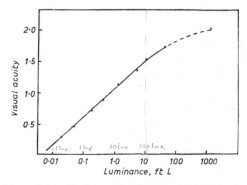

Fig. 100. The relationship between visual acuity and luminance (after Lythgoe, 1932).

significant decrease in errors when the illumination was increased from 5 to 50 lm/ft². When the illumination was further increased to 150 lm/ft² there was a slight but non-significant decrease in errors. It must be realized that these figures were obtained with objects of high contrast and it must not therefore be assumed that, when objects are very small and contrasts very poor, levels as low as these will be acceptable. On the other hand, Tinker (1949) stated quite emphatically that if visual acuity is taken as a criterion 'there is no justification for suggesting that more than 40 or 50 foot candles are necessary for adequate illumination, even for tasks that approach a threshold discrimination'. Recent thinking, however, does not agree with Tinker in this and current recommendations of the American Illuminating Engineering Society have suggested lighting levels as high as

2,500 lm/ft² on the delivery table of the obstetrical ward in a hospital, 2,000 lm/ft² for cloth inspection in the textile industry and 1,000 lm/ft² for a variety of industrial tasks, including extra fine bench work, fine inspection, and precision arc-welding. It is still an open question whether levels of illumination as high as this are really necessary and whether they are economically worth while.

The relationship between visual factors and the amount of light required has been investigated over the years by many workers, notably by Luckiesh and Tinker in America, by Weston in this country and in the last decade by Blackwell in America. The methods which each has used have been somewhat different and the results of their research have led to the recommendation of lighting levels which have increased with the years. Luckiesh and Weston carried out their work mainly by studying the effects of the various variables on visual acuity, but Weston supplemented his laboratory work with extensive studies of performance in industry, studies which have few parallels elsewhere. Luckiesh used parallel bars as his test object, each bar occupying one third of the width of a square. The task was to say whether the bars were vertical or horizontal. There were thus only two possible alternatives. Weston, on the other hand, used the Landholt Ring, a circle with a small gap. Eight common orientations are used in this test. Blackwell used a disc projected on to a translucent screen.

Using the method of threshold of detection, Luckiesh and his collaborators established a series of relationships between the size of the object, the contrast and the brightness required. However, it is undesirable that industrial work should be carried out at the threshold of detection, so that results of this kind must be extrapolated to an indefinite level above threshold at which normal work will be done. The method used by Weston, the cancellation of Landholt Rings, enables performance to be measured in relation to size, illumination and contrast at levels above threshold (Weston, 1949).

The work of Blackwell has been described by Crouch (1958). This work, which commenced in 1950, was based on the idea that the Landholt Ring used by other workers was too complex a test object for deciding the basic parameters of vision. Blackwell used a simple disc projected on a translucent screen with different sizes, contrast and time of seeing. He also used a wheel on which there were a number of discs and the subject had to pick out discs with circular defects in the centre as the wheel indexed. Quoting from the paper: 'The result of this study showed that the difference between a static laboratory condition and a dynamic moving-eye, field condition is a factor of 15, not in foot candles, but in contrast. Between a circular disc in the field as contrasted with the laboratory, one would need the light necessary for that same sized test object with a contrast 1/15th of that of the static laboratory task'. Blackwell has also introduced the idea of Visual Capacity in 'assimilations per second', that is, if the eye is permitted to see the object for one-fifth of a second, it has the capacity of assimilating five items of information per second ('items of information' is *not* used here in the same sense as the 'bits' of information theory). In order to relate these experimental results to real life tasks, the Visual Task Evaluator was developed (Blackwell, 1959). In this, the contrast of an unknown task was reduced by placing over it a veiling brightness which brought the task to the threshold of vision. When this has been done, a circular disc of four minutes visual angle is reduced to threshold under the same conditions as the unknown task. In this way the task can be compared with known parameters of performance for the four minute disc, and the amount of luminance required for the task calculated, when a range of field factors and rates of discrimination have been taken into account. Blackwell, however, admits that he has no factual basis for assigning these factors to tasks outside the laboratory. Nevertheless, using this technique Blackwell evaluated 56 tasks submitted to him and on the basis of these tasks new foot-candle tables have been drawn up by the Illuminating Engineering Society, U.S.A. (1959). It is difficult to see how the jump has been made from the laboratory results to the figures given in these tables; Weston (1961) suggests that it was 'intuitive'. For instance, for 8 point type on good quality paper, the required illumination would be 1·87 lm/ft². Yet for reading high-contrast or well-printed material in areas not involved in critical or prolonged seeing, such as conferring, interviewing or inactive files, the recommended illumination is 10 lm/ft². Similarly, samples of hard pencil writing on poor paper were found experimentally to require an illumination of 63 lm/ft², but the recommended illumination is 100 lm/ft² for regular office work, reading good reproductions, reading or transcribing handwriting in hard pencil on poor paper, active filing, and note sorting. The comment is made that in all locations involving visual tasks, levels of less than 30 foot candles do not appear desirable. These figures are a great deal higher than those in use in the United Kingdom, and for comparison the 1961 Illuminating Engineering Society's code for good interior lighting (I.E.S. London, 1961) suggests 30 lm/ft² for book-keeping, typing, accounting, filing and general office work. In machine workshops and fitting shops, very fine bench work requires 150 lm/ft² under the 1961 code and 1,000 lm/ft² under the new American code. Fine bench work and machine work require 70 lm/ft² and 500 lm/ft², rough grinding, medium bench work and machine work 30 lm/ft² and 100 lm/ft² and rough bench work and machine work 15 lm/ft² and 50 lm/ft² under the I.E.S. 1961 code and the American code respectively. As Crouch has said, 'In early days, the limitation of what one could get, established the level to be currently recommended. Light sources were feeble at their best and equipment had its limitations also'. It would seem not impossible that the same principle is still acting now in America, since the work of Weston and others suggests that the return in improved performance obtained for increasing the amount of light on a task is a diminishing function above a certain level. What does not seem to have been clearly established is whether present recommended illumination levels, British or American, are below, at, or above this level. The reader may be excused for wondering just what validity the various recommendations may have while their whole bases are still in doubt. (Blackwell (1954) and others are critical of the methods of research used by Weston, while Weston (1961) is equally critical of the work of Blackwell.) Commenting on the apparent irrelevance of the research results to the greatly increased recommended lighting levels Weston says, 'In passing, it may be remarked that substantial increases could have been proposed at any time within the past 15 years with as much – or as little – justification from available British and other data'. Whatever the merits of the case, it seems clear that Blackwell, having rejected the validity of earlier work, has failed to produce a convincing substitute.

Fortunately, the situation is not nearly as unsatisfactory as it might seem at first sight. In the first place, people can adapt themselves to a variety of conditions so that they can work quite satisfactorily in conditions worse than those thought to be ideal. Moreover, except under special conditions specific to industries such as transport or mining, the lighting provided is usually well above threshold, and in any case, sufficient general principles have been established for useful practical results to be obtained in industry.

Illuminating engineering is an established profession covering a very wide range of activities including the lighting of work-places of all kinds. It may, therefore, be wondered why it is thought to be necessary to include lighting in a discussion on Ergonomics. While there may be exceptions, illuminating engineers will normally be concerned with general problems of lighting installations but not with specific problems which arise when new methods or products are introduced, when machinery is installed or moved, nor with conditions in older factories or with design faults in the equipment. For example, in a press shop general lighting may be more than adequate but the overhang of a press and position of the operator in relation to a luminaire may be such that the light actually falling on the task may be quite inadequate. Further, illumination is only one of several factors affecting visual efficiency and the ergonomist or designer should be conversant with the requirements for good and efficient seeing and this requires a study of the whole visual environment. For instance, it was found that on a certain type of composing machine the general and local lighting were inadequate, but in addition the contrast between the copy and the body of the machine was too high and there were many troublesome reflections from bright parts of the machine – all matters which could be put right without much difficulty.

In the discussions which follow, general working rules will be given which will enable the practical man to establish reasonable conditions for efficient vision.

Factors Influencing the Amount of Light Required for a Task

The amount of light required for the performance of a visual task is influenced by four factors which are interdependent. These are, (a) the size of the object, (b) the contrast between the object and its immediate surround, (c) the reflectivity of the immediate surround, and (d) time allowed for seeing. The designer will be able to have control over all of these to a greater or lesser extent, except for the size of the object. The amount of light which is required for the task cannot be determined until all these factors have been established. Research has shown that, with illumination constant, visual performance varies with the logarithm of the size of the object up to a visual angle of about 4 minutes of arc, but from then on the logarithmic relationship diminishes. With the size constant, performance increases with the logarithm of the illumination up to about 50 lm/ft² (or foot lamberts) and thereafter increases relatively little as the illumination is increased to 1,000 foot lamberts (Lythgoe, 1932). Visual acuity as measured by Luckiesh and visual performance as measured by Weston both improve with increase of contrast but the relationship is not linear. This lack of relationship is not perhaps surprising when it is realized that the contrast is an arbitrary physical concept which may be unrelated to sensations of brightness difference, a matter which will be covered in greater detail later. It appears, however, that if the size and reflectivity of an object is kept constant, the acuity will vary as the logarithm of the reflectivity of the lighter background. There appears also to be a logarithmic relationship between acuity and time allowed for seeing and acuity and angular velocity of a moving object.

The Size of the Object

Fundamentally, ability to see small objects depends on visual acuity but this is not the whole story since the measures of visual acuity which are used in the laboratory and which differentiate between different types of acuity, do not in general bear a great relationship to tasks in industry. It will be remembered that in laboratory experiments three types of acuity are defined, line acuity, vernier acuity and space acuity, and that these have a ratio under laboratory conditions of approximately 1:4:24. (In passing it may be noted that the value of these acuities will be given differently by different writers, for instance, Weston (1949) gives them as 1:4:20.) Since work objects may by their nature be made up of forms which include one, two or all these types of acuity in their detail, these acuity figures can be taken as giving only a general indication of the size of objects which can be seen under different conditions.

Weston (1935) has shown that the average worker is able to carry out a job requiring the perception of detail which subtends an angle of 10 minutes of arc at the eye quite efficiently under normal lighting conditions provided the contrast is good. He found, moreover, that there was no appreciable difference in performance until the size is reduced to 6 minutes of arc. This would therefore appear to give the top limit of size above which a performance will improve little if the size is increased. At the other end of the scale 24 seconds of arc can be taken as being the smallest detail below which vision will become very difficult indeed, and the intervening space can conveniently be divided into four size categories, giving six categories in all. From the practical point of view, however, visual angle is not easy to work with, so it is convenient to convert it into dimensions seen at a normal viewing distance of approximately 18 in. For viewing distances which differ from 18 in., the size categories will require to be adjusted: for instance, if the viewing distance is 9 in., category II will become category I and if the viewing distance is doubled to 36 in. then category II will become category III.

In Table 26 the six categories are given together with their appropriate

Table 26. Size categories related to 18 in. viewing distance

Category	Visual Angle	Size in Inches	Example
I	less than 23″	< 0·002	Fine wires (about 0·001 in. diameter)
II	23″-45″	0·002-0·004	Fine scribed line or human hair
III	45″-1′ 30″	0·004-0·008	Lines on slide rule or micrometer
IV	1′ 30″-3′	0·008-0·015	Size 40 sewing cotton: thickness of lines of 6-pt. print
V	3′-6′	0·015-0·030	Thickness of 12-pt. print
VI	Over 6′	> 0·030	Any large object

visual angles, the approximate size of the detail in thousandths of an inch and examples of familiar articles of this size.

From this it will be seen that any objects with important detail over approximately 30 thousandths of an inch in size can be seen under normal general lighting conditions of say, 15 lm/ft². But that when the detail is smaller, somewhat stronger lighting will be required. This assumes that there is high contrast between the object and its background, but even objects of 30 thousandths of an inch will be difficult to see at the illumination of 15 lm/ft² if the contrast is very low, and more light will be required. The use of these six size categories in determining the amount of light required for a task is explained more fully below.

Brightness Differences – Contrast

In discussing 'contrast' we will deal first with the brightness relationships of two non-specular surfaces adjacent to each other when the illumination of both the object in regard and the immediate surround is the same. Where one or more specular surfaces are involved we may have the case of materials being illuminated from different light sources with different amounts of light and this requires separate treatment.

In considering 'brightness' we can think of non-specular materials as being diffuse reflectors of a known amount of incident light and express their luminance in terms of the lambert. On the other hand we can equally consider a reflecting surface as a secondary light source emitting luminous flux and express its intensity in candelas. It is usually convenient to use the former method of notation when dealing with non-specular materials which enable brightness differences to be expressed also in terms of reflectivity. But with specular (and some semi-specular) materials the latter method is often used and brightness differences are expressed in terms of intensity.

In Chapter 1, it was stated that various ways of describing differences in brightness between two adjacent surfaces have been used. The principal notations use either $\frac{B_D}{B_L}$, that is the ratio of the luminances of the darker and lighter surfaces, or $B_L - B_D$ which gives the luminance difference. A third notation $\frac{B_L - B_D}{B_L}$ (equivalent to $\frac{\Delta B}{B}$) is the relative brightness difference and expresses a logarithmic relationship between physical brightness and apparent brightness. It is this measure in the form $\frac{R_L - R_D}{R_L} \times 100$ which is commonly used to express the 'contrast' between two non-specular surfaces illuminated to the same intensity. These measures have been criticized by Hopkinson *et al.* (1941) on the grounds that subjective brightness (luminosity) can differ from luminance yet these contrast measures are expressed in terms of luminance and in addition they do not take into account the level of adaptation of the eye. Thus if an observer is in a brightly lit room he is quite unable to see detail outside the window at dusk, details which would show quite high contrast when expressed as $\dfrac{\Delta B}{B}$ and can clearly be seen by someone whose eyes are adapted to twilight. Hopkinson *et al.* propose instead a measure of 'apparent brightness' for which they have assembled a family of curves for adaptation levels between 10^{-5} and 10^3 effective foot candles (ftL) which is shown in Fig. 101. The luminance of the object and its background are measured, and from the curve for the appropriate adaptation level the apparent brightnesses are obtained in arbitrary units, contrast being expressed as the difference between these two measures. They applied this technique to data obtained by Dunbar (1938) who measured object and background luminance for constant thresholds on the road. Their calculations are given

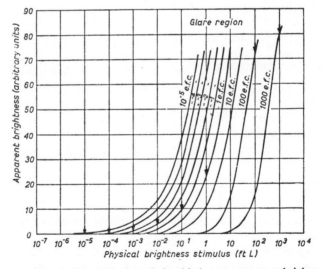

Fig. 101. Averaged data showing relationship between apparent brightness and physical brightness (from Hopkinson *et al.*, 1941).

in Table 27, which shows that the contrast values obtained by this method are reasonably constant. The curves in Fig. 101 also show that when the eye is adapted to a luminance level of about 10 ftL there is approximately a linear relationship between objective and subjective brightness both below (and above) the adaptation level. But when adaptation is to very low

Table 27

Background Brightness (ftL)	Object Brightness (ftL) (from Dunbar)	Apparent Brightness		Contrast
		Background	Object	
1·0	0·7	23	18	5·0
0·3	0·19	17	12	5·0
0·1	0·06	11	6·5	4·5
0·03	0·01	7·3	3·3	4·0
0·01	0	5	0	5·0

luminances, changes in luminance below these levels will cause relatively little change in apparent brightness, whereas those above will cause a disproportionately high sensation.

Two conclusions can be drawn from this work: first that in situations of low adaptation such as might be found in night operation of transport the expression $\frac{\Delta B}{B}$ cannot be used to express contrast, and secondly that under the normal conditions to be found in factories this expression *can* be used, provided there are no wide differences between the general illumination (which may determine the adaptation level) and the illumination on the task; care should in fact be taken to ensure that this situation does not arise. And so in order to be quite specific, the expression *relative luminance difference* (r.l.d.) will be used rather than contrast, although the latter term is commonly used in the literature. For practical purposes in determining lighting requirements in the factory, it is possible to use the expression

$$r.l.d. = \frac{R_L - R_B}{R_L} \times 100 \qquad (27)$$

The reflectivities of the task and its immediate surround can either be estimated from known materials or they can be measured. Estimation can be assisted by acquiring data such as those shown in Table 28. The figures in this Table are approximate, but they are probably near enough for most practical purposes. If measurements are to be made, an instrument with a very narrow angle of acceptance will be required and such an instrument is the S.E.I. photometer (Fig. 3, page 7) which can give readings of luminance directly in foot lamberts. However, to make a measurement of the reflectivity of materials it is necessary to measure the incident light in lm/ft² with a foot candle meter as well as the light reflected from the

Table 28. Reflection factors of some common materials
(adapted from Weston, 1949)

White plaster	About 95
Good quality white paper, white tiles	About 85
White plastic or porcelain enamel	About 75
Medium quality white paper	About 75
Bright brass	About 75
Aluminium	About 75
Bright copper	About 65
White cloth	About 65
Newsprint	About 55
Concrete	About 55
Plain white wood	About 45
Dull brass	About 35
Dull copper	About 35
Bright steel	About 25
Cast or galvanized iron	About 25
Black cloth, if not matt	About 15
Printers' ink of good quality	About 15
Matt black paper	About 5

material in foot Lamberts (both measures can be made in candelas if suitably graduated instruments are available). The reflectivity is then obtained from the formula

$$R = \frac{B}{E} \times 100 \qquad (28)$$

where R = reflectivity, B = luminance and E = illumination. When the reflectivities of the task and of its immediate surround have been estimated or measured, the r.l.d. can be obtained by the use of Fig. 102.

Although most non-specular materials are sufficiently matt to be treated as perfect diffusers, under certain conditions specular reflection may have to be taken into account when computing r.l.d. A particular condition under which this may occur is when a concentrated light source is in such a position that it produces a measure of specular reflection into the eye – the more the material departs from being perfectly matt the nearer will this condition approach disability glare. The effect of this added specular reflection can be to reduce the effective contrast. It is well known that when reading material printed on a semi-matt paper in a bright light, care has to be taken to avoid allowing the image of the light to be reflected in the paper. Under these circumstances Crouch (1945a) has suggested that the formula for r.l.d. should be modified:

$$\text{r.l.d.} = \frac{(B_{LN} + B_{LS}) - (B_{DN} + B_{DS})}{(B_{LN} + B_{LS})} \qquad (29)$$

where $B_{LN} + B_{DN}$ = non-specular luminance
and $B_{LS} + B_{DS}$ = specular luminance, of lighter and darker surfaces respectively.

When specular material forms part of the visual task, the light it reflects into the eye will usually come from a surface which is illuminated to a different level from that of a juxtaposed matt surface. Thus the r.l.d. cannot be computed from the reflectivities in the manner already described. Instead, the luminance of each surface will have to be measured separately.

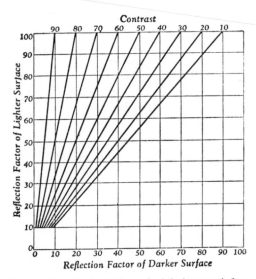

Fig. 102. Diagram for determination of r.l.d. (contrast) from reflection factors of two surfaces (after Murrell, 1957).

With reasonably large surfaces this measurement will be relatively simple, but it may not be so easy when quite small bright parts are involved. In this case an indirect method could be used which, while it is an approximation, is sufficiently accurate to produce practically useful results at distances found in normal industrial work.

Consider a mirror A (Fig. 103) in juxtaposition to a white surface B with a reflectivity of 80%. Both are seen from a position O such that a white surface C, also of 80% reflectivity, is seen in A. Both B and C are illuminated by D, the illumination on C being 20 lm/ft² and on B, 5 lm/ft² (B is 2 ft from D, and C only 1 ft from D). The luminance of C is therefore

16 ftL and that of B, 4 ftL. If A is a perfect reflector, the r.l.d. between A and B will be $\dfrac{16-4}{16} \times 100 = 75$. But suppose that A is not a perfect reflector but is of bright copper which itself has a reflectivity of 65%, its luminance will be reduced to 10·5 ftL and the r.l.d. will fall to 62. From this it will be seen that if the illuminations on and reflectivities of the non-specular surface and of the area seen in the specular surface, as well as the reflectivity of the specular material itself are known, the r.l.d. can (when the specular surface is the brighter) be calculated from the formula

$$\text{r.l.d.} = \frac{E_r \times R_r \times R_s - 100(E_m \times R_m)}{E_r \times R_r \times R_s} \tag{30}$$

where
E_r and E_m are the illuminations on the reflected and matt surfaces respectively,

and
R_r, R_m and R_s are the reflectivities of the reflected, matt and specular surfaces respectively.

In the examples given above, measurements have been given in lm/ft^2 and ftL because they are the units most likely to be used in the factory. More properly the measurements should be expressed in candelas, but the resulting r.l.d., because it expresses a relationship, would be unaltered.

Manipulation of E_r and R_r could produce very high levels of luminance in the specular material and as a general rule it would seem that this luminance should not be more than 100 times that of the background if glare is to be avoided.

In addition to brightness contrast, a second form of contrast may at times be found, namely colour contrast. It is possible to have two or more colours which reflect the same amount of incident light but which are clearly dis-

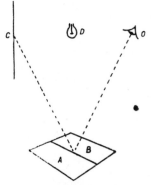

Fig. 103. Determinants of 'contrast' between a specular and non-specular surface.

tinguishable from each other by their difference in hue. For instance, four British Standard colours (B.S. 2660:1955) have a reflectivity of approximately 59% (to C.I.E. standard illuminant A which approximates to the light from a tungsten filament lamp at 2854° K). These are: 0-002 (warm yellow), 0-008 (yellowish green), 2-030 (pinky beige) and

5-062 (slightly bluish green). All these except the two greens would show colour contrast although the brightness contrast is nil. This is not intended to imply that contrast satisfactory for ease of seeing small objects can be obtained by the use of colour contrast alone. When colour is involved, contrast should be enhanced by choosing colours which are not only widely separated on the colour circle, but which have widely different reflectivities. For instance, 7-079 (blue green) and 1-016 (red) have reflectivities of 39·6

Table 29. Reflectivities of certain British Standard colours for buildings and decorative paints corresponding approximately to major hues on the Munsell system (derived from B.S. 2660:1955 and supplement No. 1 (1961))

Reflectivity	B.S. Number	Colour	Approximate Munsell Reference
80+	4-046	Yellow	5Y 9·25/1
75+	4-052	Yellow	5Y 9·25/4
70+	7-081	Blue	5B 9/2
65+	7-075	Blue-Green	5G 9/2
60+			
55+	8-087	Purple-Blue	5PB 8/2
	4-047	Yellow	5Y 8/2
	2-030	Yellow-Red	5YR 8/4
50+	7-082	Blue	5B 8/4
45+	1-021	Red	5R 7/8
40+	4-056	Yellow	5Y 7/6
	7-083	Blue	5B 7/4
	4-048	Yellow	5Y 7/2
	1-016	Red	5R 7/2
	7-076	Blue-Green	5G 7/1
35+	7-079	Blue-Green	5BG 7/4
30+	4-049	Yellow	5Y 6/2
	7-078	Blue-Green	5G 6/1
25+	2-028	Yellow-Red	5YR 6/3
20+			
15+	5-064	Green-Yellow	5GY 5/6
	3-044	Yellow-Red	5YR 4/10
10+	5-065	Green-Yellow	5GY 4/6
	4-050	Yellow	5Y 4/4
	0-010	Green	5G 5/7
	0-005	Red	7·5R 5/16
5+	0-014	Purple	5P 3/3
	1-025	Red	5R 2·5/12
0+	6-068	Green	5G 2/2

and 44·2% respectively, and it would be difficult to distinguish detail in one of these colours on a background of the other: the situation would be entirely different if 1-025 (red) (6·0%) were used with 7-081 (blue) (72·1%). The reflectivities of some B.S. colours are given in Table 29.

When the Munsell reference of a colour is known, an approximate value for its reflectivity, which is probably sufficiently accurate for most practical purposes can be obtained from

$$R = V(V - 1) \qquad (31)$$

where V is the Munsell value. Using this method the reflectivity of 1-025 is $2 \cdot 5 \times 1 \cdot 5 = 3 \cdot 75$ and that of 7-081 is $9 \times 8 = 72$, which gives a value

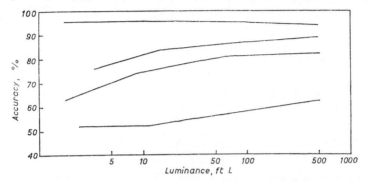

Fig. 104. The three lower curves illustrate the effect of luminance on accuracy with a 1 minute test object at r.l.d. 91, 68 and 36. The upper curve is for objects of 3 minutes and 6 minutes at all three r.l.ds (after Weston, 1945).

for r.l.d. of 94·8 compared with 91·7 obtained from the true reflectivities, a difference which is of no practical significance.

The interaction between size and r.l.d., and the interaction of both with luminance may now be considered. Weston (1945) studied these interactions using as the task the cancellation of Landholt rings. His results were measured in two ways: accuracy (the percentage of rings cancelled correctly) and performance (the product of the accuracy and the reciprocal

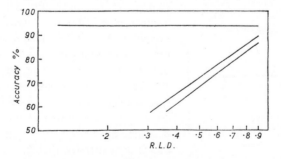

Fig. 105. The two lower curves illustrate the effect of r.l.d. on accuracy with a 1 minute test object at luminances of 500 ftL and 100 ftL. The upper curve is for 3 minute and 6 minute test objects at both luminances (after Weston, 1945).

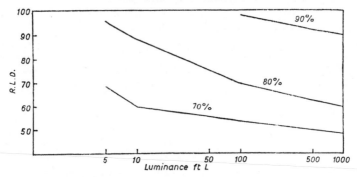

Fig. 106. The relationship between luminance and r.l.d. to achieve accuracies of 90%, 80%, 70% with a 1 minute test object (after Weston, 1945).

of the time). Relationships between accuracy, size, r.l.d. and luminance are shown in Fig. 104, for test objects 6, 3 and 1 minutes of arc, and three r.l.ds of 91, 68, and 36. For the two larger sizes, the values are so close together for all three r.l.ds that since they cover unsystematically a range of about 5%, they can be represented as a single line, r.l.d. and luminance having marginal influence on accuracy. On the other hand, both these

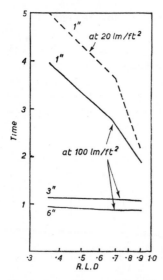

Fig. 107. The relationship between r.l.d. and time for identification of three test objects (after Weston, 1945).

factors have an effect when the size is 1 minute, that of r.l.d. being the greater. This is clearly shown when accuracy is plotted against r.l.d. in Fig. 105, with luminance constant at 500 ftL; it can be seen that even at this high level, accuracy with size 1 minute can never quite equal that with sizes 3 and 6 minutes even when r.l.d. is at its maximum. The effect of luminance and r.l.d. on accuracy is shown in another way in Fig. 106, from which it can be seen that with an object of 1 minute of arc an accuracy of 90% can be achieved only when both luminance and r.l.d. are very high. Weston points out that accuracy gives only part of the story and that time must also be taken into account; this is demonstrated in Fig. 107 in which time is plotted against r.l.d. at 100 lm/ft. It will be seen that recognition takes slightly longer with size 3 minutes than with size 6 minutes, and that these times are substantially independent of r.l.d. With size I minute, however, the time is greatly increased as r.l.d. gets less, this effect being more pronounced as the luminance is reduced as is demonstrated by the additional dotted line for 20 lm/ft².

These relationships taken together demonstrate that with very small objects accuracy cannot approach that found with larger objects unless r.l.d. is very high, and that at levels above 10 ftL, the effect of increasing the luminance with r.l.d. constant is less than the effect of increasing the r.l.d. (see Fig. 108). This puts a premium on achieving high contrasts and shows that lower contrasts cannot be compensated for by increasing the luminance even to very high levels indeed. Nor will increasing the luminance of a small object make the time required to carry out a task approach that required for larger objects even when the r.l.d. is maximal.

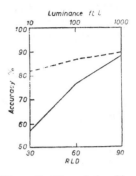

Fig. 108. The relationship between accuracy and luminance for 1 minute test object of r.l.d. = 90 (broken line) and of accuracy and r.l.d. when the luminance is 500 ftL (solid line) (after Weston, 1945).

From the practical point of view this means that to achieve the greatest accuracy with very small objects, both r.l.d. and luminance must be high, but that even then substantially more time must be allowed for the performance of a task than would be required were the task larger. Exactly what is the minimum size for both accuracy and time, this work of Weston does not show, but it would appear to be in the region of 3 minutes, although it may be less. What is clear is that the use of optical aids to increase the apparent size to about this level will be of great economic benefit.

Weston also carried out a number of shop floor investigations to supplement his laboratory work (Weston, 1949) which confirm that with very small sizes there is a performance limit imposed by contrast which does not appear to the same extent in acuity studies at threshold. Although these limitations are of obvious practical importance, Weston's work has been criticized by Blackwell (1952) on the grounds that speed and accuracy are not additive and that increments in his index of performance can come from both accuracy and speed in 'indeterminate amounts'. This is to some extent Weston's own fault, because in his 1945 paper he gives accuracy figures for only one of his experiments, his remaining results being expressed as 'performance' only. He gives no times at all and for the figures given above these have had to be reconstructed. Blackwell's own work as reported in his 1952 paper are studies of discrimination of brightness differences $\dfrac{\Delta B}{B}$ with two subjects only, using as his threshold 50% detection. He then used a mathematical transformation to extrapolate to accuracies up to 99%. The later development of his work as summarized by Crouch (1958) has already been discussed, but it seems doubtful whether his results are of such practical importance as Weston's for vision well above the threshold.

Time Allowed for Seeing

The time allowed for seeing may be determined either by a brief exposure of the object in view, which may be related to the threshold of vision, or

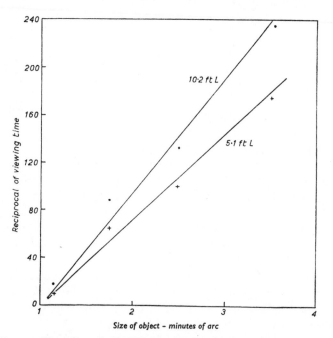

Fig. 109. The effect of visual size on visual acuity at luminances of 10·2 (above), and 5·1 ftL (after Ferree and Rand, 1922).

by rapid movement of the object across the field of view; in the latter circumstance an additional factor is introduced – the ability of the eyes to track a moving object during a brief period of time. Both conditions become of importance only when one of the parameters of size, time, illumination or speed of movement approaches a limiting value.

When an object is reasonably large and the r.l.d. is high, it is possible to see it in quite a short time. This principle is used in a number of experimental conditions involving the use of a tachistoscope. If, however, the object is small and contrast is low, the time required for seeing may be greatly increased. For instance, with a small object, if the contrast is reduced from about 70 to 50%, the time required for seeing will be increased by about four times. This point may be important if operators

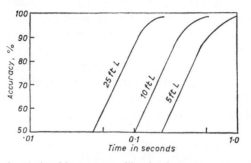

Fig. 110. A relationship between illumination, accuracy and time of viewing (after Blackwell, 1952).

have to see small defects on parts that are moving rapidly past them, or if they have to watch parts moving on a machine which is indexing. The exact relationship between ability to see and time allowed is very complex since effective acuity decreases with decreasing time of seeing. It has been shown by Ferree and Rand (1922) that with luminance constant, acuity in terms of visual angle increased linearly with increasing reciprocal of the time allowed for seeing (Fig. 109). At the same time they showed that there was a log-relationship between the luminance and the time allowed for seeing with the size of the test objects kept constant. This relationship appears to break down at very low levels of luminance.

Using the method of measuring performance already described, Blackwell (1952) established relationships between 'accuracy', luminance and time (Fig. 110). These also suggest that when luminance is constant, 'accuracy' increases linearly as the logarithm of the time, but that the relationship reaches an asymptote as 100% accuracy is approached. It is not clear whether these values were established by experiment or are the result of the mathematical transformation which he employed.

In the foregoing experiments, the contrast of the test object was unvaried, but r.l.d. is also an important variable and this has been studied by Cobb and Moss (1927) in relation to visual acuity and illumination with two exposure times of 0·3 and 0·075 sec. They showed that if the time is decreased a higher r.l.d. is required at threshold for any given size; the smaller the size, the greater the increase. For instance, at 100 mL an object of 1 minute of arc would require an increase of r.l.d. from 15 to 25 when the time is reduced from 0·3 to 0·075 sec, but when the size is 2 minutes of arc, the increase required would only be from 5 to 6 for threshold seeing. On the other hand, if the luminance is reduced to 1 mL an object of 1 minute cannot be seen at all even if its r.l.d. were 100 and a 2 minute object would require an increase of r.l.d. from 20 to above 30. While this work endorses the view that the longest time possible should be allowed, it shows also that when the time and object size are fixed, an object which otherwise could not be seen would be rendered visible by increasing either the r.l.d. or the illumination. The relationship between our size categories, r.l.d. and luminance, for a time of 0·3 sec is given in Table 30 and it must be emphasized that these are threshold values and should *not* be taken as working values.

Table 30. Approximate values of r.l.d. for threshold of seeing objects of different size categories with exposure of 0·3 sec at three levels of illumination (adapted from Cobb and Moss, 1927)

Upper Limit of Size Category	Luminance in Millilamberts		
	1	10	100
I	NV	NV	NV
II	NV	NV	35
III	40	17	7
IV	10	5·5	3
V	4	2·5	1·5

Burg and Hulbert (1961) have been studying the relationship between ability to see an object when it is stationary and when it is moving. They have repeated the work of Ludwig and Miller (whose work has been published in United States Naval Reports) who found that there was no correlation between an individual's ability to see an object and his ability to see the same object when it was moving, and that an individual's ability to see decreases as the angular velocity of the target increases, whether the target is moving when the subject is stationary or vice versa. Burg and Hulbert, however, did find some correlation between the two different types of acuity, and they think this may be due to the use of a larger sample of subjects of all ages with Snellen ratings between 20/40 and 20/13. They reached the conclusion that 'a person's ability to discriminate a moving target cannot be predicted from his static acuity and that the adequacy of prediction decreases as the speed of the moving target increases'.

Burg and Hulbert carried out their tests under two different conditions. In one the head was held fixed by means of a bite-board, and in the other condition the subject was free to turn his head to follow the target while it moved across a screen through an angle of 180° from left to right. Unfortunately, actual scores for the fixed test are not given, but the scores when the head was free show that there is a logarithmic relationship between acuity expressed as visual angle and the angular velocity of the target (Fig. 111). It may not perhaps be justifiable to extrapolate from this work, but for conditions where the head is free to follow in order to discriminate a target moving over a fairly wide area for each increase in angular velocity of 90° per second the size category of an object, as given in Table 26, might well have to be reduced by one. That is to say, if the

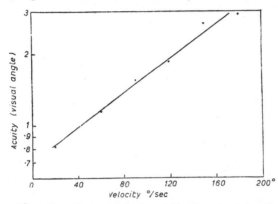

Fig. 111. The effect of target velocity on visual acuity (after Burg and Hulbert, 1961).

average individual can see an object of size II under stationary conditions, then if the object is moving at a velocity of 90° per second it might have to be treated as if it were Category I. This is the only work which gives an indication of the change which may be necessary in order to make allowance for the movement of the object that is in regard. It will clearly require further verification.

From all these results it would seem that benefit would accrue from increasing the time; it follows therefore that the longest time practicable should be allowed for seeing small objects if they are moving. If the time is being controlled by movement, the movement should preferably be intermittent, since the observation of moving objects introduces additional complications.

Luminance

It must not be assumed that because there seems to be adequate incident light as measured by the conventional footcandle-meter that the luminance is adequate for the task in hand. Several factors may influence the situation. The reflectivity of the material comprising the task may be very low – for instance, it may be necessary to detect black threads against black cloth – or the machinery surrounding the task or the operator himself may cast shadows which would reduce the amount of light thrown on the task. This fact has been well recognized by those associated with illumination, and it seems probable that the principal reason why the use of the footcandle-meter has been so well established is that it is only quite recently that relatively inexpensive portable photometers have become available which have

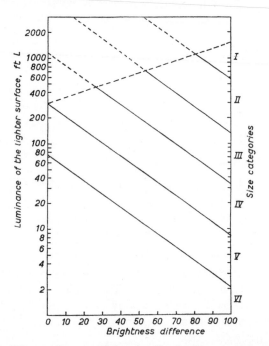

Fig. 112. Diagram for ascertaining the illumination required on a task.

a small angle of acceptance and will measure the luminance giving the answer directly in foot Lamberts.

We have seen that the gain in facility of seeing as luminance is increased is a steadily decreasing function and therefore it is a moot question up to what point it is economically worth while increasing the light on a task. Based on the experimental information available, Fig. 112 gives a relationship between illumination levels, size and luminance *difference*, the latter is standardized by expressing it as reflectivity difference. The upper limiting line represents the level of combinations of size and luminance difference beyond which visual performance is likely to be rather poor, however much light is provided. This is based on the findings of Weston (1945) which were discussed earlier. This does not mean that no improvement in visual performance is possible if illumination is increased above the values indicated, but rather that vast quantities of light cannot compensate completely for very adverse conditions of size and contrast. It must be emphasized that there is no finality about lighting levels obtained from this diagram, but it is suggested that if the levels obtained from it are used, adequate conditions for seeing will be established without too great a capital outlay or running costs.

The method of using the diagram is simple. The size of the detail is first determined and a size category is allocated to it. The reflectivities in the task are determined and the difference between them is calculated. The diagram is entered at the figure so obtained and at the point where the line cuts the appropriate size category, the required luminance of the lighter surface will be found. From the reflectivity of the lighter surface the illumination in lm/ft^2 can be readily calculated.

When the specular and non-specular materials are involved, Fig. 112 cannot be used in this way. In this instance there are two illumination variables which will together determine the r.l.d. (assuming the reflectivities to be fixed). Thus when the specular material is the brighter it is necessary to choose first a suitable illumination for the reflected surface (E_r) by taking a desirable reflectivity difference (r.d.) and finding from Fig. 112 the required luminance of the specular surface (B_s) corresponding to the size category involved. The illumination on the reflected surface is calculated from:

$$E_r = B_s \times \frac{10,000}{R_r \times R_s} \tag{32}$$

where R_r and R_s are the reflectivities of the reflected and specular surfaces respectively. The illumination of the matt surface (E_m) can then be calculated from:

$$E_m = \frac{100B_s}{\text{r.d.} + R_m} \tag{33}$$

where R_m is the reflectivity of the matt surface.

Should the specular surface be required to be darker than the matt surface, as when a material with low reflectivity such as steel is used, it will usually be possible to make the reflected surface black and to treat the object as if it were matt with a low reflectivity. Where this is not possible the luminance and illumination of the matt surface should be determined from Fig. 112 as already described, and the illumination on the reflected surface is calculated from:

$$E_r = \frac{100\,[100B_m - (\text{r.d.} \times E_m)]}{R_r \times R_s} \tag{34}$$

where B_m is the luminance of the matt surface.

By way of example, let us assume that we have a bright brass part which has a reflectivity of 75% and is about 0·015 in. in size (lower limit of category 5) which is to be seen against a matt black fixture having a reflectivity $R_m = 15\%$. Taking a reflectivity difference of r.d. = 90 as appropriate, we find from Fig. 112 that $B_s = 12$ ftL. The background reflected in the brass is white paper of $R_r = 80\%$, so, from formula 32:

$$E_r = \frac{12}{80 \times 75} \times 10{,}000 = 20 \text{ lm/ft}^2$$

and by formula 33:

$$E_m = \frac{12 \times 100}{90 + 15} = 11·4 \text{ lm/ft}^2$$

Thus the paper reflected in the brass part should be illuminated by 20 lm/ft² and the fixture by 11·4 lm/ft².

Alternatively, if the metal is dark and is to be seen against a light background with $R_m = 80\%$ it will no longer be possible to have an r.d. of 90, since under these conditions r.d. must be less than R_m. It will therefore be taken as 75 and from Fig. 112 $B_m = 21$ ftL and $E_m = 26$ lm/ft². Substituting these values in formula 34 we obtain:

$$E_r = \frac{100(100 \times 21 - 75 \times 26)}{75 \times 80} = 2·5 \text{ lm/ft}^2$$

Thus the surround should be illuminated with 26 lm/ft² and the reflected background with only 2·5 lm/ft². In practice the background would probably be as dark as possible and the specular surface treated as if it were matt black of very low reflectance. These examples show that it is possible to get the greatest luminance difference when the specular surface is the lighter. When the specular surface is the darker, the luminance difference is limited by the reflectivity of the matt surface.

In its latest and entirely new lighting recommendations (I.E.S. London, 1961) the Illuminating Engineering Society (London) has adopted the line that good lighting should be sufficient to produce 'near maximum visual performance' which they define as a range of 90 to 100% of performance under ideal conditions; 'performance' being in the sense used by Weston.[*] Following his finding that contrast appears to be relatively unimportant

*The basis of this new code has been explained by Weston (1961).

when sizes are large, and that visual performance is poor except with high contrast when sizes are small, a standard luminance is defined only in relation to the size of the detail:

$$\text{Standard luminance} = \frac{180}{S^{1.5}} \text{ ftL} \tag{35}$$

where S is the angle subtended at the eye by the detail in minutes of arc. A nomogram is given which will enable the illumination to be determined, if S and the reflectivity of the lightest part of the detail are known, in accordance with the formula:

$$E = \frac{18,000}{R \times S^{1.5}} \text{ lm/ft}^2 \tag{36}$$

The value of 180 is empirical, based on the best estimate of illumination levels required for various tasks. We have already noted that these levels will be lower than those recommended in America. While the approach adopted by the I.E.S. is an eminently practical one and a notable advance in specifying lighting levels, it must not be taken to mean that no improvement at all will follow from using higher levels than would be obtained from the above formulae if exacting visual tasks or unusual conditions are found.

CONTRAST BETWEEN A TASK AND ITS IMMEDIATE SURROUNDINGS

So far, in discussing contrast we have considered the contrasts in the task itself, that is, between the object in regard and its surroundings. We now have to consider the effect of contrast between the task and the more immediate surfaces which surround it. It has been established that the eye will tend to move towards the brightest part of the visual field. This tendency is called the phototropic effect and is well known to car drivers as the compelling tendency to look at the headlights of an oncoming car at night rather than at the road in front of one's own vehicle. The phototropic effect plays an important part in outdoor advertising and in window dressing.

If we consider as our task black print on white paper, it will be in high contrast with its surroundings if it is placed on a black desk top and illuminated with a desk lamp and leaving the remainder of the room in darkness. Under these conditions the eye will tend to be drawn to the white paper and maximum concentration or attention to the task will be obtained. On the other hand it may be obtained only at the cost of some visual discomfort and the extent of this discomfort will depend on the amount of light thrown upon the book. On the other hand if the book is placed on a pale desk, say of cream colour, and the room is illuminated with general lighting, we have established a condition for maximum visual comfort but at the cost of reduced visual attention. Thus all contrast manipulation in the field of view must be a compromise between the need to avoid visual distraction and the need to avoid visual discomfort or fatigue. Lythgoe (1932) studied the effect on visual acuity of varying the brightness of the immediate surrounding to the task. His results are shown in Fig. 113 in which the interacting effects of surround brightness and luminance of the task on visual acuity are plotted. The surround brightness in one case was equal to the task brightness up to 38·1 ftL, after which the task was maintained at this level while the surround luminance was further

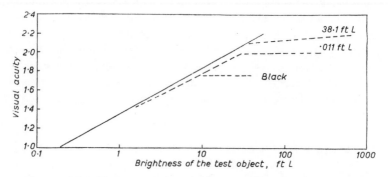

Fig. 113. The effect on acuity of varying the brightness of a test object against three backgrounds of constant brightness (broken lines) and a background of equal brightness (solid line) (after Lythgoe, 1932).

increased; in the other two cases the surround brightness was maintained at 0·011 ftL and complete darkness respectively while the task brightness was increased. Under conditions of equality, acuity was linearly related to the luminance, but under the other two conditions there appears to be a maximum level above which further increase in luminance will produce no increase in acuity. (The experiments were not carried far enough to show what happens at high levels of task luminance.) This level is about 9 ftL when the surround is dark and about 30 ftL when the surround is 0·011 ftL. It is difficult to quantify these results because in two of the three conditions the task/surround brightness ratio was rather indeterminate; in other work, however, he kept the luminance of the task constant at 12·6 ftL and varied the luminance of the background to give varying ratios. He found that the acuity was at its optimum when the surround brightness was equal to or slightly less than the task brightness. The fall in acuity was sharpest when the ratio was 1:10. When the surround was three times brighter than the task, the loss of acuity was equivalent to reducing the illumination by 60%.

Brightness ratios were also controlled in a study by Biesele (1950) who maintained task brightness constant at 20 ftL and who had his subjects mark Landholt rings under task/surround brightness ratios between 1:0·05 and 1:10. The average score over four hours is shown in Fig. 114 together with Lythgoe's results. It will be seen that for optimum performance (or acuity) the task should be lighter than the surround up to a brightness ratio of about 1:0·3.

Lythgoe also studied pupil size in his research and he points out that the surround brightness will determine the level of adaptation of the eye as revealed by pupil size (see Fig. 114). The importance of surround brightness in determining the adaptation level, it will be remembered, plays an important part in determining the brightness differences which can be

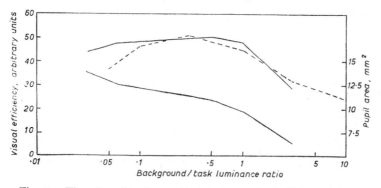

Fig. 114. The effect of background/task ratio on visual efficiency (upper curves) and on pupil area (lower curve) (after Lythgoe, 1932 – solid line – and Biesele, 1950 – broken line).

discriminated. From this it follows that surround brightness is of equal importance to illumination in determining visual efficiency, and that surround brightnesses which result in too great a contrast between the task and its surroundings will seriously impair efficiency. From these results we can conclude that for maximum visual comfort and efficiency the task should be of high reflectance and the immediate surroundings should be rather darker than the task, and the more general surroundings should, ideally, be rather darker still. The luminance of the task should not differ by a ratio of more than three to one from that of the immediate surroundings (I.E.S. London, 1961) and these in turn should not differ by more than three times from the reflectance of the general surroundings. If a task of low luminance is unavoidable, that is, suppose fine white lines have to be seen on a dark plate, it is important that the plate should not be placed on a white surface, since under the high illumination which may be necessary in order to see the fine lines the amount of light reflected into the eye by the white surface may be such as to produce an adaptation level at which it is difficult or impossible to see the fine lines. Rather the task should be placed on a dark surface large enough to cover the major area of the visual field: beyond this lighter surfaces can be introduced.

Experimental evidence of the effect of the luminance ratio on a task was obtained by Luckiesh and Moss (1932). They made their subjects set one pointer against another by means of a control. The two pointers were black on a white background and the surroundings of the task were either illuminated to the same extent as the background or were not illuminated at all. Performance with the dark surroundings was approximately 10% lower than that with the light surroundings.

The luminance ratios obtained in these experiments represent contrast limits which should not normally be exceeded and in practice reflectances giving luminance ratios lower than 3:1 may very well be used with advantage to produce a working situation which will give maximum visual comfort and efficiency. It is important to avoid areas away from the task which have a luminance near to or greater than that of the task itself so that the eye is drawn away from the task. If polished objects fall into the field of view and these reflect light directly from the luminaire, bright areas may result which will cause visual discomfort and may well detract from visual efficiency. Thus control panels consisting of black shiny material with chromium plated bezels can cause acute visual discomfort and loss of acuity and are to be avoided at all cost.

The luminance of the surroundings of the task may be manipulated by the balancing of local and general lighting, by the use of materials of different reflectance or by the use of paint or colour. As a rule it will be necessary to consider all these together. For instance, in a press it is no use increasing the local lighting alone in order to obtain adequate contrast within the task if consideration is not also given to painting the dark metal of the sides and bed of the press a colour which will produce visual efficiency and comfort. This is especially true if the task comprises bright metal parts which will reflect light specularly from either the luminaire or the distant surroundings and will be in too high a contrast with the dark metal of the machine. In general, therefore, it may be said that quite apart from any aesthetic qualities, the important use of colour in industry is in the manipulation of contrast to obtain visual efficiency.

GLARE

With the steady increase in the number and brightness of lighting fittings, the problem of glare is becoming of concern. Glare is generally recognized as being of two types: discomfort glare and disability glare, and both will arise from the inability of the eye to adapt itself to a very wide range of brightnesses. Discomfort glare, as its name implies, may produce a feeling of strain due to the worrying effects of sources of brightness in the visual field, which may come either from the luminaires themselves or from reflections of luminaries on bright surfaces. Disability glare can cause a diminished ability to see, especially when fine detail is involved, and this also may arise either from the light sources being very close to the line of sight or from reflections on the task. The effect of both types of glare may be to cause distraction by drawing the eyes away from the visual task. These two types of glare will be discussed separately.

Discomfort Glare

Discomfort glare has been studied by a number of workers for many years in an attempt to quantify what is in effect a subjective assessment. Early work in this field was that of Holladay (1926), who produced a formula for a glare constant – G, which he defined as:

$$G = \log B + 0 \cdot 25 \log Q - 0 \cdot 3 \log F \qquad (37)$$

where B = the brightness of the source, Q = the visual angle of the source in steradians, and F = the 'ftL factor', that is the brightness of the surround. The method of experiment used by Holladay was a 'shock' exposure in which the observers were subjected to a glaring source for only short periods of time. In this, his method differs from later workers who expose their subjects to a constant glare source. For the purpose of convenience of comparison with other work we can re-state this formula in the form:

$$G = \frac{B_s \, \omega^{0.25}}{B_b^{0.3}} = \frac{B_s^{3.3} \, \omega^{0.83}}{B_b} \qquad (38)$$

where B_s = the source brightness, B_b = surround brightness, and ω = the apparent area of the source in steradians.

Holladay's work was mainly directed at studies of the disability effect of glare, but Harrison (1945) developed the concept further to produce a modified formula which was intended to extend glare ratings to the area of discomfort. Harrison's formula, given originally in a slightly different form, may be modified for comparison with Holladay's formula:

$$G = \frac{B_s^2 \omega}{B_b^{0.6}} = \frac{B_s^{3.3} \, \omega^{1.7}}{B_b} \qquad (39)$$

Comparison of the two formulae will show that Harrison attributes a greater influence to the area of the source than does Holladay, but the high exponent of the source brightness is retained. Parsons, in the discussion of Harrison's paper, criticizes this and points out that it is probably unjustified because of the different type of glare studied by Holladay and the effect of the direction of the glare source. The formula was later modified by Harrison and Meaker (1947) and in a subsequent paper, Meaker (1949) put forward a tentative interpretation of a scale of the Harrison-Meaker glare constant which is given in Table 31 and also suggested that the exponent of the source brightness should be between 2 and 2·15.

Table 31.　Harrison-Meaker glare constants

0–10	Not likely to be criticized because of comfort.
10–15	Few people will find this uncomfortable. Fifteen is the upper limit of a completely comfortable range for people working in one position for long periods.
15–25	Some people find this annoying.
25–40	This range should not be exceeded in rooms where critical work is being done. This range may be satisfactory for bench work in industry.
40–100	Many factory installations fall within this range and it is assumed to be not unacceptable provided that people are moving around from place to place.
Over 100	Is unacceptable.

Working on street lighting in the United Kingdom, Hopkinson (1940) developed a third version of this formula, which applies specifically to conditions where there is a high source brightness and a small apparent area of the source. His formula is:

$$G = \frac{B_s^{1.3}\omega}{B_b} \qquad (40)$$

from which it will be seen that he attributes slightly more effect to the source area and slightly less effect to the source brightness than does Holladay. Subsequent work conducted in the laboratory (Petherbridge and Hopkinson, 1950) working with sources of lower brightnesses and larger area led to the modified formula:

$$G = \frac{B_s^{1.6}\,\omega^{0.8}}{B_b} \qquad (41)$$

The authors point out that a simple formula of this kind will apply only over a very limited range of the straight line portions of the equal glare curves and that it is at the best a working approximation; they give a warning of the danger of extrapolating beyond these conditions. In conducting their experiments the multiple criterion technique was used

(Hopkinson, 1950) and this systematic approach to the use of subjective judgements will probably have led to a more accurate assessment of the situation than the American work, bearing in mind that in the Harrison-Meaker formula the exponent B_8 was 'borrowed' from Holladay and was subsequently modified in the light of experience.

Formula 41 is applicable to sources which are located 10° from the direction of viewing, and the glare constants which result from its application can be defined in terms of subjective appraisal as shown in Table 32.

Table 32. Values of the Petherbridge-Hopkinson glare constant

Degree of Glare	Glare Constant
Just intolerable	600
Just uncomfortable	150
Just acceptable	15
Just imperceptible	8

In America the Holladay-Harrison-Meaker approach has come under fire from a number of workers who have suggested instead a measure known as the borderline between comfort and discomfort (b.c.d.). Crouch (1945b) suggests that when the size of the source is 0·01 steradians this threshold is obtained when the brightnesses are adjusted so that the glare factor calculated in terms of Holladay's formula is 75. Luckiesh and Guth (1949) have quantified this further in the formula:

$$B_8 = 108F^{0.44} (Q^{-0.21} - 1.28) \tag{42}$$

where B_8 = the brightness (ftL) of a circular source in the line of sight,

Q = its solid angle in steradians, and

F = the brightness of a uniform surround (ftL)

This formulation was obtained with a flashing light as was the original formulation of Holladay. In spite of the controversy which had raged over these rival methods of presentation, it does not appear that the latter has received general support.

The formulae given above do not take into account the extent of the displacement of the glare source from the line of vision. This important factor was dealt with by Harrison and Meaker (1947) based on later work by Holladay and by Luckiesh and Guth (1946) whose data fit the expression:

$$\frac{B_\theta}{B_c} = \frac{\theta^{0.3}}{\cos \theta} \tag{43}$$

where θ = the angle of the displacement of the source,

B_θ = the tolerable brightness of the displaced point, and

B_c = the tolerable brightness of the fixation point.

This formula, again based on Holladay's work, is related to the shock effect of an intermittent light and therefore should not be used for continuous glare sources.

Petherbridge and Hopkinson (1950) also studied the effects of displacing the glare source from the line of sight and found that their results could be expressed, over the range $\theta = 5°$ to $50°$, in hyperbolic form as:

$$\log_{10} \frac{B_\theta}{B_c} = \frac{0 \cdot 137}{1 \cdot 85 - \log \theta} \tag{44}$$

where $\dfrac{B_\theta}{B_c}$ represents the ratio of the acceptable brightness of a source displaced $\theta°$ from the line of sight, to the brightness of this source (for the same degree of glare) when in the line of sight. Graphic representation of this relationship suggests that it is not until the source is between $45°$ and $50°$ above eye level that any appreciable difference is noted in the permissible brightness of a glare source. At $40°$ the brightness of the source can only be three times what it is at $0°$, and at $50°$ only seven times, but at $60°$ it can be 20 times to produce the same degree of glare. In other words, if at $10°$ the amount of glare is assessed as uncomfortable it will not be assessed as just acceptable until it is displaced $50°$, and just imperceptible until $60°$ from the line of sight. From this it follows that the proper way of dealing with glare is not to try to reduce the brightness of sources close to the line of sight, but to remove them from the visual field altogether.

A practical assessment of the Harrison-Meaker glare formula was carried out by Lowson *et al.* (1954). As subjects they used 14 lighting engineers with an average experience of 14 years, and made them rate one incandescent source and two types of fluorescent sources used both sideways and crossways, that is five assessments in all on a 5-point scale. (They found in the event that all the assessments fell within a 4-point scale.) The scale used was defined as (1) comfortable, (2) just comfortable, (3) slightly uncomfortable, (4) definitely uncomfortable, and (5) very uncomfortable. The various ratings given by the observers were averaged and are shown (Fig. 115) plotted against the Harrison-Meaker glare factors, which are taken as 6 = comfortable, 20 = almost comfortable, 60 = slightly uncomfortable, and 200 = very uncomfortable. The line of correspondence between the two methods of assessment together with the best fitting regression line (which has been calculated by the present author) and the limit lines shown by the authors of the paper, are also given in the Figure. It will be seen that while there is agreement between the two assessments at the higher degrees of discomfort, there was a substantial discrepancy at the 'comfortable' end of the scale, which suggests that the assessments in this experiment were high compared with the Harrison-Meaker scale. No practice was given to the subjects and no attempt appears to have been made to validate the results.

While the glare formulae of Petherbridge and Hopkinson, given above, may be used to calculate glaring effect of a single light source without much difficulty, if the glare from a whole lighting installation is to be computed

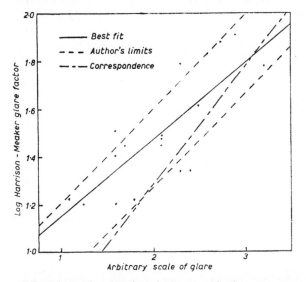

Fig. 115. Experimental comparison between subjective assessments of glare and the Harrison-Meaker glare factor (after Lowson *et al.*, 1954).

the process would be extremely tedious. Using these formulae, the Illuminating Engineers Society in the 1961 Code (I.E.S., 1961) has done the hard work and gives the results in a series of tables which enable the glare index of any particular installation to be calculated. Together with the recommended levels of illumination for particular situations, the limiting value for the glare index thus obtained is given in the Code; it is suggested that any installation which will produce a glare index above the permitted value should be modified to reduce the effect of glare.

Disability Glare

Disability glare can be caused by an excessive amount of light entering the eye from objects close to the task, this may come either from a light source or from light reflected from the immediate surround or from the task itself. The immediate effect of disability glare is to reduce contrast and so the threshold of acuity, and this will be equivalent to reducing the effective lighting level. For instance, a glaring source of 2 lm/ft² displaced 5° from the line of vision will reduce the ability to see by about the same amount as it would be reduced by lowering the lighting level from 100 lm/ft² to 1 lm/ft². From this it will be appreciated that luminaires placed on the machine in the almost direct line of sight can cause great difficulty to operators. Disability glare can also occur when very bright tasks are in a very dark surround, so that the eye is adapted to a very low level of illumination, with the result that the detail of the task cannot be seen.

The effects of disability glare have also received a good deal of attention, in fact it will be remembered that the original Holladay work was in this connection. Research was pursued particularly in the United Kingdom by

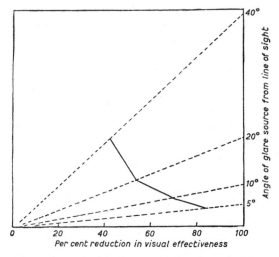

Fig. 116. The effect of a glare source in the line of sight on visual effectiveness (after Luckiesh and Moss, 1932).

Stiles and Crawford (e.g. Crawford and Stiles, 1935) who were especially interested in the effect of the scatter within the eye. This scattering theory was critically appraised by Fry (1954). Studies of performance under glare conditions have been carried out by Luckiesh and Moss (1932) who used their usual test object of two parallel bars under an illumination of 10 lm/ft². A glare source producing an illumination of 5 lm/ft² at the eye was used at different angles in relation to the line of vision. The reduction in 'visual effectiveness' with the glare source in four positions is shown in Fig. 116. Disability glare of another kind was studied by Sharp and Parsons (1951). With the large increase in the number and size of luminaires consequent upon the introduction of the fluorescent tube, it has been suggested that the reflection from the surface of printed and other papers could cause trouble through glare. Using formula 29 of Crouch (1945a) for computing the r.l.d. when some specular reflection is present from a glare source, they computed the loss of visibility on the various quality papers due to reflection from a concentrated light source of 1,200 ftL compared with a diffused source of 200 ftL when the papers were illuminated to general values of 30 lm/ft^2 and 50 lm/ft^2. Their results show that reducing the brightness of the source by 1,000 ftL by increasing its area so that it ceased to be a cause of glare, produced an increase of visibility equivalent to that which would have been produced by increasing illumination by almost 50%.

It seems from this work that the importance of avoiding reflections from bright luminaires cannot be over-emphasized. Reflection from paper surfaces, for instance, may be a cause of the dislike often expressed in drawing offices for the use of general fluorescent illumination in place of local lighting. Other sources of reflection which may cause difficulty are to be found in control rooms from the cover glasses over dials or from shiny bezels or panel surfaces.

An example of difficulty due to disability glare of this kind was found in one of H.M. ships where a control panel was constructed of shiny black material with chromium plated bezels and chromium plated fillets between the different sections of the panel. Due to the shape of the compartment it was impossible to place the lights in positions such that reflections from the surface of the panels and from the chromium plating could be avoided. In addition to trouble from the lighting, the contrast between the white dial faces and the black panels was too high for visual comfort and as a result there were substantial complaints of visual fatigue from those who had to watch this panel. However, when a recommendation was made that the panel should be painted grey, it was said that this could not be done because this particular panel was the Captain's pride and joy and he was accustomed to showing it to visitors. When the Captain left the ship, and before his relief arrived, the panel *was* painted grey and complaints of eyestrain from the watchkeepers ceased forthwith. This story illustrates not only the importance of controlling contrast and reflection to avoid disability glare, but also the importance of designing a task for the benefit of those who had to do it and not for the benefit of visitors.

Although strictly speaking it should not be included under disability, the effect of distraction by glare sources must be considered. If the glare sources with their phototropic effect attract the eye away from the task they may have an effect on performance. To this extent then they can be said to cause disability. This subject has been relatively little studied, but a pilot experiment by Hopkinson and Longmore (1959) has been reported. Their tentative conclusion seems to suggest that the gaze is fixed for longer periods on a large source of low luminance than on a smaller source of high luminance. This is an interesting result, since attempts to reduce discomfort glare, as a result of the work already quoted, have led to re-design of lighting equipment to produce larger surfaces of lower brightness; it would now seem that these may tend to draw the eye away from the work for longer periods than did the older type of smaller high brightness sources. The authors, however, make the point that their results do not show that the low brightness sources are more distracting than high brightness sources; the latter provoke responses in the form of a series of jerky darts towards and away from the source which may be more fatiguing than the slower and longer movement towards the low brightness sources. Clearly this is a factor in the visual situation which requires more attention.

COLOUR

Colour in factories has already been briefly mentioned, and it was suggested that its primary use should be to improve conditions for visual comfort by reducing sharp contrasts in the immediate surroundings to the task. This may be carried further into the more general surroundings by making sure, for instance, that ceilings and walls adjacent to windows should have a reflectance of not less than 60%. This may well mean painting the walls which have windows in them a lighter shade of the main colour of the room itself. Internal and external reveals and window frames can also with advantage be painted a light colour. Distracting contrasting features such as roof trusses should also be light, but vertical pillars which are liable to be a hazard may well be painted a colour which will call attention to their existence. Just as low contrasts may obscure unwanted detail, vivid contrast may be used to call attention to situations such as gears, moving parts, dangerous cutting edges and levers or control buttons. Both colour and brightness contrast may be used here to call attention to these features, which may be a hazard to the operator. The external parts of machine guards should be considered as part of the machine itself and should therefore form part of the machine's own colour scheme, but the danger area behind the guard which is revealed when the guard is removed should be painted in a vivid colour. So far as possible the inner surfaces of the machine casing and the underside of overhanging parts should be kept light, which will aid visual comfort (as already discussed) and cleanliness as well.

Within the above limitations the aesthetic quality of colour or colour combinations which may be used is very largely a matter of opinion. It is often said that warm saturated colours such as yellows and red tend to give a sense of warmth and an advancing effect. Conversely, cool shades of blues, greens and greys tend to give a sense of coldness and recession. It is a common belief that people in a room decorated in warm colours will tend to feel comfortable at a lower temperature than those in a room which is decorated in a cool colour. For instance, Gloag (1951) says . . . 'it is well known that an adverse indoor climate . . . can be relieved to some extent by the use of colours associated with "warm" or "cold" effects'. But he does not produce supporting evidence. This popular belief was challenged by Berry (1961) who gave 25 subjects a synthetic driving task in an illumination of four colours, two 'cool', two 'warm', and then in white illumination. The subjects, who were unaware of the purpose of the experiment, were told that the lighting installation would produce heat which might interfere with the experiment. They were, therefore, to signal when they felt too hot. Actually the heat was provided by blower-heaters and was carefully controlled. The temperature and humidity at the moment each subject indicated discomfort was recorded. Analysis of variance shows that the effect of colour is almost exactly equal to that expected by chance alone ($p = 0.50$). At the close of the tests, subjects were asked to rank the four colours according to the amount of heat transmitted; the rank difference between yellow and amber as 'warm', and green and blue as 'cold' was significant at $p < 0.01$. There was no relationship between the ranking of the colours and points at which individual subjects signalled discomfort. From this it seems that the notion that it may be an advantage to have a cold colour scheme in hot working areas or in rooms with a southern aspect, or to use warm colours in rooms with a northern aspect is a fallacy.

Long narrow rooms may be made to appear wider by using cool colours on the side walls and warm colours on the ends. Horizontal breaks, such as dados, especially when they are in a cool colour, will tend to reduce the apparent height of a room. Conversely, if the upper part of the wall and the ceiling are in the same colour, a feeling of spaciousness will be produced in a room which may be rather low. A room with a low ceiling will appear higher if it is lighted fairly brightly and a high room will appear lower if the major part of the lighting is thrown downwards and the ceiling is allowed to be fairly dark. Most of these effects just described are, of course, subjective, but they seem to be accepted as common practice in interior decoration. Within these limitations, the actual hue used must be largely a matter of personal opinion for which it is difficult to establish a norm. It is probably true to say that provided the general principles which have been laid down in the foregoing are followed, it does not matter very much one way or the other what particular colours are used in order to achieve the particular end in view, provided they do not conflict too outrageously with the accepted views of the people who will have to live with them, and do not violate the primary object of obtaining visual comfort by the avoidance of high contrast where this will be distracting or the use of high contrast where compelling attention is needed. For a discussion of the use of colour in interiors see Gloag and Keyte (1957).

THE QUALITY OF THE ILLUMINATION

In the preceding discussions we have been dealing with the phenomena associated with good vision on the assumption that the incident light has been white or approximately white, that is, the light source has emitted radiation in reasonably equal quantities over the entire visual spectrum. Surfaces which we have described as coloured have therefore had available to them for selective absorption the whole range of wavelengths. For instance, a red material will reflect most of the wavelengths in the red band which fall upon it and will absorb the rest. It follows therefore that if the incident light is deficient in red, the surface will be graded towards grey. The quality of the illumination must therefore be considered together with other factors affecting visual efficiency.

North daylight is approximately white within the normally accepted meaning of this term. Artificial sources are, however, far from having even distribution over the visible spectrum. In many instances this does not matter very much, provided that the use of colour to manipulate contrast is not upset by the quality of the illumination used or provided that colour discriminations do not have to be made. On the other hand, there may be instances where coloured light can be used with advantage in order to enhance colour contrasts which may otherwise be rather low. The range of possibilities in this situation is so great that specific recommendations cannot be made. As a general principle it may be accepted that if it is desired to increase the contrast between two coloured surfaces, light which is deficient in the wavelength of the darker of the two may be used. For example, should it be necessary to distinguish a pale blue object on a yellow background, both object and background having similar reflectivity and comparatively low brightness and colour contrast, it is possible to improve visibility by viewing the task under an orange light.

Special uses of light of this kind will, however, be comparatively rare compared with the use of 'white' light and in industry the commonly used sources may be incandescent, fluorescent, mercury vapour and sodium vapour. These will vary greatly among themselves in their emissions, there being within the fluorescent range alone a wide variety of phosphors which can be used to give light of different spectral qualities. There is as yet no evidence that, provided the amount of light available on the task is the same from different types of light source, there is any demonstrable effect on efficiency on tasks where a change of colour is not affected. Complaints about lighting installations may well arise from other causes. The amount of light may not be the same in two installations, or the use of fluorescent lighting to give a flat or almost shadowless effect may prejudice tasks where the operator depends on shadows for seeing the position of one part relative to another. This may occur more frequently than is generally realized, and special care should be taken to ensure that when general lighting is installed important cues which aid the operator in doing his job are not thereby removed.

In some fluorescent installations there may be trouble from flicker, and this may occur particularly when tubes begin to get old. It has been shown that, even when tubes are mounted out of phase, superimposed frequencies may develop when the tubes have been in use for some time which become noticeable to the operator and may cause annoyance (Collins and Hopkinson, 1954). Certain types of light which are deficient in the red end of the spectrum may have an unfortunate effect on the appearance of the complexion and, while it is easy to say that this should be ignored, the likes and dislikes of the workers have to be taken into account in a happy factory. A condition which older people, at any rate, have come to associate with incandescent lighting is a feeling of warmth which may well be absent from certain types of fluorescent tube, and this also may have an effect on subjective feelings of comfort. It is too early yet to say whether the increased use of fluorescent lighting will cause the younger generation to become so accustomed to it that this effect will disappear in time.

Summing up it can be said that, provided there are no objections on subjective grounds, it does not seem to matter what kind of light source is used for normal work so long as the quantity of illumination is adequate for the task. Light of special colour may with advantage be used to improve visibility by a selective darkening of parts of the task. Special care should be taken in lighting installations to ensure that shadows that are used by operators are not destroyed by overall lighting. The levels of general lighting that are now being recommended would be difficult to achieve economically with incandescent lighting, so that the choice seems to lie between the various types of discharge lamps and there is no evidence that these lamps have any deleterious effect on the operator or on his eyes. There is no evidence either that the amount of ultra-violet emitted by these lamps is liable to cause any harm. In any event, it is substantially less than the amount of ultra-violet found in natural daylight.

Effects of Flickering Light on Human Beings

The effects of flicker in discharge tubes has been mentioned. This is a subject which has received some attention from experimentalists both from the fundamental and practical aspects. Amongst the former, Bach *et al.* (1956) report that unpleasant subjective effects are consistently reported following exposure to diffuse flickering light. They include interference with consciousness, sensations involving the eyes such as irritation and watering, sensations of muscle twitching such as eye blinking and facial twitching, sensations relating to other parts of the body such as queasy feelings in the stomach, headache, nausea, chill and tense muscles. The most consistently effective frequency for production of these effects is 9 flashes/sec, though any particular frequency is not critical between the limits of 7 and 20 flashes/sec. The effects are not cumulative after an exposure of about five minutes, while maximum effects occur with high brightness in the field of view. Monochromatic light seems to hold no advantages over white light. Some degree of drowsiness was reported in all cases when the light was modulated by a spontaneous 3 c/s EEG activity of the subject. Hand-eye coordination was significantly impaired by flickering light in a simple tapping task but not in a more complex tapping task. There is an increase in apparent brightness of a flickering light which appears to be greatest in the neighbourhood of 10 flashes/sec.

The second, practical approach has been taken by Collins and Hopkinson (1954) who used the multiple criterion method of subjective judgement to obtain estimates of flicker on a four-point scale. They examine the effect of various factors such as light/dark ratio, wave form and modulation on assessments of the effect of flicker, and they relate the sensations of flicker to a Flicker Index. This work is closely parallel to their studies of glare which have already been discussed and it can result in useful relationships being established between subjective sensations and physical characteristics of lighting.

PLANNING FOR EFFICIENT VISUAL PERFORMANCE

Ideally, every task should be studied separately and special conditions established. In most instances where a factory lighting installation is planned, the exact position of the machines and the shadows cast by the operator may not be known. After a while machines may be moved and new and different tasks installed in the area, but the lighting, which may be entirely adequate for general work, may be quite inadequate for a number of specialized tasks. It is not good enough to leave the operator to demand local lighting which he can adjust as he thinks fit. The complex interaction of the various factors we have discussed makes it clear that the provision of luminaires is only one part of the job of planning for good seeing.

If no control over the size of the task is possible, the first step is to determine this size and the contrast involved in the task and to take steps to increase the contrast as far as possible. This may be done by the use of shadow from oblique lighting or by providing the lightest possible background to define dark detail by painting jigs white or by using ceramic or enamel backgrounds. If the task is light, the background should be, conversely, as dark as possible. It must be remembered that parts which are made of bright metal will reflect light into the eyes from a surface in the reflected line of sight and this should be lightened or darkened as necessary and be adequately lighted since the light falling directly on the task will not, in this instance, influence brightness. In other instances it may be necessary to use diffused lighting in order to reduce the brightness of specular surfaces. Oblique light, besides being valuable in providing additional contrast by means of shadow, can also be used to enhance detail on black surfaces and oblique back lighting can with advantage be used to emphasize detail on transparent objects. Diffused oblique light may be used either to enhance or diminish the brightness of inscribed lines on both specular and non-specular surfaces.

Many nominally non-specular surfaces have a certain shine, and care should be taken to avoid placing these in a position where luminaires will be reflected directly from the surface of the material. For instance, black print on white art paper becomes impossible to read if it is placed in front of a window due to specular reflection from the paper's surface. In certain circumstances polarized light may be used in order to overcome difficulties with reflections of this kind. Where accurate positioning movements are required, it should be ensured that the light is placed so that full use is made of shadow in order to give the operator information about the position of one part relative to another. At the same time the shadows should not obscure the detail which is required for accurate positioning. This is especially important if the surrounding levels of illumination are very high and if sources of reflected light in the shadowed areas are small. The contrast between the bright surfaces and those in shadow can be such that all detail in the shadow areas could be lost due to the eyes being adapted to too high a level of illumination.

If a job requires very high concentration which may be obtained at the expense of some visual comfort, the task can be isolated from the surroundings by increasing the brightness ratio between the task and its surroundings beyond the level of 3:1 which is considered the maximum for visual comfort. If on the other hand tasks do not require a very high concentration, the contrast may be low, but it has been claimed that if contrasts are too low the visual presentation may become highly monotonous.

Weston (1953) makes the interesting suggestion that there may be such a thing as 'visual boredom'. He suggests that what is fashionably called brightness engineering, in which there is substantially a uniformly bright field, is insupportable on psychological grounds and may rapidly lead to inattention or even to sleep. He makes the valid point that this should not be confused with muscular or neural fatigue induced by excessive use of the eyes. Going to the other extreme however, very bright contrast in the surroundings causing glare will definitely appear to cause muscular fatigue in the eye, and this can rapidly produce feelings of weariness and may again lead to an inability to maintain visual alertness in a task. A happy medium must be struck between complete evenness of lighting and lighting which is over-contrasted or even glaring, while visual monotony can be reduced to a greater or lesser extent by the use of colour.

Special attention should be paid to the avoidance of glare, both from light sources, natural or artificial, and from reflections from specular objects in the field of view. Particular attention should be paid to the ceilings against which the light fittings may be seen, to the windows, and to fittings provided for supplementary lighting. If the sources of glare cannot be moved from the field of view, their effect must be minimized or even removed altogether by the increase in luminance of the surface behind the luminaire, by either additional lighting or colour. Glare sources, whether they are luminaires or reflections, become more troublesome the closer they are to the line of sight, and therefore such sources of glare as chromium-plated bezels on dials should always be avoided.

Planning for visual efficiency should start at an early stage in job planning. If the positioning of adequate local lighting is left until machinery has been built and installed, it may be very difficult to achieve the ideal situation simply because it is impossible to put lighting fittings in the proper place due to the structural shape of the machine in question. If the need for adequate lighting is considered at the design stage, provision for lighting fixtures can be made and the practice of putting the lighting fittings in a press opposite the operator so that the light shines straight into his eyes would be avoided.

These then are the factors which must be taken into account as a matter of routine when a new job is installed or when any old job is to be improved. In many instances, the size of the detail is such that special lighting will not be required. This does not mean that other factors relating to visual comfort should not be looked at, since even in gross tasks, contrast and glare may be such as to cause visual difficulty. Nor need this be confined to the shop floor. The managing director who attempts to do concentrated reading on a black desk top may be suffering unnecessary visual fatigue without being aware of it. Offices with many dark surfaces provided by the fashionable olive green steel furniture may be visually unsatisfactory. It is not unlikely that the lack of adequate visual planning may contribute to a feeling that a particular factory or office is not a very satisfactory place in which to work. It is likely that under these circumstances the cause of any discontent will not be realized and it may be rationalized in other ways. A great deal can be done by good lighting, the use of colour, by the manipulation of contrasts to make a factory or office a pleasant place where people will be happy to work and it is probably well worth while giving a little time and thought, at no great expense, to achieving this end.

CHAPTER FIFTEEN

Environmental Factors IV.
Vibration

During his work a man will frequently be subjected to vibration from various sources. He may be riding in the cab of a railway locomotive or a road vehicle; he may be operating a pneumatic drill or he may receive vibration from a machine tool. Vibration will be transmitted to him through those parts of his body in contact with the source of vibration, usually the buttocks, the hands and arms, or the feet. Gross periodic movement may also be present and it is convenient to distinguish between periodic movement and vibration by defining the former as having a frequency greater than 1 c/s. The effects of movement and vibration upon man may range from motion sickness through slight discomfort to physical damage.

Vibration is characterized by having a wave form which may be a simple or a complex harmonic motion, but most of what is known about the effect of vibration on man comes primarily from research with simple wave forms. Movement and vibration, since they are wave forms, are specified by their frequency and amplitude but, in addition, the direction in which the vibration acts on the man is very important in assessing its effect. Movement with frequencies greater than 1 c/s can normally be found in ships and vehicles and may have an amplitude up to several feet. Its main effect is to produce motion sickness. In addition, if the movement is sudden and of large amplitude there may be danger of injury through a person being thrown about. It must be borne in mind that frequencies greater than about 30 c/s will produce vibrations which, as well as being transmitted through solid objects, are transmitted through the air as sound and which may play a part in transmitting vibration to a man quite apart from the fact that he will perceive them as sound in the ear.

Research on vibration suggests that it may be conveniently divided into three main areas, though the boundaries of these areas are slightly indefinite and differ from author to author. In general, it seems to be convenient to consider low frequency as being 1 to 6 c/s, medium frequency 6 to 60 c/s and high frequency above 60 c/s. A number of investigations have been carried out in order to determine subjectively either the tolerable frequency and amplitude limits of vibration or the threshold of sensation of vibration. These investigations are in reasonable agreement and suggest that at the tolerance limit for comfort there is a not quite linear relationship between log frequency and log amplitude from a frequency of about 20 to 30 c/s at an amplitude of 0·0008 in., to a frequency of 1 c/s at an amplitude of approximately 2 in. Summarizing various research results Janeway suggested, in an unpublished report, that between 1 and 6 c/s the sensation of vibration depends on the rate of change of acceleration, so that at the tolerable limit the relationship between the amplitude (A) and frequency (F) of a vibration can be expressed by the formula:

$$AF^3 = 2 \qquad (45)$$

He goes on to suggest that between 6 and 20 c/s the limit is dependent upon the acceleration and that the relationship between amplitude and frequency is then

$$AF^2 = 1/3 \qquad (46)$$

Above 20 c/s the limit appears to be directly proportional to the velocity of the movement, so that

$$AF = 1/60 \qquad (47)$$

These limits are based on subjective judgements of vibration which can just be tolerated in a vertical direction. Similar relationships representing the onset of severe discomfort and so on have also been determined, but measurement does not extend to the conditions under which actual physical damage is likely to appear. Similar reviews have been made by Goldman (1948) and (1961).

Vibration which is at or near the tolerable comfort limit may be experienced for long periods without causing any harm, although they may interfere with a man's efficiency as an operator and produce sensations of fatigue.

Useful though the subjective approach may be, the alternative method, involving measurement of the effects of vibration, has been followed by a number of workers. Dieckmann (1958) studied the effects of vibration in both the vertical and the horizontal directions. In particular he investigated movements transmitted to the head of a man standing on a vibrating platform and measured a number of physiological functions such as change in skin resistance. He then combined these measurements with a subjective assessment of discomfort, in order to develop what he described as 'scale of vibration strain' using the symbol K to represent the degree of strain. He gives in his paper a series of curves showing K values from 0·01 to 100. The lower value represents the lowest limit of the perception of the movement and the higher value the upper limit of strain which could be tolerated by the average man. Dieckmann differs from Janeway in suggesting that strain is best measured by the peak acceleration of the vibration over the frequency range 1 to 5 c/s, by the velocity from 5 to 40 c/s, by the amplitude for frequencies above 40 c/s. It will be remembered that Janeway suggested that from 1 to 6 c/s the degree of strain was indicated by the rate of change of acceleration, whereas Dieckmann suggests that over this frequency range strain is indicated by acceleration only.

It has been demonstrated that when a man, seated on a hard seat, is subjected to vertical vibration, his head will vibrate with a frequency similar to that of the seat from 0 to 10 c/s (Müller, 1939; Latham, 1957 and Dieckmann, 1958). As the frequency increases so does the amplitude

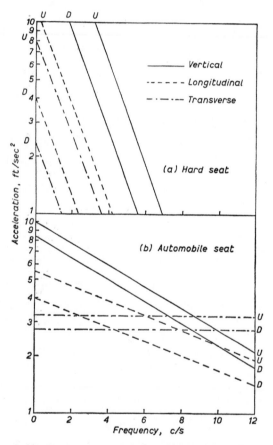

Fig. 117. Subjective assessments of uncomfortable (u) and disturbing (d) vibration (after Jacklin, 1936).

of the head movement, which reaches a peak at seat frequencies of 3 to 6 c/s, when it is between $1\frac{1}{2}$ and 3 times the amplitude of the seat. Above these frequencies the amplitude of head movement falls until at about 10 c/s it is again approximately equal to the seat amplitude and further increases in frequency are accompanied by diminishing amplitude of head movement. A similar effect has been shown by Coermann (1938). There is a very rapid decrease in the amount of vertical vibration transmitted to the head of a standing man between 20 and 40 c/s, and beyond 60 c/s the amount of vibration transmitted is almost negligible. For the horizontal vibration of a seated man, Dieckmann found that head amplitude is greatest at about 2 c/s, but the diagrammatic presentation of his results makes it impossible to give the numerical relationship between the head/ seat amplitude. At 5 c/s the head amplitude is quite small with a vertical component equal to the horizontal component.

Of specific design problems related to vibration, vehicle seating has received the most attention. Jacklin (1936) used subjective judgements of seated people on three degrees of vibration: (1) perceptible – when movement is just felt, (2) disturbing – movement of some organs of the body is felt and is counteracted by muscular effort; (3) uncomfortable – tolerable for only a limited time. Hard seats were first investigated using 100 young women, an unspecified number of young males, and a 'considerable number' of older people, to give, it is claimed, results representative of the average 'normal' human. The results can be expressed in the form:

$$K = Ae^{0.6f} \tag{48}$$

where $K =$ a constant (not to be confused with Dieckmann's K)

$A =$ peak acceleration in ft/sec^2

$f =$ frequency in c/s.

Limiting values of K were determined for each criterion and for three directions of vibration; the resulting relationships, shown in Fig. 117 (a), indicate that tolerance is greatest for vertical and least for transverse vibration. The vector sum of the Ks in the three directions can indicate the total effect of simultaneous vibration.

The investigation was extended to soft automobile seats in road tests, which gave groups of data depending on whether movements in two or three directions were in resonance. An accelerometer ('the dummy') was mounted on a pad on the seat beside the subjects, while the vehicles were driven at varying speeds. From the results (Fig. 117 (b)) a combined index of 'riding comfort', Kc, was computed to give

$$Kc = 11\cdot9 \text{ ft/sec}^2 \text{ for 'uncomfortable'}$$
$$\text{and } Kc = 9\cdot78 \text{ ft/sec}^2 \text{ for 'disturbing'.}$$

Jacklin goes on to evaluate a number of production cars against these criteria and finds several of them unsatisfactory. The general applicability of these results seems to depend on whether the equipment used to measure acceleration has characteristics similar to the 'dummy' used by Jacklin, since no measurements were made upon the subjects themselves.

Studies with padded car seats have also been reported by Wisner (1961), but unlike Jacklin he placed instruments to measure vertical acceleration on the right hand side of the head of his subjects at the level of the semi-circular-canal, and on the mid-line of the chest. Each subject sat on a plastic cup, which was individually moulded to the shape of his buttocks and placed between him and the padded seat. A third accelerometer was located in this cup to record movement of the pelvis (Fig. 118). Wisner

Fig. 118. A subject fitted with three vertical accelerometers (from Wisner, 1961).

found that the natural frequency of the head was 1·1 to 2 c/s and suggests that this coupled with the natural frequency of the vestibular apparatus of 0·5 to 1·0 c/s is a reason why vehicles in which the suspension has a low natural frequency of, say, 1 to 2 c/s will cause car sickness. A further finding was that at about 4 c/s the pelvis and thorax show maximum phase difference causing stretching and compression of the lumbar region. This may account for the 'rough ride' experienced on tractors which often have a natural frequency of about 3·5 c/s.

Field studies of tractor seats have been summarized by Lehmann (1958a) in which energy expenditure in Kcal/min was measured. The difference in phase of upper and lower parts of the body reported by Wisner no doubt contributes to the quite wide differences in energy expenditure of 0·5 to 1·65 Kcal/min found with different types of seat suspension examined by Lehmann (these figures do not include basal metabolism).

In addition to measuring energy expenditure, Lehmann also measured the 'electrical capacity of the skin', which he considered to represent nervous stress. Tractors fitted with nine different seats were driven over a standard test course and it was found that a seat suspension based on a parallelogram linkage (so that the seat remains parallel to the base of the tractor) was much superior to a hinged type of suspension. Vibration characteristics of these seats are not given in the paper. The use of energy expenditure measurement to assess the effects of vibration is not very common and would appear to be quite useful. Discussing seating Guignard (1959) suggests that pronounced total body movement will occur only at frequencies below 6 c/s, while at frequencies above 10 c/s vibration appears to be absorbed by the tissue if the individual is seated, or by the insulating properties of the seat pad. Unfortunately, it is these lower frequencies which are most usually found in transport. Wisner (1961) has pointed out that the system made up of body, seat and suspension is so complex that it cannot be expressed mathematically. In field investigations of vibration conditions, a man should be used if possible, but if a dummy is used, it must have the same 'system of suspended masses' as a man, that is it must have dynamic similarity.

It should be noted that most of the experiments on subjective tolerance to vibration have been of comparatively short exposure durations, rarely more than about half an hour. Little is known therefore about the effects of prolonged vibration, even at levels which are considered tolerable for shorter periods. Unauthenticated reports from individuals who have spent long periods in vibrating vehicles suggest that they find this vibration fatiguing. No definite scientific evidence can be drawn from this kind of report, however, because 'fatigue' is an extremely elusive condition and the term is used in a variety of ways.

It will now be clear that the effects of vibration on man depend on the structural complexity of the human body and in particular, on the way in which the different dynamically suspended masses will respond to vibration of different frequencies. This subject is too complex to be dealt with in detail here, but an extensive treatment is given by Goldman (1957) and Cope (1960). Vibration may have the effect of producing a displacement of the internal organs of the body and under extreme conditions damage may occur. Large organs may pull their supporting ligaments and cause injury, due to crushing of the soft tissues. Men who travel in vehicles such as tractors, frequenty complain of traces of blood in the urine or lumbar and abdominal pain. It is also known that lorry drivers may suffer from sacro-iliac strain, but it has not been determined with any great certainty whether this is due to vibration. One of the difficulties of research on these effects is that the experimenter cannot deliberately inflict injury on his subjects. While experiments may be carried out on animals, it is not by any means certain that effects found in animals will occur, under similar conditions in a man, since the vibration characteristics of animals differ very greatly from those of men. Excessive vibration of amplitudes of the order of 0·2 in. at frequencies of about 10 to 20 c/s has been shown to cause damage to the lungs in monkeys, probably due to the vibration of the heart (Riopelle et al., 1958), and in man to cause rectal bleeding, blood in the urine and constipation (Mozell and White, 1958). Mechanical injury to the auditory system has been described and under certain conditions heart failure may occur in animals.

The effects so far discussed have been due to vibration of the whole body. In addition, by vibrating parts of the body only damage may be caused such as that resulting from the operation of hand-held vibrating tools or pneumatic drills (Bart, 1946; Giullemin and Wechsberg, 1953). Agate and Druett (1947) have shown that the vibrations set up by handtools can be progressively damped by passage up the arm, so that no vibration can be detected in the shoulder. Thus it seems likely that damage is caused by a succession of rapid sharp blows, rather than by the continuous oscillation, which occurs when the whole body is vibrating. Although there is an extensive clinical literature on injury from vibrating tools, relatively little is known about the way in which the damage is caused. In some instances actual bone damage has been observed, but a more common condition produced by vibrating tools has been described as 'white fingers', 'dead hand', or 'Reynaud's phenomenon' and is characterized by cyanosis of the hands when exposed to cold, by pain and by numbness. It seems likely that these effects are caused by changes in the blood vessels and capillaries near the surface of the skin, which influence the ability of the affected skin surface to respond to heat and cold in the normal way.

Very little is known about the effect of vibration on efficiency. It seems that efficiency is most likely to be affected when the body is vibrating relative to the task. It has been shown, for instance, that visual acuity can be impaired and the effect is attributed to movements of the eyeballs, which are at a maximum for frequencies between 40 to 80 c/s. It seems clear that when an operator is sitting on a seat which is vibrating at a different frequency from an instrument panel, difficulty will be experienced in reading instruments. Anxiety may be experienced, especially by pilots, though some of this may be due to fear of losing control of the aircraft (Cope, 1960).

It is impossible to make specific design recommendations which will ensure that the effects of vibration on efficiency are minimal. This is because vibration characteristics vary enormously so that results from one particular situation can rarely be extrapolated to a different situation. The main point is to ensure that the amplitude and frequency are confined within the limits of tolerance of the operator. Goldman (1961) suggests that under certain conditions peak accelerations above 1 g will be dangerous and above 0.03 g will cause discomfort. Biological effects of greatest practical interest are likely to be found between 2 to 20 c/s and at peak accelerations above 0.001 g. Damping or other means of control may be introduced where appropriate. Another difficulty in dealing with the effects of vibration on performance is that individual attitudes to vibration vary considerably. Some people may be prepared to accept it, some may actively dislike it, while others may not notice it. The existence of this subjective variable makes it difficult to predict the effect of even a modest degree of vibration on the performance of any particular individual or whether he will report experiences of 'fatigue' induced by vibration. In industry this problem is not made any easier by the fact that many people believe that they are paid, in part at least, to tolerate discomfort (which can include that due to vibration) and it would appear that most workmen will accept vibration provided that it does not cause actual physical damage. Thus there is an obvious need for more extensive data on individual differences in attitude towards vibration on the job.

CHAPTER SIXTEEN

Organizational Factors I.
Methods of Investigating Work

Before any change is made in equipment or working methods it is necessary to discover where change is needed. At times it will be only too obvious, on other occasions the only clue may be vague complaints, while quite often inefficiency may be unrealized or be put down to human failure. On the whole industry is equipped to deal with only the most obvious faults which are revealed by Method Study, while the sole criterion of evaluating a situation will be output, based on time studies. It is true that output is the ultimate arbiter of success, but time studies, although they may show which of two solutions is the better, do not usually indicate whether either is the best that can be achieved.

Occupational physiologists and psychologists are also interested in isolating, solving and evaluating problems of human performance and have developed a number of methods mainly for laboratory use which could with advantage be further developed for use by industry. These methods will be briefly described in this Chapter, with some examples of their application to specific work situations. Established methods of industrial engineering such as motion study, activity sampling and the charting of movements of men or materials, can also be useful in isolating some problems, but they are too well known to require description here.

ISOLATING ERGONOMIC PROBLEMS

User Opinions

Design faults may sometimes be revealed by obtaining the views of people using a piece of equipment; misleading results are, however, probable unless certain precautions are taken. In the first place it is not easy to frame unbiassed questions which will give no clue to the kind of answer which it is hoped to obtain. A user may not realize the design faults in which the questioner is interested and attempts to direct his attention to these points may result in answers being given which are influenced by this direction. The use of 'open ended' questions may also prove to be unsatisfactory for the same reason: the faults may exist but the user may have adapted himself to such an extent that he will see no reason to complain unless they are very severe – when they may be very obvious anyway.

A second difficulty is that complaints may be unrelated to the true cause of the trouble. The user may feel that there is something wrong, but because he does not have the expert knowledge to realize what it is, he will pick on something obvious. Thus, if a man has difficulty in reading a scale he may complain about the lighting when in reality it is the scale design which is poor.

One way of overcoming these difficulties is to attempt to determine the likely effects of the faults under investigation and to question the user on these effects in a way which will leave him in ignorance of the true purpose of the questions. For instance, if users are questioned on seat height they will usually be satisfied with what they have, but if they are asked about the extent to which they use the back-rest, how often they need to get up during a working period and whether their backs or shoulders ache at the end of the day, useful results may be obtained.

Another approach is to question learners about the difficulties which they are experiencing. They should not by then have had time to adapt themselves to unsatisfactory conditions and have come to have accepted them. For instance, drivers of diesel trains will usually be satisfied with their driving positions but if carefully questioned will often admit that at the outset they suffered physical discomfort which gradually disappeared as they got used to the job.

Accidents

Data from accidents may sometimes be used to reveal design faults. Unfortunately most accident reports are intended to show who was to blame rather than to reveal the true cause, so that if data from accidents are to be of any value, special report forms will have to be used which are designed to pinpoint design factors (if any) which may have contributed to the accident.

The Critical Incident Technique

Fortunately, accidents are comparatively rare. More rewarding can be the study of near-accidents or *critical incidents* and a technique for doing this has been developed by Flanagan and his collaborators. This can best be described in Flanagan's own words (Flanagan, 1954):

'The Critical Incidents Technique consists of a set of procedures for collecting direct observations of human behaviour in such a way as to facilitate their potential usefulness in solving practical problems . . . (it) outlines procedures for collecting observed incidents, having a special significance in meeting systematically defined criteria. By an incident is meant any observable human activity that is sufficiently complete in itself to permit inferences and predictions to be made about the person performing the action. To be critical, an incident must occur in a situation where the purpose or intent of the act seems fairly clear to the observer and where its consequences are sufficiently clear to leave little doubt concerning its effects.'

The basic idea underlying the technique is that if sufficient data are obtained on a number of incidents in a design or a task which may have led to either an accident or a near-accident, the cumulative evidence may be sufficient to warrant further investigation. Thus if a process worker reports mis-reading a dial, thereby causing a mal-functioning of the equipment, it may, on a single occasion, be put down to human error, but if this occurs on a number of occasions and with a number of different process workers there would seem to be a very good case for assuming that there is something wrong with the instrument. The technique can be carried out in two ways: observers can make daily, weekly or monthly reports of incidents which have come to their notice or a number of individuals may be questioned about incidents which occurred in the past and which they remember. When on-going information is obtained from a number of individuals, it is important that only very simple types of judgement should be called for, that the observations should be as objective as possible and that there should be agreement between observers on the types of judgement to be made and the type of event to be reported. This procedure should result in the collection of a series of objective facts which have a minimum of inference or interpretation. Subjective evaluation of the data should be avoided.

The earliest and most frequently quoted example of this technique is that of Fitts and Jones (1947b) who collected information on what is often called 'pilot error'. They questioned a large number of pilots on whether they had ever themselves made or had ever seen anyone else make an 'error in reading or in interpreting an aircraft instrument, detecting a signal or understanding instructions'. As a result, they collected a number of incidents involving errors of interpretation of instruments which had more than one pointer, in particular the three-pointer altimeter which was frequently mis-read by a thousand feet. In addition, errors of misreading instruments such as the artificial horizon, or errors in interpreting direction and heading from compasses were reported.

Following the work of Fitts and Jones, an extensive programme of research into instrument design was initiated and it is from this that part of the material given in Chapter 9 has come.

Perception Analysis

Motion study takes into account only the physical activity of an individual, but it may at the same time be necessary to attempt to study the information which an individual has to receive, its sources, and the sequence in which it is gained. Crossman (1956) has proposed 'perception study' similar to motion study but the results of the use of this technique have not apparently been published.

An analysis of the sources from which an individual obtains information while doing specific tasks can sometimes be obtained by making studies of eye movements. This method has been used, for instance, in the study of pilot activity, when making blind landing approaches, the objective being to evaluate the layout of the instrument panel (Jones et al., 1949). From this kind of study the sources of the information required by an individual in doing the task and the relationships between the sources can be obtained.

Work Measurement

In most rates an allowance called 'compensating rest' is incorporated which is intended, in part, to compensate for unsatisfactory or stressful features in the task. Whenever an allowance is given which is in excess of perhaps 10%, there are likely to be unsatisfactory features in the job which warrant investigation. For instance, if a 25% allowance were given for heat, steps should be taken to remove the heat stress. When there is a heavy perceptual load, no compensating rest would normally be given, so that it can only be used to show which physical features of the job are not satisfactory.

METHODS OF MEASURING WORK AND ACTIVITY

Energy Expenditure

The amount of energy expended in doing physical work is closely related to the amount of oxygen consumed, so that methods of measurement attempt to assess this amount either directly or indirectly. Methods of direct measurement involve estimating the volume and composition of the expired air while indirect methods are based upon variations in the heart rate. The first practical apparatus for measurement of oxygen (O_2) consumption was developed by Douglas in 1911 who produced the famous 'Douglas Bag', a large impermeable envelope carried on the back of the subject into which the whole of the expired air was breathed through a mouthpiece. This bag is however somewhat cumbersome, and has the disadvantage that its capacity is limited to about 100 litres so that it can only be used for comparatively short periods before it is full, which would take perhaps five minutes of moderately heavy work. Thus if the ventilation rate is high or the measurements are continued for any length of time, a method of continuously measuring the volume of expired air must be used.

An early attempt to do this was made in 1859 by Edward Smith who used a valve facepiece and a gas meter. He passed the whole of expired air through a gas analysis apparatus to measure the carbon dioxide (CO_2) but did not measure oxygen consumption. In the early 1900s several pieces of equipment were constructed using gas meters, but most of them were not portable. Apparatus of this kind is still used in many laboratories

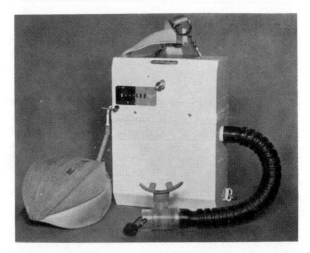

Fig. 119. Kofranyi Michaëlis apparatus for measuring volume of expired air (by courtesy of the Max Planck Institut für Arbeitsphysiologie).

where the 'captive' subject works, for instance, on a bicycle ergonometer.

The first practical portable apparatus was produced by Kofranyi and Michaëlis of Dortmund (1941) which measures directly the volume of the expired air and takes a small fraction, approximately 0·05%, into a rubber bladder for analysis later. It consists of a small gasmeter weighing about 4 kilogrammes and is carried on the back; the expired air is passed into it by means of a facepiece and a connecting rubber tube (Fig. 119). It cannot be used for measuring very high work rates, since if the ventilation rises to about 50 litres a minute the meter tends to under-record. It also has the disadvantage that there is a definite resistance to respiration which may prove unpleasant to the wearer.

To overcome these difficulties the IMP (integrating motor pneumotachograph) has been designed by Wolff (1958). This instrument (Fig. 120) is sensitive to variations in pressure of the expired air and produces a voltage output which is proportional to its instantaneous flow. This voltage

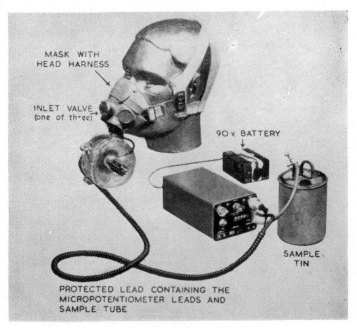

MASK WITH
HEAD HARNESS

INLET VALVE
(one of three)

90 v. BATTERY

SAMPLE
TIN

PROTECTED LEAD CONTAINING THE
MICROPOTENTIOMETER LEADS AND
SAMPLE TUBE

Fig. 120. The IMP.

is integrated mechanically to give a direct reading of volume on a mechanical counter. As with the Kofranyi and Michaëlis instrument a small sample of expired air is taken at intervals and passed into a plastic bag.

In its simplest form, the method of pulse count requires no apparatus at all, the wrist being palpated for approximately 15 seconds at fixed intervals; this has the disadvantage of interrupting work for what are virtually tiny rest pauses. Furthermore it tells nothing about the fluctuations in pulse rate which may occur between the taking of readings. Nevertheless, the method has been used successfully by workers in Scandinavia. Continuous readings of pulse rate can be taken with a photoelectric pulse counter which is carried on the back and receives signals from an attachment on the lobe of the ear which contains a tiny photoelectric cell and an electric lamp (Müller and Himmelmann, 1957). As the blood flows through the ear lobe, its density increases thereby causing variations in the current from the photocell. The signal from the earpiece is fed to a transistorized amplifier which operates a counter. The equipment can be fitted with a transmitter so that the pulse rate can be recorded remotely. This instrument is available commercially (Figs. 121 and 122).

A second method of electrically recording the heart rate is an adaptation of the electrocardiograph. Two electrodes pick up electrical activity resulting from the action of the heart. This can either be recorded on an apparatus carried by the subject or data can be transmitted to a remote recorder

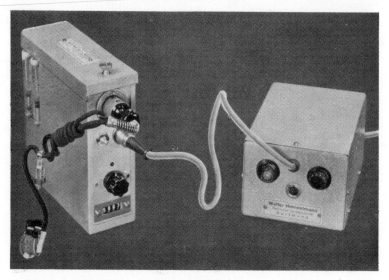

Fig. 121. The Müller Photo-electric Pulse Counter (by courtesy of Prof. E. A. Müller).

(Metz *et al.*, 1961). The equipment is available commercially in America and Holland. Heart rate can be related to oxygen consumption or energy expenditure: for the normal average man a heart rate of 75 beats/min seems to be equivalent to an oxygen consumption of about 0·5 l/min or a calorie expenditure of 2·5 Kcal/min, and each increase above this of 25 beats is equivalent to an increase of 0·5 l./min of oxygen or of 2·5 Kcal/min. The resting pulse rate is usually taken to be about 62 beats/min when the O_2 consumption is 250 ml/min and the calorie expenditure is 1·25 Kcal/min (Christensen, 1953). The heart rate for women may be up to 10 beats/min higher than that for men of equivalent aerobic capacity (Åstrand, 1952).

This relationship between pulse rate and O_2 consumption has been confirmed by Wyndham *et al.* (1959) working with Bantu subjects. In particular they have shown that it reaches an asymptote which defines a maximum level of O_2 consumption and pulse rate for each individual.

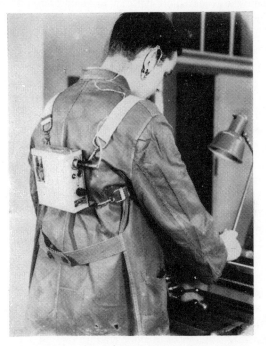

Fig. 122. An operative wearing a Müller Pulse Counter (by courtesy of Prof. E. A. Müller).

'Electro-' Methods of Measurement

Electrical potentials are produced by muscles when they contract and in Electromyography (EMG) these are picked up by suitable electrodes and fed through amplifiers to a suitable pen recorder or to a cathode ray tube. In theory the greater the activity of a muscle the greater will be the potential. Thus it is argued that a movement which displays greater activity may be less satisfactory than one which displays less, provided both achieve the same end. It is also thought that a fatigued muscle will show greater activity than one which is not fatigued.

There are some difficulties in the way of this simple interpretation. In the first place, surface electrodes are rather unselective in the potentials which they pick up, so that it is not always possible to be sure that the muscle groups in which one is interested are actually those whose activity is being recorded. This difficulty can to some extent be overcome by the use of co-axial needle electrodes, but these may interfere with the work being done. A second difficulty is that the records are difficult to interpret. Changes in muscle potential are not easy to quantify and it cannot be said with any certainty that the changes which are observed are of any significance. Ryan (1953) has described a method of recording in which the amplified potentials are rectified and fed into a condenser which when charged will fire a thyatron tube whose output could be either recorded on an ink-writer or fed into a counter. He attempted to use electromyography as an indicator of effort in visual tasks without great success.

In spite of these difficulties, useful research has been undertaken with electromyography on sitting (Floyd and Silver, 1955); on postural work (Carlsöö, 1961); on women – with special reference to high-heeled shoes (Joseph and Nightingale, 1956); on muscular fatigue (Lippold et al., 1960); on typewriting (Lundervold, 1958); on visual tasks (Alphin, 1951) and on glare (Ryan et al., 1950).

Electroencephalography (EEG) is similar to EMG, in fact the same apparatus can be used for both. The difference lies in the potentials being measured. Although EEG is essentially a technique extensively used in research on the brain, attempts have been made to adapt it to the study of performance. Kennedy (1953) has described his attempts to develop an Alertness Indicator in which he first recorded the alpha rhythm of subjects in a vigilance situation. The idea was that a slowing of the alpha rhythm might indicate decreased alertness. It was found that there was such a wide range of individual differences in alpha rhythm that no useful result could accrue. He then tried sponge rubber electrodes placed on the forehead just above the supraorbital ridges and the resulting potentials were integrated electronically into a moving average over one second. An automatic warning device would activate when the tension level fell below a predetermined threshold which could be set to act as a useful predictor of very long reaction times or response failures. Pen records showed that some subjects would go for several hours without a serious drop in tension level whereas the records of others would start to show oscillations soon after the start of the experimental session, these were particularly marked when a subject was deprived of sleep. No later use of this method in a practical situation seems to have been reported.

Electro-oculography (EOG) is a method of electrical recording of eye positions. There is a standing potential difference between the back and the front of the eyeball whose field moves as the eye rotates. The use of this field movement is complicated, but the practical problems have been largely solved (Shackel, 1960). Prior to the development of this method, eye movement was recorded by reflection from a mirror mounted on a contact lens, by corneal reflection or by direct photography of eye position. This last method was used by Fitts and his co-workers, at the Psychology Branch of the U.S.A.F. Aero-Medical Laboratory, to study the fixations on blind-flying instruments in order to arrive at the optimum layout and by the present author to check scanning patterns on panels of dials during check reading. The photographic method is, however, somewhat crude and very time consuming so that the use of EOG will greatly facilitate studies of this kind.

The Force Platform

A device for measuring the forces which result from either the making of movement or the application of static forces has been developed by Lauru (1957). The principle of these platforms is that the forces which are set up through the muscular activity of the subject are measured by piezo-electric quartz crystals mounted at suitable points on the platform. Lauru's platform is triangular and has crystals mounted at its three corners to measure vertical, transverse and frontal forces. It is portable and can, if necessary, be taken into a factory. While the force platform is mainly of use to physiologists in studying muscle activity it can be used to a limited extent to determine the forces which are applied when work is carried out by various methods.

Flicker Fusion Frequency

Flicker fusion frequency (also known as critical flicker fusion) is the rate at which successive stimuli cease to appear to flicker and become steady or continuous. This has been used in studies of vision for a long time and it was usually taken for granted that it was a function of the retina. More recently, however, it has been suggested that flicker fusion frequency is more likely to be a cortical function and it can be used as a measure of the fatigue in the central nervous system (Simonson and Enzer, 1941), or 'mental fatigue' (Brozek and Keys, 1944). Recently the phenomenon has been investigated in relation to a number of physiological activities, these studies are important because it is difficult at times to say whether 'fatigue' is entirely mental and may not have some physical component. Both visual and auditory flicker have been studied but the bulk of the research effort has been concentrated on the former.

Visual Flicker Fusion

An extensive summary of the work on VFF was made in 1952 by Simonson and Brozek but a more recent brief summary has been given by Grandjean and Perret (1961). Different workers have used different methods of measuring VFF and for this reason the results are often not strictly comparable. Most workers, however, express their results in terms of flashes or cycles per second. One variable of importance is whether the flicker is presented to the subject continuously or discontinuously. In the former method the rate of flash is increased or decreased until the subject either sees the light as continuous or ceases to see it as a flicker, whereas in the discontinuous method, different rates of flicker are presented for varying intervals of time and with this method it appears that the length of exposure may be a variable.

When the continuous method is used, the most common procedure is to start from a low rate and increase up to the point of fusion. The alternative method of starting with a high rate and decreasing produces slightly higher results, the extent of the difference between the two appearing to depend upon the speed with which the rate of flicker is changed. The duration of exposure has also been shown to be an important variable (Grandjean and Perret, 1961), increasing exposure from 10 to 16 sec reduced the VFF by about 3 flashes/sec, while VFF seems to be influenced by the starting point for an ascending rate. Thus Ryan and Bitterman (1951) obtained a VFF of 34·4 flashes/sec when starting from a rate of 20 flashes/sec but VFF of 38·3 flashes/sec when starting from 30 flashes/sec, a difference which is greater than the differences supposed to have been found due to the variable which is being investigated.

Two main methods of producing flicker have been used, the interruption of a beam of light by a disc and the generation of flicker by electronic means. The shape of the light pulse will depend to an extent on the method being used and it would seem that the nearest approach to a square pulse is obtained when a rotating shutter is used in conjunction with a known distance. Approximately square-wave light forms can also be obtained with some types of electronic device and one of these has been described by Fritze and Simonson (1951) in which both the light/dark ratio and the brightness can be varied within wide limits. The area of the test patch, the intensity of illumination and the light/dark ratio are all variables which can influence the results, from which it is quite evident that standardization of methods of testing is an important prerequisite if this method is to be used to evaluate changes which are intended to reduce fatigue.

The effects of these different variables on VFF have been widely investigated and the following variables have been noted:

(1) VFF increases linearly with the logarithm of illumination intensity over a wide range of intensities of the light stimulus. The range of this relationship appears to depend on the area of the test pitch.

(2) When the brightness of the test patch and of the surrounding area are the same, VFF is at its highest and falls off as the brightness of one exceeds the other.

(3) At very small visual angles, the VFF increases linearly with the logarithm of the area of the test patch.

(4) Colours of equal brightness appear to have no effect on VFF.

(5) An increase in light/dark ratio appears to cause a decrease in VFF. There appear, however, to be a number of discrepancies in the findings of different authors on this point and age appears to be a variable (Misiak, 1951; McFarland et al., 1958). Most experimenters who have not been investigating the effect of light/dark ratio appear to use a 50:50 ratio.

(6) VFF decreases during dark adaptation.

(7) VFF is influenced by the size of the background (Foley, 1956).

(8) There is wide individual variability in VFF but individuals are fairly consistent within themselves.

(9) Effects of practice are almost negligible.

Auditory Flicker Fusion

The analogous phenomenon to VFF, auditory flicker fusion (AFF), has received very little attention from researchers. It is measured by changing the rate of interruption of white noise until the subjects report the disappearance of flutter. An early study by Miller and Taylor (1948) suggested that AFF might be as sensitive as visual flicker to any of the changes which the latter purported to measure. Moreover, it seemed to be a simpler technique with fewer variables. This method was further investigated by Davis (1955) who made his subjects carry out mental multiplication for periods of 2 hours and 1 hour, the latter period having two levels of work load. He measured both visual and auditory fusion frequencies before and after the tasks were performed. He found a decrease of AFF of 10% after the 2 hour task and 4·5% and 2·4% after the 1 hour tasks, all the differences being significant at $p = 0.01$. With VFF, on the other hand, he found a decrease of only 2·2% after 2 hours ($p = 0.05$) and 1·7% and 0·3% after 1 hour, differences which were not significant. These results suggest that AFF is likely to be a more sensitive measure of the effect of work than is VFF, but so far as can be traced, no attempt has been made to use AFF as a measure of 'fatigue' in an industrial task.

To sum up, therefore, it seems that visual fusion frequency may be a useful technique to evaluate 'mental fatigue' in industry, but in the present state of knowledge it is not nearly reliable enough to be used as a standard of measurement of 'fatigue' (itself an indefinite term in this context). This does not mean that it is a method which should not be used, and a great deal of useful evidence might be obtained if more attempts were made to validate it in industrial situations. That changes of VFF do occur under certain circumstances cannot be denied, but it appears to be less sensitive than AFF under similar conditions. It seems, therefore, that AFF and to a lesser extent VFF may be promising methods for evaluating the result of changes which may be made in the working environment, but more research will have to be done, particularly on AFF, before they can have any useful validity.

'Spare Mental Capacity'

The idea behind the measurement of 'spare mental capacity' is that when a man is doing a task which is not very demanding he has something in reserve with which he can undertake a secondary task. As the primary task takes more of his attention, the secondary task may be progressively neglected and is therefore a measure of the demand made by the primary task.

This method was initiated by Bornemann (1942) who studied the interaction of mental arithmetic as a subsidiary task with a number of primary tasks. Other workers who have developed the method are Bahrick *et al.* (1954) on performance as a result of practice, Broadbent (1956) on relative difficulty of listening tasks and Garvey and Taylor (1959) to compare tracking tasks.

A systematic study which is continuing has been reported by Schouten *et al.* (1962) who used as a primary task the pressing of pedals in response to high or low tones presented in unpaced or paced conditions at different rates representing various percentages of maximum performance. Secondary tasks included putting washers and nuts on screws, solving arithmetical problems and writing. The results show that simple tasks are relatively unaffected by raising the demand made by the primary task, while complex tasks may disintegrate completely.

These are laboratory studies designed for specific research purposes, work of a more practical nature has been reported by Poulton (1958) who compared two arrays of dials in order to assess their relative readability and by Poulton and Brown (1961) on car driving.

As it stands at the moment this method of studying mental demand has been showing promise in the laboratory and on the road. It needs, however, further development for industrial use and, in particular, calibration against known standards of performance.

Subjective Tests

Subjective assessments have been used in a number of ways for many years. Their use is usually associated with factors related to personal feelings which cannot be measured directly, for instance subjective judgement has been used in setting up the effective temperature scales. Subjective assessments of remarkable consistency have been used as a rating procedure in time study, but as with all subjective judgements, this rating procedure is relative rather than absolute. Another area of research, the study of glare, has led to the most formal development of the use of subjective assessments in the 'multiple criterion technique of subjective appraisal' (Hopkinson, 1950). This technique is based on the idea that observers should not be made to assess the degree of a function in terms of pre-arranged categories, but rather that they should be asked to make settings of one variable to correspond to one of a limited number of related criteria. Hopkinson found that observers who were unwilling to make attempts to perform under the first condition were quite happy performing under the second condition.

There are several steps in the procedure which must be considered to be essential. First, there must be a pilot experiment in order to discover which of the variables in the situation could best be put under the observers' control and whether the criteria which had been chosen were clearly understood by the observers. The form of words used to describe the criteria could be very important and failure to choose the best wording would probably make itself known during the pilot experiment through the observers' questioning exactly what the experimenter required.

The second essential feature of the technique is the use of a control experiment which has to continue throughout the main experiment. In it each observer makes assessments, say twice daily, under a set of standard conditions. This not only enables the observers to gain confidence in their ability to make the required assessments, but it also enables the experimenter to detect when an observer has an 'off day'; it further enables him to set up confidence limits from the scatter of each observer's reading under the standard condition. Hopkinson showed that an experienced observer was markedly consistent under these control conditions, but occasionally produced results which deviated substantially. When this occurred it was not considered advisable for him to take part in an experimental series.

As a result of readings taken in a control experiment, it was possible to make a selection of observers. Not everybody is equally good at this kind of task and Hopkinson believes that it is better to work with a smaller number of consistent observers, rather than a larger number who are inconsistent. Just what degree of inconsistency can be accepted will depend on the function being tested in the experimental series. Confidence can be built up in a subject by giving him knowledge of results or by inducing him to guess, and in this way an apparently unsatisfactory subject can be 'trained' until he becomes a useful observer.

The method by which the experimental series will be run will depend upon the variable tested. It would seem that the judgements on the different criteria should be made in ascending and descending order, each subject being allowed as much time as he wishes to make his judgement. Several precautions would obviously have to be taken if the variable under review involves vision: the subject must be allowed time for his eyes to become adapted to the prevailing brightness; if control movements are made to adjust a function, the control knob must give no clue to the subject about the level which he has set, nor if it is a visual task, should the mechanism of the apparatus give him a clue as to what is occurring, for instance by an increase of noise from a motor.

The only point which remains in doubt about subjective judgements is whether a small number of specially trained observers could give results representative of the population as a whole. By working in this way, Hopkinson appears to differ somewhat from the majority of other investigators who have set up various criteria of comfort or discomfort. The effective temperature scale, for instance, was built up on the assessments of a large number of individuals and much of the work on vibration was carried out in the same manner. To meet this point, Hopkinson gives details of the applications of laboratory findings to field conditions, particularly in relation to studies of radar and street lighting; the results which he presents suggest that a method using a small number of highly trained observers will give results as acceptable as the alternative method of using a substantial untrained cross-section of the population.

Miscellaneous Methods

The device known as UNOPAR (universal operator performance analyser and recorder) is described by Nadler and Goldman (1958). A small loudspeaker is attached to the limb whose movement is to be studied, which emits a signal at a frequency of 20,000 c/s. The Doppler shifts of this frequency caused by movement are picked up in three dimensions, amplified and fed into a pen recorder. Although this is an ingenious method of measuring the velocity of movement in three dimensions, rather extravagant claims are made for the device as a universal solver of work problems. In its present form it can only measure one movement at a time. A

UNOPAR is installed in the Department of Engineering Production, University of Birmingham.

A number of other methods of measuring work have been described in the literature, for example, the 'universal motion analyser' (Smith and Wehrkamp, 1951) but these are primarily refinements in methods of timing rather than developments of new techniques of measurement.

THE EVALUATION OF WORKING METHODS

It is often necessary to determine which method of work is likely to be the best or to decide upon the correct design of some equipment. Views may differ on what is meant by 'best', but it should imply not only that the method adopted produces a satisfactory output but that this should be achieved with the minimum strain on the worker.

When physical work is involved, energy expenditure measurements may be used. It is assumed that the method requiring the lowest expenditure of energy is the most efficient, provided it does not take any longer. For instance, in a study by Bonjer (1959) men were required to move cases of beer by four different methods. He measured energy expenditure, total heart beats and time and calculated the theoretical mechanical work based on the distance through which the case was lifted. The results are given in Table 33.

Table 33. Comparison of methods of moving eight cases (after Bonjer in Murrell, 1958)

	Tilting on Short Edges	Tilting on Long Edges	Tilting on Long Edges with Hook	Turning on Angular Points
	I	II	III	IV
Mean time (in minutes)	4·24	4·64	4·72	3·62
Energy expended (Kcal)	17·7	19·9	17·4	9·4
Total heart beats	1,079	1,087	1,214	981
Theoretical mechanical work per metre	0·214	0·207	0·207	0·138

All methods of measurement agree that Method IV, turning on the points of the case, is superior; on the basis of time and heart beats Method I would be placed second, but on the basis of theoretical mechanical work it would be placed last. On the other hand Method III is placed second on energy expenditure but took the longest time. This emphasizes the importance of making a decision on more than one criterion whenever possible.

In this example the rate of work was varied but in other instances the rate and/or the quantity of work is unvaried so that time or output cannot be used as a measure. One such study was an investigation by Bedale (1924) into different methods of carrying loads. In all she carried loads varying between 20 and 60 lb by no less than eight different methods. She carried each load 100 yards at a pace of 2·8 m.p.h. This routine was continued for a period of 1 hour.

The results show that the method of carrying with a yoke involved the least energy expenditure and that even when 60 lb was carried in this way the gross energy expenditure was only 3·7 Kcal/min. For the other methods a maximum economic load was 40 lb, but when 50 lb was carried the energy expenditure exceeded 4 Kcal/min. It would appear, therefore, that for this subject the energy expenditure of 4 Kcal/min represents the level at which further increases of load commence to be uneconomical. Bedale points out that so long as the loads to be carried do not cause too high an energy expenditure it is probably most economical to carry fewer heavier loads. For instance, the additional cost of carrying two 20 lb bundles in the hands instead of two 10 lb bundles is only 0·37 Kcal/min. If, however, these are increased to 25 lb each, the increase in calorie cost is a further 0·67 Kcal/min. If the same total weight has to be moved in the same time, the worker will have to walk twice as fast carrying only 20 lb at a time as carrying 40 lb, and merely walking at 5·6 m.p.h. would require an expenditure of at least 7 Kcal/min, which would rapidly lead to exhaustion.

This finding is confirmed by Müller et al. (1958) who studied the stacking of stones of different weights. They found that if 1 long ton of stones were stacked from ground level, the energy expended when the stones weighed 4 kilograms each was double that which was expended when the stones weighed 28 kilograms each. The difference was somewhat less if the stones were taken from a platform either ½ metre or 1 metre above the ground, and it seems that the additional energy expended was largely due to the additional work done in bending the body much more frequently when the lighter stones were lifted.

The force platform has been used to compare methods of work. Lauru (1957) studied bricklaying by the conventional method and by methods using both arms simultaneously and one after the other, proposed by Gilbreth. The results suggest that the last method requires the least muscular effort, in particular the vertical forces are greatly reduced. Methods of manual lifting have been compared on the force platform by Whitney (1958) who doubts whether the accepted method of lifting with the knees bent is better than keeping the knees straight and bending the trunk.

The jobs just described have been performed without equipment but equally useful results can be obtained when work involving equipment is studied. Crowden (1928) investigated the physiological cost of wheeling bricks from kilns. Each worker moved about 25 tons over 9 hours taking 70 bricks at a time over a distance of between 20 and 60 metres. The barrow has a large wheel of approximately 24 in. diameter with the axle approximately 1 inch in front of the loading platform. The bricks were stacked on either side of the wheel, seven deep by five high (Fig. 123). Measurements were made using a Douglas Bag. The results of this investigation showed that over a standard distance of 50 metres about 8% of the energy was expended in raising and lowering the handles, about 21% in

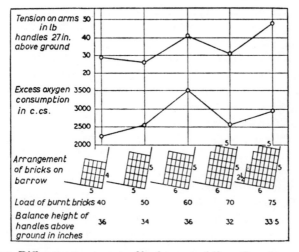

Fig. 123. Different arrangements of load compared (from Crowden, 1928; by courtesy of the Controller of Her Majesty's Stationery Office).

attaining walking speed and stopping, and the remaining 71% on maintaining speed over the distance. The proportion of the energy expended in starting and stopping must clearly depend on the distance travelled and will be greater the shorter the distance; over a distance of 10 metres, lifting, starting and stopping will account for approximately two thirds of the energy expended. From this it follows that once the journey has been commenced the worker should have an uninterrupted run.

Crowden also investigated the size and arrangement of the load, the height of the handles and the centre of gravity of the load which in turn influences the handle height for load balance. The nearer the load is to this balance point, the less will be the weight resting on the arms, but once the balance point is reached the load will tend to tip over forward and it appeared that the optimum position was reached when the load on the hand was about 15 lb. With the arrangement in use at the time this balance was achieved when the handles were 36 in. above the ground and the optimum load on the arms could be achieved only by men of 6 ft or more in height, who had a palm height of 34 to 35 in. Smaller men whose palm height might be as low as 28 in. would have an excessive weight resting on the arms. The balance was altered by re-arranging the load to give a balance height of 32 in. so making the load suitable for men with a palm height of 26 to 31 in.

Other results of the investigation showed that the optimum length of the leg of the barrow was 18 in. rather than 14 in., that the normal brisk walking pace adopted and the load of 70 bricks were the most satisfactory. When these points were established, the only variable remaining was the height of the workers themselves and it was recommended that the arrangement of the load should be adapted to the palm height of the workers.

The force platform has been used by Lauru (1957) to examine five different types of pedals for operating stamping presses. He found that the minimum forces are exerted when the pivot of the pedal is in line with the tibia: in this position the weight of the leg can be rested on the pedal when it is not in operation. A further study of the pedal of a welding machine showed a substantial reduction in the force required when a fixed support was provided for the heel.

Brouha (1960) used the same method to compare forces developed when using manual and electric typewriters and found that the effort was reduced by about 50% when the electric typewriter was used. Typewriting has also been studied by Lundervold (1958) using electromyography. He investigated a number of variables including the height of the seat, the use of the back-rest of the chair, the height of the typewriter in relation both to the elbow height of the typist and the rate of depressing the keys. He found that there was an optimum seat height and when the seat was too low so that the thighs formed an acute angle, or too high so that the feet were barely touching the floor, the muscle potentials increased substantially. Further he found that there was least activity when the relaxed posture was adopted using the back-rest: a more rigid posture would increase the muscle activity. The height of the typewriter in relation to the typist was found to require the least muscle activity when it was as low as possible with the forearms sloping down slightly.

Yllö (1962) used electromyography in a practical study of card punching on IBM 0-24 machines at the Volvo Skövde Works, Göteborg, which showed that the keyboard was too high in relation to the working arm and was in a position which caused the forearm to be twisted, with the wrist in an unnatural position which was the probable cause of hyperfunctional myalgia which troubled the operators. The cards were also too far from the eyes. These faults were ameliorated by fitting the keyboard lower, to the side of the operator and tilted outward, while the whole machine was raised on blocks and tilted forward. In addition both the seating and lighting were improved. Two years after the changes were made performance had doubled.

Visual flicker fusion has been used in an investigation into a process in which needles which have to be straightened are projected onto a screen and are manipulated by the worker until they correspond to a standard. The contrasts in this job were rather high and Grandjean (see Murrell, 1958, p. 125) found that the VFF of women on this work was lowered at the end of the working day, a change not shown by other women in the same shop on different work. This he believed was evidence of fatigue. Changes were made in the visual environment to reduce contrast and as a result there was a smaller drop in VFF, and at the same time output increased.

In addition to comparisons of different ways of doing the same job, the techniques described may be used to study different jobs in order to decide which, if any, require modification, or to investigate complaints. Energy expenditure measurement will clearly indicate when an excessive physical effort is being demanded and pulse rate can show when there is heat stress (Christensen, 1953).

This brief review of some of the research into working methods using biological techniques shows that if they were more widely applied and developed, quite substantial reductions in the demands made on the operator and increase in production could be achieved. As things stand at present, the biological methods have been almost entirely used by academic research workers on research projects; what is now needed is further development work on the shop floor in order to perfect these techniques for industrial use and to define the areas in which they can most usefully be applied.

CHAPTER SEVENTEEN

Organizational Factors II.
The Organization of Work

The problem of how much work a man should be called upon to do and how much rest he should take has been an ever-present problem since 'scientific management' was introduced by Taylor in the closing years of the nineteenth century. In an effort to reach a practical solution to this problem, time-study was developed, by means of which the 'rate' for the job and appropriate fatigue allowances were determined. An extension from time-study has been the pre-determined motion-time systems which enable 'rates' to be built up synthetically. When applied to physical activity these systems work reasonably well, but they have no absolute validity, since time is rather a poor measure of the physical and mental effort required to do a job and of the requirements for recovery. That for practical purposes the results of measurement must be expressed in terms of time does not mean that time is the only or even the best method of measurement. When relatively little physical activity is involved in a job, the stop-watch is useless. Perception and other 'mental' functions cannot be expressed in terms of time alone, so that other techniques will have to be developed if any advance is to be made towards a clearer understanding of factors affecting performance under modern conditions.

Methods used by physiologists and psychologists have already been described, and before discussing the ways in which these methods of measurement can be applied, it is desirable to distinguish different kinds of work: heavy physical work and light work, and repetitive and non-repetitive work. Heavy work can be measured by means of energy expenditure but light work, because it requires relatively little energy, will require other methods.

HEAVY WORK

Energy Expenditure during Work

Before use can be made of measurements of energy expenditure, it is necessary to decide how much energy a man should have available for his diurnal activity and how much of this total should be used for his work.

The first may be indicated from dietary surveys and the second from accumulated evidence on the level at which work can be maintained indefinitely by an average man. If the calorie expenditure for basal metabolism and for non-work activities is known, the amount of energy available for expenditure during work can be determined.

Dietary studies indicate that the calorie intake of food will vary substantially from individual to individual and between groups of individuals within the same occupation. Average calorie intakes for large groups may therefore be somewhat misleading though they may indicate some general trends. For instance, Edholm et al. (1955) studying the training of army cadets, found that young men of similar age offered the same amount of food and carrying out the same amount of activity had a food intake varying from 1,500 to 7,000 Kcal/day, while energy expenditure was in the range 2,000 to 5,000 Kcal/day (Edholm, 1957).

In spite of these difficulties it is probably possible to arrive at a reasonable estimate for the calorie expenditure of that rather unsatisfactory creature the 'average man' employed in different types of occupation. On this basis the F.A.O. (1949) produced a reference man who is estimated to have a daily calorie expenditure over an extended period of 3,200 Kcal per day. His activity would be that involved in occupation in light industry, general laboratory work or driving a truck. By way of comparison, Garry et al. (1955) found that clerks expended approximately 2,800 Kcal per day, whilst miners in the Fyfe coalfield expended 3,600.

The figure given by the F.A.O. relates to individuals doing relatively light work. If heavier work is being undertaken, the daily expenditure will rise and it is generally accepted now amongst work physiologists that over a period of time the calorie output of a normal man in good health should not exceed 4,800 Kcal/day (Lehmann, 1958b). Edholm (1957) thinks this is a little high and suggests that only exceptional individuals can maintain a level above 4,000 Kcal/day indefinitely. These two figures represent an average expenditure of 3·3 and 2·8 Kcal/min over the 24 hours respectively.

Daily averages are, however, not of very great use in deciding how much energy may be expended in work. In order to obtain a value for work alone the 'average man's' daily activity must be divided broadly into three stages. Time spent in bed, time spent in work activity and time spent in non-work activity. Passmore (1956) has suggested that for the period spent in bed (on the average about 8 hours in 24) 500 Kcal are expended, and that 1,500 Kcal should be allowed for non-work activity. This will leave about 2,500 Kcal for work. This is based on his estimate of 4,500 Kcal/day.

Lehmann (1958b), on the other hand, suggests that the requirements for leisure and basal metabolism will be 2,300 Kcal/day (allowing for basal metabolism this is 180 Kcal/day less than Passmore's figure). Taking this from his suggested 4,800 Kcal leaves a *maximum* of 2,500 Kcal of which 2,000 Kcal should be used for a *normal* day's work. Over an 8 hour day this works out at 4·17 Kcal/min to which must be added 1 Kcal/min for basal metabolism to give 5·17 Kcal/min. (A similar value, 5·21 Kcal/min is reached on the basis of Passmore's estimate.) Lehmann, therefore proposed 5 Kcal/min, and this seems to have met with acceptance by many workers: it represents a reasonable maximum which, on physiological grounds, should be regarded as a figure which would not be exceeded when averaged over a long period of time.

In work which remains relatively the same throughout the year, the energy expenditure week by week may remain at roughly the same level but in some occupations, particularly farming or building, the energy expenditure during peak periods may be maintained at a very much higher rate than 5 Kcal/min for weeks at a time. This will do no harm, and it is probable that the food intake will rise to compensate for the extra energy used. These occupations, however, are characterized by periods of lower energy output in the winter, and although we have no direct evidence of this, it seems likely that taken year by year the overall expenditure is unlikely to exceed that expended by individuals who remain on the same sort of task throughout the twelve months.

An alternative approach is developed from the fact that the lactic acid formed during exercise is oxidized by means of oxygen carried in the bloodstream, and further, that muscle itself has a reserve of oxygen which can be called upon when the rate of supply of oxygen to the muscle is insufficient for the removal of all the lactic acid which is being formed. When this happens the reserve of oxygen will gradually be depleted and an oxygen debt will be built up. Under these conditions some of the work is said to be anaerobic. When the oxygen supply is sufficient the work is said to be aerobic.

It has been suggested that the rate of energy expenditure which corresponds to the level at which some of the work ceases to be aerobic is about 5 Kcal/min. At the commencement of work which does not exceed this level, oxygen consumption and pulse rate will rise rapidly and will then reach a steady state, and recovery after the cessation of activity will be quite rapid. However, if the work is above the 5 Kcal/min level, the pulse rate and the oxygen consumption may not reach a steady state for some time or can continue to rise throughout the whole period of activity, and the time for recovery after work has ceased will increase disproportionately.

Five Kcal/min is a *gross* value including basal metabolism and in the discussion which follows all values will be given in gross Kcal/min. Some authors, particularly Germans, give the net calorie expenditure or 'work Kcal' of different tasks, having deducted the basal metabolism. The literature on basal metabolism is vast and those interested should refer to Robertson and Reid (1952). For practical purposes a figure of 1 Kcal/min may be used.

When he is working a man must move his body so that the energy expended doing similar tasks by two individuals of different weight will be greater for the heavier individual, the difference between the values depending upon the amount of bodily movement involved. In a static job the difference is likely to be small, but when heavy exercise is taken the difference is likely to be quite great, as is shown in Table 34. For this reason many experimenters give their results in terms of Kcal/kilo body weight, or in Kcal/metre^2 of the body surface area. To convert the figures given in this form to Kcal/min for the 'average' man, his body weight may be taken as 65 kilos and his surface area as 1·77 sq. metres. The surface area of an 'average' woman is approximately 10% less than that of a man. Other experimenters give their results directly as oxygen consumption in litres per minute and it is normally taken that 1 litre of oxygen is equivalent to 4·8 Kcal.

Table 34. Energy expenditure in walking, related to men of various weights

Speed (m.p.h.)	Energy expended in Kcal/min		
	80 lb	144 lb	200 lb
2·0	1·9	2·9	3·8
2·5	2·3	3·5	4·5
3·0	2·7	4·0	5·3
3·5	3·1	4·6	6·1
4·0	3·5	5·2	7·0

So far we have discussed the total energy expended in 24 hours or during the working day. This total is made up of a variety of levels of energy expenditure as work varies throughout a shift. To obtain optimum conditions two criteria have to be satisfied: first, energy expenditure should not average more than 5 Kcal/min, and secondly, work should continue at a rate exceeding 5 Kcal/min for an optimum length of time after which rest should be taken.

The methods of measuring energy expenditure through O_2 consumption have already been described, but these require specialized equipment and expert knowledge so that the alternative method of measuring pulse rate is more likely to be used for routine measurement in industry. The relationship between pulse rate, oxygen consumption and calorie expenditure suggested by Christensen is given in Table 35.

A working figure of 65 beats/min can be taken as the resting pulse rate, approximately 15 beats above the basal rate. It is to this datum that increases in pulse rate must be referred. Thus, if we take 5 Kcal/min as the maximum expenditure, we would allow an increase of 30 to 35

Table 35. Relationship between oxygen consumption, energy expenditure and pulse rate (after Åstrand, 1952 and Christensen, 1953)

O_2 L/min	Kcal/min	Pulse Rate (beats/min) Male	Female
2·5	12·5	175	195
2·0	10·0	150	165
1·5	7·5	125	140
1·0	5·0	100	110
0·5	2·5	75	85
0·2	1·0	60	63

beats/min in the pulse rate, on the average, over the working day before any recovery time would need to be allowed.

To clearly understand what is going on it is necessary to realize that the effects of muscular exertion are delayed. It will be remembered that when we were discussing the action of the muscles it was pointed out that they contain quite a substantial reserve of energy and that this energy is replaced by oxidation after the activity has taken place. Thus if you make a sudden sprint after a bus, although you may start breathing rapidly while you are sprinting, you continue with a high pulse rate, rapid respiration and, possibly, sweating to get rid of the heat, for some time after you are sitting comfortably in the bus. From this it follows that when activity starts the pulse rate does not rise instantaneously, nor does it drop instantaneously when the activity ceases. A typical curve for pulse rate against time during activity and rest is given in Fig. 124.

If energy expenditure itself is used to determine the amount of work and rest required, the expenditure on all the component activities must be determined, a detailed activity study must be made and the average expenditure for a typical phase or cycle of activity must be determined. The *total* rest required can then be computed from the formula:

$$a = \frac{w(b - s)}{b - 1·5} \tag{49}$$

where a = recovery time in minutes, w = total working day in minutes, b = average calorie expenditure per minute and s = level of energy expenditure adopted as standard. The value of s may be taken as 5 Kcal/min for an 'average' man, or more precise results may be obtained by adjusting this value to the surface area of the individual when his height and body weight are known (Sendroy and Cecchini, 1954) and/or to his aerobic capacity (Åstrand and Ryhming, 1954).

This calculation will give only the total time during the working day which should be taken as rest. The optimum arrangement for work and rest is that work should cease at the point in time at which lactic acid

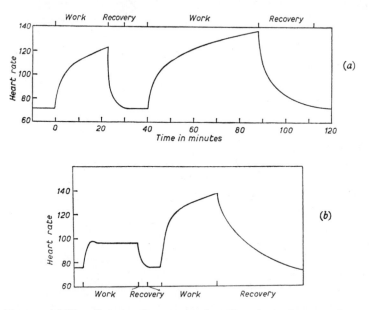

Fig. 124. (a) The effect on pulse recovery time of continuing heavy work beyond the optimum duration; (b) Pulse recovery time with moderate and with heavy work of equal duration.

starts to accumulate in the body; rest taken at this time will be minimal. This point cannot, however, be determined subjectively because it will take some time for sufficient lactic acid to accumulate to give a subjective sensation of 'fatigue' – by then work will have continued too long and a relatively much longer rest will be required. In practice, this probably means that under the present system a longer recovery time (C.R.) is allowed than would be necessary if work were organized into properly determined periods of activity and rest. This is an idealized situation, since many industrial jobs vary greatly in work effort from moment to moment, but when a substantial part of the work is carried out at a rate exceeding 5 Kcal/min, the duration of periods of work and rest can be estimated.

In order to do this Müller (1953) has suggested that the energy reserve of the 'average' man can be taken as being about 25 Kcal and this reserve will not be drawn upon as long as the work remains below the 5 Kcal/min level. When the work exceeds this amount, however, the energy reserve will be steadily depleted at a rate depending upon the amount at which the energy expenditure exceeds 5 Kcal/min. Thus it follows that if excess work of 2·5 Kcal/min is done, the reserve will last for 10 minutes. At the end of this time the supply of oxygen will drop by the equivalent of 2·5 Kcal/min and the lactic acid will start to build up in the muscle causing the feeling of muscular fatigue. Therefore, the activity should continue for just as long as there is an energy reserve to be drawn upon, and rest should then commence.

The duration of the period of rest will be determined by subtracting the energy expenditure during rest (say 1·5 Kcal/min), from the base of 5 Kcal/min to give 3·5 Kcal/min to replace the reserve of 24 Kcal. This will take 7 minutes. Thus a cycle of work at an average of 7·5 Kcal/min should be organized into periods of about 10 minutes activity and 7 minutes rest if the optimum output and minimum fatigue are to result.

This is, perhaps, an oversimplification of a rather complex and still somewhat obscure situation, which would not be acceptable in any accurate physiological study of work and recovery. Nevertheless, it would seem that in our present state of knowledge it could give sufficiently realistic results for it to be used as a basis for work organisation in industry.

If pulse rate measurements are made, a somewhat different approach will be adopted. If, during work, the pulse rate rises to about 105 to 110 beats/min and then levels off, recovery will be very short, perhaps 3 min. If work is more severe, pulse rate will rise steeply at first and then will continue to rise more slowly until exhaustion is reached. The time taken for the pulse to return to normal will get progressively longer the greater the duration of the work (Fig. 124). The method of determining the optimum periods of work and rest is therefore to experiment with varying work schedules and to measure the resulting pulse recovery *times* (Brouha, 1960). An alternative approach is to use the pulse recovery *sum* as a measure of the rest required (Müller, 1959-60); the principle involved in the two methods is, however, the same: that work should continue only for a length of time which will not unduly prolong recovery.

LIGHT WORK

Under 'light work' is included all activities which have not previously been discussed. For reasons which will soon be apparent, it is not proposed to distinguish between mental work and physical work with a very small mental component. The theory which will be developed will be applicable to all types of work other than heavy physical work which can be measured by the methods already described.

After an activity has continued for some time, most people require a short break even when highly motivated. If prolonged activity is studied closely, it will be found that before an individual has realized that his concentration is slipping there will have been evidence of deterioration in his performance. This has been known for many years and has been studied by psychologists under a number of names, including 'attention', 'alertness' and 'vigilance'. In using these terms there are semantic difficulties, since over the years they have had specialized meanings attributed to them which conflict either with the original meaning or with the commonsense everyday use of the term. For instance, 'span of attention' has commonly been used as a synonym for 'span of apprehension' or 'understanding', and apprehension is not used in the sense of being afraid of something which might be coming in the future. As a result, a series of specific theories, each covering a fairly narrow field, have developed round these particular words. With the foregoing semantic difficulties in mind, the term 'actile period' has been proposed by Murrell (1962b), defined as 'a period during which there is a state of preparedness to respond optimally to stimulation either discretely or continuously' – the period during which a worker can maintain concentration upon the task in hand.

Murrell has further proposed that the 'actile period' will have a finite length which will depend upon the individual involved and the task being undertaken. When the end of the actile period is reached, performance will start to deteriorate, the manner in which the deterioration manifests itself being dependent upon the nature of the task. In a continuous task, it may mean the appearance of discontinuities in the performance which will cause longer cycle times to appear. In other instances, it may mean that signals which should have been seen will be missed. Thus there may be pauses in the work, misinterpreted signals, missed signals or longer cycle times which are characteristic of the end of the actile period. Just as there is in heavy physical work an optimum period of activity before there should be a rest, in the same way the actile period represents the optimum length of time during which light physical or mental work should continue. If activity is continued beyond the actile period, the deterioration of performance will become serious and in the case of repetitive work the output will begin to fall. This fall, which is well recognized in industrial studies, has been attributed to either 'fatigue' or 'boredom', according to the circumstances. These decrements of performance are in fact due to the continuation of an activity beyond the end of the actile period. There is a good deal of experimental and factual evidence to support the view that if a break in the nature of the activity or a rest pause is taken at appropriate times, a fresh period of optimum activity will result. With the actile period in mind, work should be organized into a series of periods of work and rest arranged to give the optimum result as for heavy physical work.

The actile period has been put forward as an operational description which can be tested by observing behaviour in various industrial occupations. It is not intended to supplant the various psychological concepts which explain *why* this behaviour occurs. It has the advantage that it has no common meaning and can be used to describe behaviour on all kinds of light work from repetitive assembly to vigilance tasks in the laboratory. Moreover, unlike 'fatigue' it is a positive concept.

Disintegration of performance at the end of an actile period may affect performance in two main situations. First, an operator may be engaged upon a continuous, repetitive task during which no new signals for action are likely to be presented. Secondly, an operator may have to respond to signals which are unexpected in time or place. He may be doing another task, either major or subsidiary, or he may be waiting for a signal to occur. Breaks in attention in repetitive work are likely to lead only to a reduction in output, but in the second situation they may cause a breakdown of performance which, in adverse circumstances, may lead to accident. Thus when discussing the applicability of the actile period to industrial tasks it is convenient to divide them into five categories. First, *repetitive work* in which the same sequence of operations is expected to be repeated continuously without a break. Secondly, *monitoring*, which is the circumstance under which an inactive individual is expected to watch for longer or shorter periods of time for signals which may occur at infrequent intervals. Monitoring takes place in some control rooms, flying certain types of aircraft and in a continuous inspection of moving products. The third type of activity may be called *driving* and this would include such activities as the driving of railway locomotives, road vehicles, cranes or rolling mills. Fourthly, operators may be *machine minding*, that is they will be watching machines, automatic or semi-automatic, in order to ensure that they are operating correctly. Finally, in *inspection* the actile period is only one aspect of the task, and inspection will be dealt with in a separate chapter.

It will be noticed that repetitive work is unique in being alone in the first of the two situations presented above. All the other non-repetitive activities are related to the second situation, since they have in common the need to be watchful for signals coming from the surroundings which may be unexpected in time or in place. The efficiency with which an operator responds to these signals may be of greater or lesser importance according to the nature of the task. For instance, in inspection it may not matter very much if a defective part is missed, but if a driver fails to see a halt sign and thereby causes a bad accident at a cross roads, the result of failure is extremely serious. We have a good deal of industrial evidence on repetitive work but most of our knowledge of the remaining fields of activity comes from laboratory experiments. If mishaps in industry or failures of performance were analysed in terms of an actile period, a clearer picture might be obtained of the part this may play in industrial activity.

Repetitive Work

In this section we are concerned primarily with jobs which are continuous and repetitive and in which the perceptual demand may be low to high, in industrial terms from light assembly to adjustment of electrical equipment.

Since the war, decrement of performance in repetitive tasks has been studied almost exclusively in the laboratory in relation to tasks making a high perceptual demand; that high demand jobs should have been chosen for study is not surprising since it can be unrewarding and is certainly very time consuming to do otherwise. However, the majority of industrial tasks are low-demand and so we must depend for our evidence mainly on field studies by the Industrial Fatigue Research Board, most of which were undertaken between 1920 and 1939.

The Duration of Work

The working hours which existed in industry before 1920 were frequently quite long. (Taylor (1915) reports on girls who were *inspecting* ball bearings for $10\frac{1}{2}$ hours and this was made up of spells of work lasting, without a break, for up to $5\frac{1}{2}$ hours.) Early investigations were, therefore, concentrated on discovering whether these long periods should be broken by a single rest pause; many years elapsed before the possibility of more frequent pauses was discussed. The practice of giving a mid-period break is now reasonably established, but we are still a long way from wide acceptance of more frequent pauses resulting in an even greater increase in output.

If the work of the I.F.R.B. merely produced evidence that a mid-period break was desirable, there would be little object in quoting it here. But a good deal of our knowledge of long term variability of performance comes from this work, which remains an almost unique collection of studies of the 'natural history' of industrial work.

In repetitive work, the approach of the end of an actile period may be shown by the appearance of longer cycle times representing the start of momentary disintegration in performance. These longer cycle times (or brief pauses) would cause a decrement in performance when they are frequent enough (Davis and Josselyn, 1953). At the outset, however, the extra long times may well be compensated by some extra short times (Bartlett, 1953), so that it is not until longer cycle times become frequent that their effect will make itself felt on the output; this is shown in Fig. 125. It will be seen, therefore, that a sensitive way of determining when a pause is required might be to study changes in the standard deviation of cycle

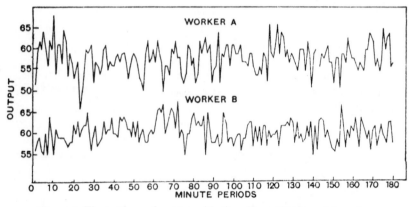

Fig. 125. Fluctuations of output with time (from Wyatt and Langdon, 1938; by courtesy of the Controller of Her Majesty's Stationery Office).

times. Unfortunately, it has not been possible to trace any published results giving actual cycle times against clock time. There are, however, a number of output curves in the I.F.R.B. reports which give either the time taken to complete a small unit or the number completed in a relatively short unit of time. In the absence of better evidence, we must make do with these.

Wyatt and Fraser (1925) carried out a study of the folding of handkerchiefs and they gave curves showing the time taken to complete one dozen for several workers. From two of these curves the basic data have been reconstructed. Work lasted for about 3½ hours and Fig. 126 shows histograms for workers A and B for the first 40 dozen and the last 40 dozen. Table 36 gives the means and standard deviations of the data in these histograms which show that somewhere between the beginning and the end of the period of work a very marked variability has crept into the performance.

The report also gives the rate of work for these operatives based on three-week averages for each 15 minutes of the working day and from

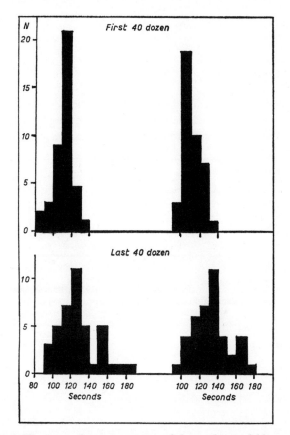

Fig. 126. Histograms for two operators of times taken to fold one dozen handkerchiefs (from Murrell, 1962b, based on data from Wyatt and Fraser, 1925).

these data it is possible to plot output curves against clock time. These curves, given in Figs. 127 and 128, show that the output began to fall after $2\frac{1}{4}$ hours and it was at this point that Wyatt and Fraser introduced their rest pause, on the basis that it should be introduced at the point at which output begins to deteriorate. To look into the matter more closely, Table 37 is exhibited in which the mean times taken to fold 1 dozen handkerchiefs are given for each 20 dozen with their standard deviations. It will be seen that for worker A the standard deviation begins to increase after the 60th dozen and with worker B after the 80th dozen.

Table 36.   Changes in mean and standard deviation in handkerchief folding for
$3\frac{1}{2}$ hours
(after Murrell, 1962b)

Operator	First 40		Last 40		Percentage Increase	
	M	S.D.	M	S.D.	M	S.D.
A	106·25	10·05	127·75	21·33	20	108
B	105·75	8·95	132·5	18·97	24	112

Before the performance has disintegrated sufficiently to make a marked difference to the standard deviation, there may be long cycle times which may be a more sensitive indicator than the standard deviation itself. What

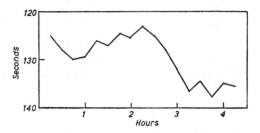

Fig. 127. Fluctuations in time taken by operative A to fold one dozen handkerchiefs (after Wyatt and Fraser, 1925).

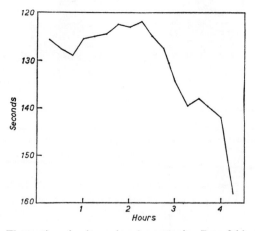

Fig. 128. Fluctuations in time taken by operative B to fold one dozen handkerchiefs (after Wyatt and Fraser, 1925).

should be taken as a long time must be a matter of opinion, however for the purpose of this argument cycles which are longer by two standard deviations from the level optimum performance have been taken as being 'long times'. In this case the limit has been set at 125 seconds per dozen

Table 37. Means and standard deviations for three operators engaged in handkerchief folding (after Murrell, 1962b)

Operator	Dozens Completed	20	40	50	60	70	80	90	100	120
A	M	110·0	102·5		102·0		104·5		121·5	144·0
	SD	7·75	9·95		9·80		11·53		17·38	19·97
B	M	106·0	106·0		103·0		107·0		129·5	133·1
	SD	8·60	10·20		10·05		9·54		15·87	22·34
C	M	142·5		112·0		132·0		135·0		
	SD	14·35		11·66		14·0		22·14		
A	Time	0·36	1·10		1·44		2·19		2·59	3·22
B	in	0·36	1·10		1·45		2·20		3·10	3·35
C	hours	0·48		1·44		2·28		3·13		

and the number of long times occurring during each 20 dozen are given in Table 38. Both operatives show one long time in the first 20 dozen and this is probably a characteristic of the warming-up period. It will be noted that the mean times for folding the first 20 dozen also show this tendency. Operative A begins to show long times after the completion of the first 40 dozen and operative B after the first 50 dozen. Taking the two operatives together it might be suggested that the time for a pause would be after 1¾ hours, which is the approximate time which they took to fold the first 60 dozen; this is half an hour earlier than the pause given in the experiment. If it were decided that the pause should be given before deterioration has become evident, it should be given after 1 hour 10 minutes, the time taken to fold 40 dozen.

In the investigation under review, production data were obtained altogether from eight operatives. As might be expected, they showed great

Table 38. 'Long times' for three workers engaged in handkerchief folding (after Murrell, 1962b)

Operator	Dozens Completed										
	10	20	30	40	50	60	70	80	90	100	110
A	0	1	0	0	1	1	2	5	5	5	5
B	0	1	0	0	0	1	1	4	4	6	6
C	1	1	0	0	0	8	3	5	7		

individual differences. The output curves of Workers A and B are the only two which display what Dudley (1958) has called a normal 'saddleback' shape. The curves of four of the others show a steady decline from the outset of the working period, while that of a seventh worker shows a steady increase up to 3 hours after the commencement of work. It would be impossible from output curves of this kind to determine the point at which a pause might be given. The eighth worker, whom we will call C, has an output curve which is exhibited in Fig. 129. She was clearly a very slow starter and achieved a peak output after the first $1\frac{1}{4}$ hours. After this,

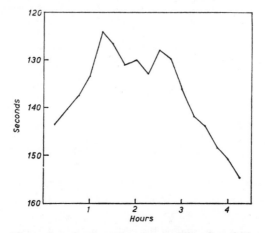

Fig. 129. Fluctuations in time taken by operative C to fold one dozen handkerchiefs (after Wyatt and Fraser, 1925).

her output remained substantially level until $2\frac{1}{2}$ hours had passed and then fell off very rapidly. If a decision had to be made from this curve on the introduction of a rest pause, should it be introduced at $1\frac{1}{4}$ hours or at $2\frac{1}{2}$ hours, or would two pauses be required (as they may well be)? A detailed analysis of the variations in her speed of work are also given in Table 37. Up to the completion of the first 22 dozen, her work was very irregular which is shown by the rather high standard deviation of 14·65 sec for the first 20. For the next 30 dozen her performance remained substantially level with a slight decrease in the standard deviation, but on the 51st dozen she started to go to pieces, her output dropped and her standard deviation increased. Taking again two standard deviations above the mean between 20 and 50 dozen, the number of her long times is also exhibited in Table 38. There was a high number of long times in the first 30 dozen. These are, in a sense, an artifact because of the low working speed at the outset of the period. If the two standard deviations are taken from a mean time of 142 sec for the first 20 dozen, the number of long times is reduced to 1, 1 and 0 for the first three spells in this table. The long times show that once the 50th dozen had passed her performance went to pieces, with eight long times out of ten in the 51st to the 60th dozen. It took her approximately 1 hour 44 minutes to complete 50 dozen; a period which is very close to that shown for workers A and B, and falls about half way between the two peaks of her average output curve. The reason why there may be discrepancies between output curves and the incidence of long times is fairly simply explained; long times are very frequently followed by short times with the result that the mean may vary very little although there is a substantial increase in the standard deviation. For instance, between the 51st and 70th dozen and the 71st and 90th dozen there is a difference of only 3 seconds in the mean time for worker C but her standard deviation has increased by half, from 14 to 22·14.

A further study of fluctuations in output from which data can be obtained is supplied by Wyatt and Langdon (1938). The task in this study was semi-machine paced; the operatives had to inset metal cylinders in holes in a moving belt, but since six holes were exposed at the same time the constraint on the operators was not absolute. The belt speed was greater than that at which the operators could do the task. For maximum performance 69 cylinders per minute should have been inserted, but operators were only able to insert at the best 67 and at the worst 57. The

Table 39. Average output and standard deviation for semi-paced machine feeding (after Wyatt and Langdon, 1938)

Worker	Rate per Minute	Standard Deviation (SD)
A	57·9	3·68
B	60·4	2·79
C	67·5	0·99
D	57·1	2·82
E	60·6	3·58

mean outputs with the standard deviations for the five operatives studied are given in the report and are reproduced in Table 39. Because this job is machine paced the 'normal' output curve will not be found, but variability is possible due to six holes being available at one time, though it will not be as great as if the job were self-paced. Long times have, therefore, been taken as those which deviate from the mean by one standard deviation only, and these are shown in Table 40. The variability of worker A is very evident as in only three of the 10 minute periods does she fail to have one long time. The other workers, once they have settled down, work a period of at least 40 minutes without a long time and when 80 minutes have elapsed (70 in the case of worker B) long times begin to appear. This suggests that a pause should be given after 70 or 80 minutes of work, a conclusion it would have been quite impossible to reach from output curves. This method of using standard deviations shows useful promise,

Table 40.   Long times in semi-paced machine feeding
(after Murrell, 1962b)

| Worker | Elapsed Time in Minutes | | | | | | | | | | | | | |
	10	20	30	40	50	60	70	80	90	100	110	120	130	140
A	1	2	3	–	1	1	1	1	2	1	–	3	–	1
B	–	–	–	–	–	–	–	1	1	1	2	3	1	1
C	3	2	1	1	–	–	–	–	1	–	–	–	–	1
D	–	–	–	–	–	–	–	–	2	1	–	1	2	–
E	3	–	–	1	–	–	–	–	1	2	2	3	3	3
Total	7	4	4	2	1	1	1	2	7	5	4	10	6	6

but further research to test the hypothesis is needed. Preliminary results obtained by the author do, however, confirm that 'long times' can be used in the way proposed and that rest pauses introduced at the indicated times produce satisfactory results (Murrell and Forsaith, 1963).

The Nature of a Pause

In the foregoing discussions, the term 'pause' has been used deliberately without defining what it means. It is true that in the experimental work quoted the pause has been taken as rest, but in many instances it may equally well take the form of a change of occupation. For instance one way of introducing a change of work may be to require operators, at regular intervals, to leave their place of work and carry away a tote pan of finished parts and collect new material. If the size of the stint is chosen so that this break in activity takes place at the right time, it may have all the advantages of a complete cessation of work. The use of service girls, or conveyors, to take parts to and from the operatives may therefore have the effect of reducing rather than increasing output.

As an alternative to a pause of this kind, actual changes in occupation may be introduced at regular intervals and this may be just as effective as an actual rest. This aspect of industrial organization has been investigated by a number of workers.

Miles and Skilbeck (1923) studied the effect of introducing a change of activity into an unspecified task. They followed the view of Wyatt (1924a) that the point at which output begins to fall is the correct time for the introduction of a change. They were of the opinion that output will show

signs of falling before a worker begins to feel fatigued and for this reason
a change of work is better than a complete stop since she would want to
continue work and would resent a break at this point. They introduced
the change of activity by arranging for the girls under observation to collect
material from the stores at between 10.50 and 11.15 in the morning,
instead of when they first arrived at the beginning of the day's work. An
increase in output for the remainder of the period of work until 12 o'clock
was recorded. In the afternoon they first tried a change period at 2.30 p.m.
but this was not a success, nor was a period at 3.20 p.m.: a fall in output
at 3 o'clock was still in evidence. They eventually settled for 3 o'clock and
the increase in output achieved by these changes was 14·2%.

Wyatt and Fraser (1928) investigated the folding of handkerchiefs in
two styles. Eight girls were observed over a period of 3 weeks. On some
days the two styles were folded alternately at approximately 1 hour intervals, on other days the style of folding was unchanged. The result of the
investigation showed that there was little to choose between the two
methods of work, and the authors comment that the change in the nature
of the job was comparatively slight. They therefore conducted a further
study of workers manufacturing bicycle chains. In one instance they worked
on a machine-paced job which indexed every 1·64 sec – 'theoretically the
operative was expected to keep pace with the disc but, in practice, she was
usually unable to do so and many shafts were allowed to pass unnoticed'.

The job used as an alternative to this during the first half of each
morning and afternoon up to the mid-period break was to place links on
two long parallel bars in readiness for another operation. This was a self-
paced job. The gain in efficiency due to the change of task was 5·8%, but

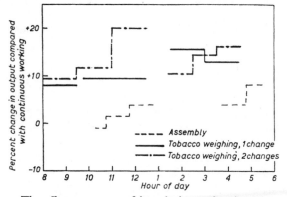

Fig. 130. The effect on output of introducing variety into the tasks of
bicycle chain assembly and tobacco weighing (after Wyatt and Fraser,
1928).

most remarkable was the progressive gain throughout the working period, the greatest gain being at the end of the working day. This is shown in Fig. 130, together with similar results obtained in a study of tobacco weighing. In this task one girl cut off and weighed tobacco while a second girl packed it. Three conditions were studied: in the first, the girls stuck to the same task continuously, in the second they changed tasks once in each spell of work and in the third the task was changed twice within a spell. The second condition showed 11·2% increase in output over the condition of continuous work, whereas in the third condition, where the task was changed twice, there was an increase of 13·8%.

Wyatt and Fraser also made a study of the hand rolling of cigarettes, an operation which is carried out in two parts; the enclosing of the tobacco in the cigarette paper and trimming the protruding end of tobacco with a pair of scissors. Five different conditions were tested:

(1) Making and cutting alternatively, according to the girls' inclination.
(2) Making for 1 hour and then cutting.
(3) Making for 1½ hours and then cutting.
(4) Making for 3 hours and then cutting.
(5) Making all day with no cutting.

Ten operators worked in the order 1 to 5 and another ten in the order 5 to 1. Results showed that the highest output was achieved under condiditions 2 and 3, i.e. with 1 and 1½ hours of making before cutting. When the making period was extended to 3 hours there was a decrease of 5·7% in output from that achieved in conditions 2 and 3, and when making continued all day there was a decrease of 11·2%.

Wyatt (1924b) reports on a laboratory experiment in which three tasks were given: simple addition of five digits, mechanical computation in which a column of digits was added by means of a comptometer and the results recorded on a typewriter (ten digits were used in this task), and muscular work in which the subject was seated and pulled against a powerful spring balance at half minute intervals with the left and right hands alternately. Three subjects were used; in one scheme they worked continuously for 2½ hours in the morning and afternoon of each day on one task, in the other they worked for three spells of 50 minutes on each task. The experiment covered a period of 6 weeks with tasks taken in rotating order. The results are summarized in Table 41.

It will be seen that the effect of doing varied work has been to increase the amount of work achieved with all three subjects and to decrease the number of errors made. As with many tests of this kind, the difference between the subjects is quite substantial but all subjects showed the effects described.

It is clear from these results that variation in the work at appropriate intervals may have a marked advantage. This method may overcome the real or apparent disadvantage of introducing frequent rests; changes of work, should not, however, be used as a substitute for rest throughout the whole working day. Changes should not be too frequent: the methods already discussed should be used to decide when changes should be made. The actual change itself may be of quite short duration – fetching material, logging data and so on – or it may involve a complete change in occupation.

Table 41. Improvement in performance when work is varied
(after Wyatt, 1924b)

Test	S	Work Done Varied – Unvaried Percentage	Errors Varied – Unvaried Percentage
Adding	A	+24·2	−26·1
	B	+ 8·8	−55·1
	C	+10·6	−25·1
Comptometer	A	+12·7	−43·4
	B	+ 4·2	− 9·2
	C	+18·2	−27·9
Muscular	A	+ 7·8	
	B	+ 2·4	
	C	+ 5·1	

The Duration of Rest

When it has been decided that an actual rest pause should be given, it will be necessary also to decide how long this should be. This may to some extent be settled on administrative grounds. If there is only one pause in the morning or afternoon and refreshment is to be taken, the time necessary to do this will clearly determine the length of the pause. If pauses are given more frequently, it may again be administratively more convenient to round off and give either 5 or 10 minutes. Virtually nothing seems to be known about this aspect of work organization, since the various experiments that have been reported seem to have been conducted with a 'shot gun' technique. For instance Wyatt and Fraser (1925) described a laboratory experiment involving the placing of bicycle chain links on a pair of steel shafts. In the early part of the experiment, with a work period of 2½ hours, they compared giving one rest pause with working continuously. They then introduced two rest pauses each of 5 minutes. 'In the course of these experiments', they say, 'they (the subjects) felt that the rest would be more advantageous if introduced earlier in the spell of work. The desire to pause often occurred after work had been in progress about 45 to 50 minutes and it was decided to try the effect of introducing two rest pauses of 5 minutes each instead of a single pause of 10 minutes as before. The work spell was accordingly divided into three equal periods of 50 minutes.' That their aim was reasonably accurate is shown by the results: the output increase was 10·1% with two rest periods as compared with 5·2% with the single longer period,[*] but it is possible that even better results might have been obtained by other arrangements and other durations of pauses. Burnett (1925) found similar results, also in an experimental situation lasting over 2 months. The results are, however, strongly suspect because she used a different pattern of work on each of 4 days of the week, and the day on which three rest pauses of 5 minutes were introduced, instead of the single pause of 15 minutes, followed a day on which no talking and no rest pauses were allowed, which she admits was heartily disliked by the subjects.

Jones (1919) gives details of an experiment in which it was found that when workers took spontaneous rests they produced 16 pieces per hour. When working for 25 minutes and resting for 5 minutes the production was 18 pieces per hour. When working 17 and resting 3, production was 22 pieces per hour and when working 10 and resting 2, production was 25 pieces per hour.

Wyatt (1927) in discussing an experiment by Graf, in which varying periods of rest were given, suggests that too long a pause may be unsatisfactory and that if rests last longer than 10 minutes rhythm may be lost. In the absence of evidence to the contrary this view might well be accepted.

Meal Breaks

In the experiments reported by the Industrial Fatigue Research Board, the workers may have taken some refreshment, but the effect of pauses with refreshment does not seem to have been compared to that of pauses when no refreshment was taken. So we do not know the extent to which the improvements in output noted were due to the effect of food intake. Hutchinson (1954) is of the opinion that decreases in output towards the end of the morning and afternoon periods of work are due to empty stomachs causing a nervous condition, particularly in sedentary workers. He therefore conducted a series of tests to determine the effects of refreshment on efficiency. In one experiment two 1·5 min tests of concentration were given in the late morning; subjects who had consumed sandwiches and tea performed better than those who had not, the difference being significant at the 1% level. Similar significant differences were found with a 3 min test administered to another group of 88 subjects, again at the 1%

*Similar results have been reported by Murrell and Forsaith (1963).

level. Using typists, Hutchinson also studied the effect of midday meals of various sizes. Fifteen minute tests were given immediately after the meal was consumed. A larger number of words was typed when the solid content of the meal was 9 to 12 oz than when more food was taken; however, with what was believed to be a highly motivated group this difference disappeared. He concludes that there may be a decrease in production if a full three-course meal is eaten during the working day.

From this evidence there seems to be a case for giving at least one pause in each working period in a form which will permit some refreshment to be taken.

Paced Work

In many industrial tasks an operator may have to keep pace with a machine. The extent of the pacing will vary, from a machine presenting only one task at a time which must be completed before the machine indexes, to semi-pacing when a sufficient number of tasks are presented for the operator to complete each task with some time in hand. The difficulty of the job will depend on the rate at which the machine is running. Clearly if a machine is running slowly enough even rigidly paced work will cause the operator no difficulty. Conversely if the machine is running very fast, then even if a job is semi-paced the operator may have difficulty in keeping up with the machine.

Paced work will be influenced by variability of the operators. In all human performance there will be variability about a mean performance and this will not disappear just because an operator is paced by a machine. In relating the performance on a paced job to the variability of the operator we must consider what sort of performance an operator would put up if he were free to do the same job in his own time. In practice of course, this is not always possible, but in order to discuss the effects of pacing we will assume that data are available from which a frequency distribution of cycle times can be constructed and a mean obtained. If the job is now paced and the machine is set at a speed equivalent to the mean, on more than half the occasions the operator will complete the task within the time allowed and on the remainder he will be forced to work faster than he would do naturally if he is to keep up. The extent to which he can do this is not known exactly, but it is suggested that, when allowance is made for any tolerance permitted by the machine, he will 'miss' on most of the occasions on which he would take longer on the unpaced condition.*

*This suggestion has now been confirmed for one type of rigid pacing by Murrell (1963b) who has proposed a mathematical model to predict the number of 'misses' at a given machine speed.

If the operation of the machine is not seriously affected by misses then it seems likely that the machine speeds would be set at or about the mean speed of the free time. But if it is important that 'misses' should not occur (and completing the processing 'off the line' is usually quite expensive and must be avoided) it is likely that machine speeds will be set lower than the free mean speed. This means that the output can be substantially reduced below that which is possible when self-paced.*

Wyatt and Langdon (1938) state that many firms deliberately run their machines at a speed which is faster than that with which the operative can normally cope. They report on a machine on which small steel chains were manufactured which was run at 37 units per minute, at which speed the average machine efficiency was 64·3%, while even during the best hour of the day it reached only 68·2%. The output curve which they show gives a fairly steady fall in output throughout the working day and the output for the last hour was nearly 20% below that of the highest level recorded. Wyatt and Langdon suggest that this kind of situation causes a good deal of strain and that the task is disliked by the operatives. They suggest, further, that running the machines at speeds faster than the operatives can cope with is fairly prevalent, and they quote 29 jobs which they studied, on none of which was the machine efficiency higher than 86%, while the lowest was 52%.

In a laboratory study on the effect of pacing, Broadbent (1953a) gave his subjects a five-choice serial reaction task under both unpaced and paced conditions. The subjects were required to respond to one of five signal lights by pressing the appropriate key. In the unpaced condition the next light came on when a key was pressed either correctly or wrongly. In the paced condition the lights were presented at regular intervals, the interval corresponding to the average rate of response during the unpaced condition. The paced group showed a fall in correct responses after the first 10 minutes but this effect was not shown by the unpaced group until the sixth 10 minute period was reached. In the paced condition the occurrence of a longer response time could mean a missed response, whereas in the unpaced condition there could be a short pause and the subject could work faster in order to catch up. During the first 10 minutes of the unpaced task there was an average of five of these pauses and in the second 10 minutes an average of eight and a half pauses but there was no significant change in the output of correct responses during these periods.

Different operators will have different variabilities, often about similar mean performance times. The effect of this will be that operators who have

*This suggestion has also been confirmed by Murrell (1963b).

a smaller variability are likely to have a smaller number of 'misses' than those who have a greater variability, and as a consequence their output will be higher. This again may influence the setting of machine speeds, since in practice individual machine speed will not normally be set to correspond to the individual variability of the various operators: they are likely to be set to correspond to the performance of the more variable operators, and the better operators may be somewhat underloaded.

An industrial study on the effect of individual variability upon performance has been made by Conrad (1955). The task investigated was the operation of an automatic weighing machine, which allowed only a set time for the operator to insert material. If she had not completed a task by the end of the fixed time, she would have to wait for the next cycle to come round. Conrad gives an example of two operatives both of whom had a cycle time of 6 sec. One of them had a coefficient of variation of 0·15, the other's was 0·30. Table 42 shows the effect on output of running these machines at various speeds.

So far we have been considering rigidly-paced tasks. These are probably much less common, in industry, than are tasks which are less rigidly paced, in that the operative may have several positions available to him

Table 42. Effect of distribution of operative cycle time on output
(after Conrad, 1955)

Machine		Operator A (C.V. = 0·15)		Operator B (C.V. = 0·30)	
Cycle Time (seconds)	Rate per Minute	Output per Minute	Percentage of Maximum	Output per Minute	Percentage of Maximum
15	4	4·0	100	4·0	100
10	6	6·0	100	5·9	98
8	7½	7·5	100	6·6	87
6	10	8·6	86	7·5	75
5	12	6·8	56	7·8	65

which will allow a larger measure of variability than will the rigid task.

Most of the jobs studied by Wyatt and Langdon (1938) appear to have been semi-paced in this way and they give details of a number of industrial investigations and special experiments which they carried out, with these tasks. Unfortunately they do not appear to have realized, as did Conrad, the importance of individual variability, so that only on one or two occasions do they give the standard deviation of the performances of the various operators. They did, however, realize that even when groups of operators had become highly experienced and apparently very efficient, there were still marked differences between their performances.

With the idea that slowing down a machine might increase output, Wyatt and Langdon conducted an experiment with three operators on a chain-making process. The subjects were three adult research workers, who were given practice periods lasting for 18 weeks, at the end of which time the learning curve showed practically no improvement. They were then given tests at machine speeds varying from 28 to 36 units/min, each working period lasting for $2\frac{1}{2}$ hours. The whole series was replicated once. The results showed that as the machine speed increased, so did the standard deviation of each subject, while the percentage efficiency decreased. From the data given the present author has correlated the standard deviations of each of the three subjects at the various speeds with the machine efficiency, using Kendall's tau, the resulting correlations are: for subject A $\tau = -0.94$, for subject B $\tau = -0.84$ and for subject C $\tau = -1.0$.

Output also increased as the speed increased, but at the higher speeds the output of two of the three subjects started to decline. Subject A reached his peak at 35 units/min, B at 34 units/min and C at 36 units/min. These speeds are in the same order as are the correlations obtained between the standard deviations and the machine efficiency.

As a result of this experiment, Wyatt and Langdon seemed very satisfied that a reduction in the machine speed could improve output and they therefore investigated the situation on the shop floor. The process investigated was wrapping chocolates. In this case it was essential that every space in the machine was filled, so that if the operative was unable to complete the task in the time, she had to stop the machine and fill the empty slots before restarting. An experiment was conducted in which the machine speed was progressively changed from 68.5 to 85.7 units/min, a relative speed increase of 25%. Two operatives were used and their relative increases in output at the highest speed tested were 2.6 and 7.1% respectively. However, they achieved their greatest increase when the machine was running at 81 units/min, when they showed 8 and 9.5% increases respectively.

An analysis was also made of the time lost due to the inability of the operators to keep the machines supplied with chocolates. Even at the slowest speeds some time was lost in this way and this showed a sharp increase when the speed reached 78 units/min in the case of one operator and a more gradual increase in the case of the other. There was a further substantial increase when the machine speed was advanced from 81 to 85.7 units/min; one operative lost 10% of the time due to stoppages, while the other lost 6.1%.

In another study of chocolate-wrapping in a machine capable of running at 240 units/min, the machine speed was increased in steps of 15 units from 150 to 210 units/min. Four workers were tested for 2 days each, at each of the machine speeds. As before, it was found that there was an optimum speed for the machine but this was different for the different workers. One worker produced her greatest output at 195 units/min; two at 180 units/min and one at 165 units/min. Unfortunately the standard deviations of the various operators are not indicated.

We must now consider the effect of an increase in variability which may take place in one individual as the day proceeds. This point was made by Conrad and Hille (1953) when they studied the packing of jars. The working day was divided almost equally by breaks into four parts and they took times of a number of cycles about the middle of each period. Table 43 is constructed from their report, in which the differences between the mean packing times in the first and the fourth periods are contrasted with the differences between the coefficients of variation. Only in one instance was there any substantial change in the rate of work – with operative D, whose mean packing time increased about 11% between the morning and

Table 43. Changes in packing time (seconds) and its coefficient of variation
over the working day
(after Conrad and Hille, 1953)

| Operator | Mean Packing Time | | | | Coefficient of Variation | | | |
| | Period | | Differences | | Period | | Differences | |
	1	4	4-1	%	1	4	4-1	%
A	7·39	7·34	−0·05	− 0·68	0·177	0·207	+0·030	+14·5
B	7·20	7·29	+0·09	+ 1·25	0·247	0·315	+0·068	+21·5
C	6·85	6·84	−0·01	− 0·15	0·211	0·222	+0·011	+ 5·0
D	7·43	8·26	+0·83	+11·2	0·146	0·184	+0·038	+21·0
E	7·71	7·59	−0·12	− 1·56	0·156	0·201	+0·045	+22·5

the afternoon. In the other instances, the differences are so small and of such variable direction that they are negligible. This is not true of the coefficient of variation. One operative who was obviously extremely consistent, had an increase of only 5%, but the coefficient of variation of three others increased by over 20%. This was a self-paced task and the results show that it is variability which will tend to change with time rather than rate of work.

It is possible to go one step further and to argue that if variability will increase throughout a working period, more satisfactory results will be obtained if the speed of machines is varied also. This line of thought occurred to Wyatt and Langdon, but for different reasons, and they conducted an experiment on a wrapping machine. The operators normally worked 4 hours at a spell and there was generally a decrease in output-throughout. Two workers took part in this experiment. In one condition the machine was maintained at the uniform speed of 220 /min, the normal speed of the machine. In the other condition the machine was run at 240 /min for the first 80 minutes, 220 /min for the second 80 minutes and 200 /min for the third 80 minutes. The results of this experiment are shown graphically in Fig. 131. It will be seen that the output declined under both conditions as time went on. That this should have occurred when the machine speed decreased is not very surprising, but it is interesting that under this condition the output of both individuals was higher than when

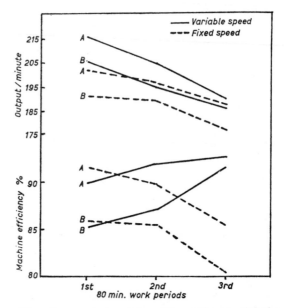

Fig. 131. The effect on output of changing machine speed during a shift (from Murrell, 1962; based on data from Wyatt and Langdon, 1938).

the machine speed was kept constant. The output of subject A was up by 4·1% and that of subject B by 5·4%. When we look at machine efficiency we see that this also declined substantially when the machine speed was unchanged. When the machine was run at a variable speed the machine efficiency increased throughout the working period.

Two conclusions can be drawn from this discussion. First that it may be an advantage if a machine running at a constant speed as part of a production line is fed by means of a buffer or queue, which should be sufficient to permit the operator to show his natural variability (for a discussion of queues see Hunt, 1956). The length of the queue will depend on the cycle time, the time available for processing, and the variability of the operative. By providing in this way for operator variability it is highly likely that it will be possible for machines to be run faster without failing to have all parts processed. From this it follows that the financial benefit of providing buffers between operatives and machines and carefully investigating the expected variability in the operatives may be quite substantial. It is only in this way that limitations imposed on output by operatives' variability can be reduced to a minimum.

The second important point is that if buffers cannot be provided, it is probable that machines could be run faster than would otherwise be the case, if the morning and afternoon work periods are broken up by pauses. These shorter work periods should end before variability has begun to increase substantially. If operatives have to work continuously or almost continuously throughout a 4 hour period, the rate of a production line should be set to take into account the substantially increased variability which develops at the end of a working period; that is, the machine will have to be run more slowly than if the speed were set in relation to the smaller variability at the beginning of the period. From this it follows that with shorter working periods the machine speeds might be increased and, far from reducing output this could result, by the introduction of pauses at the right time, in substantial increases. This is contrary to the commonly held belief that output is related only to the time a machine is running, which is clearly questionable.

The length of the working period before a pause can be determined in much the same way as when work is self-paced. As time goes on, variability will start to increase, but this will show as an increase in the number of 'misses' rather than in the incidence of long times. All that is necessary, therefore, is to record the incidence of misses in relation to clock time and to determine when they start to increase. This method has been used by the present author in a pilot experiment. Three rest pauses of 5 minutes introduced at times when there were increases in the number of missed cycles led to increased output, even though the actual working time was thereby reduced from $3\frac{1}{2}$ to $3\frac{1}{4}$ hours (Murrell and Forsaith, 1963).

Production Lines

In addition to being paced by machines, operatives working on production lines may in effect be pacing each other. Work will be passed hand to hand down a line or down a conveyor belt and the effects of individual variability may be just as marked under these circumstances, as they are when an individual works on his own. Even when the work is balanced and apparently each operative ought to be able to keep up with all the others, there will be varying differences in work rate between operatives, and these must be allowed for, otherwise difficulties due to the pressure of one operative on another are almost bound to arise. Thus it is essential to arrange for buffers between each operative to minimize the effects of individual variability. It will be necessary to calculate the probable size of these buffers in relation to the expected variability of the various operatives, so that they may be of an adequate size.* Where the work content of each station may differ, this will not always be easy, while the cost of increased space and work in progress must be balanced against increased output due to a reduction in waiting time.

Non-repetitive Work

Non-repetitive jobs may be divided broadly into two groups: first, continuous tasks with a physical content which may vary from almost nil, as in radar watchkeeping, to steady light muscular activity, as in driving a car. Common to most of these tasks is the need to concentrate on what is going on, since even momentary inattention may have unfortunate consequences. Secondly, there are discontinuous tasks which may vary from moment to moment in pace and content. There is usually variety in such tasks, so that by their very nature they provide breaks in continuity which will mask any disintegration in performance, if it occurs at all. Because of the difficulties involved, performance in discontinuous work does not seem to have been studied. Nor is it easy to see how non-specific investigations could be carried out.

In contrast, continuous tasks have received a good deal of attention, particularly that aspect which has come to be called 'vigilance'. Interest in this subject came to the fore during World War II when it became evident that the reliability of individuals on radar watchkeeping was of a strictly limited duration, so that after a period of time the chance of an operator seeing a signal quickly or seeing it at all was seriously reduced. This lead to the Clock Test of Mackworth whose first report appeared in 1944. In discussing the results of his work he used the term 'vigilance' for this particular aspect of attention rather than 'alertness' or 'attentiveness'. The term vigilance had originally been proposed by Head to describe a state of excitation of the nervous system which kept it in a state of readiness to respond to stimuli. In this sense its use by Mackworth was quite appropriate, but subsequently its sense has become narrower and it now describes a state of readiness to respond only under specific circumstances.

*An elegant example has been reported by van Beek (1961).

These have been defined by Fraser (1957b) who said that the classical vigilance situation can apply only if (a) the display consists of a series of neutral signals throughout which the significant signals are randomly interspersed; (b) the conditions of the experiment are such as to render it a stress situation in terms of speed, load, duration, etc.; (c) knowledge of results is minimal. A related definition by Jerison (1959) is that 'vigilance is defined as a probability of detecting rare and near threshold events'.

Alertness is therefore implicit in a vigilance task. In some circumstances an operator may know where and what kind of signal he is to expect, though he may not know when the signal will occur. In other instances he may know neither where nor when a signal will occur. If there is a lapse of attention a transitory, random signal may be missed altogether; if it is of longer duration it may not be seen until too long a period has elapsed from its initiation.

Not all research workers seem to agree with Fraser. For instance, the ambiguity of 'vigilance' is discussed at some length by McGrath et al. (1959) who point out that if it is used to refer to a central process determining performance on some tasks it is duplicating existing psychological concepts such as 'drive', 'inhibition' and so on. Moreover, they say, it is unsatisfactory that monitoring performance results from some subjective state called 'vigilance'. 'To say that performance declined because the vigilance of the observer declined is to follow a circular route of reasoning. Since the state of vigilance must be inferred from the observable phenomena, it cannot then be used to explain those phenomena.' From this it follows that it is more logically consistent to search for the conditions which control what can be observed and in this sense vigilance should mean performance on a vigilance task. This argument illustrates the difficulties which occur only too often when psychologists adopt, for a specialized purpose, a common word which can easily be misunderstood. To the industrialist vigilance means alertness, and he can be forgiven for failing to appreciate that performance measured under the circumstances of the vigilance situation will not necessarily be applicable to widely differing industrial tasks even though they may require vigilance in the non-psychological sense. Performance on a vigilance task, in common with the jobs we have been discussing, has an actile period which, as it is a high stress task, is quite short. As a psychological research tool studies of vigilance performance are invaluable and research results may help to explain why performance starts to deteriorate after a period of time; they do not, however, give a working description of industrial jobs which will enable executive action to be taken to improve performance. It follows, therefore, that in discussing continuous non-repetitive tasks vigilance research must be borne in mind, although performance on most of these tasks may have to be described in other terms.

Monitoring

As automation increases, more men will be required to use panels of dials to ensure that plant is functioning correctly. This can be a very inactive job. Usually it is not. That industry has very little faith in the ability of men to carry out this task efficiently by vision alone is perhaps indicated by the warning devices fitted to most of the important functions in a plant. Thus a process operator gets an auditory warning by means of a bell or buzzer and a visual indication by means of a red flashing light, indicating which function has gone out of limits. No research can be traced which indicates failure of operators to respond to deviations at the right time without an alarm system. Auditory warnings have been fitted from experience or common sense, or because it has been felt that no man can reasonably be expected to monitor visually all the indications in a large control room. That operators will respond to alarm systems there appears to be no doubt; it is less certain whether they will always respond promptly and correctly and whether the efficiency of their response is influenced by the length of time they have been at work. Failure in monitoring can thus occur in two ways: a signal for action may not be perceived, or if it is perceived the wrong action may be taken. Were it not that few control room tasks bear any resemblance to vigilance tasks in the laboratory, it might be possible to state the length of the actile period from the results of these tests. But there seems to be no evidence that process operators actually start missing signals after about $\frac{1}{2}$ to $\frac{3}{4}$ hour on duty, if they are not provided with alarms. The situation is probably more subtle than this: in the control rooms of chemical plants, the process is recorded on a variety of charts and it is the duty of the operator to make such adjustment as he thinks necessary in order to obtain maximum output under varying conditions. If there are periods when his efficiency is temporarily impaired, he may fail for a time to notice and act upon deviations from optimum with a consequent loss of product. Until more field evidence is available, this form of deterioration must remain largely speculative, but if it is accepted that changes in the traces are 'wanted signals' and that these are liable to be missed after a period of relatively inactive work, it would seem that deterioration of this kind could and will occur.

An aspect of monitoring which could be related to the vigilance situation is to be found in the transport industry. In some jet aircraft, the pilot is provided with an automatic device to fly his aircraft, but it is still necessary for him to monitor the airspeed within fairly close limits, since should it pass outside the limits, the automatic pilot may put the aircraft either into a steep dive or a steep climb. Such an incident occurred over the Atlantic in the spring of 1959 when the second officer of a Boeing 707 was left alone on the flight deck and failed to notice that the airspeed had varied. The plane dived from 35,000 ft to about 4,000 ft before he and the Captain, who had regained his seat at the controls, were able to pull the aircraft out of the dive. It might be imagined that the use of the automatic pilot would make the task of the pilot somewhat easier than before. This is, of course, true if the automatic pilot is truly automatic, but in this instance it has transferred the task of flying the aircraft from a physical activity to a perceptual activity, which may impose a much greater stress on a pilot than might ordinarily be realized. Attempts have been made to measure pilot 'fatigue' but these have not been very successful. Nevertheless as more monitoring tasks are created, some attempt should surely be made to estimate the strain imposed by this kind of task, especially when the results of only a moment's inattention can lead to very dramatic results.

Turning now to the second form of failure in monitoring – taking the wrong action – it is necessary, in the absence of experimental evidence, to speculate on what could happen when the end of the actile period is reached. If the level of activity in a control room is low (and the more efficient instrumentation becomes the lower it is likely to be), as time passes the operator may steadily become less alert; if an alarm then goes off, he may require a little time to become fully aware of what is going on and may either do the right things in the wrong order or even fail to appreciate the situation and take the wrong action. The plant may still 'fail-safe', but this is a poor consolation if a shut-down could have been avoided. This speculation is supported by some anecdotal evidence, but until better proof appears it must remain a hypothesis that the end of the actile period will be marked by a greater likelihood of error in operation. In making this hypothesis, the incidence of errors towards the end of the actile period in repetitive tasks has not been overlooked, but while this evidence is suggestive it is no more than this. It might be possible to use 'flicker fusion' or a development of the 'alertness indicator' or the concept of 'spare mental capacity' to indicate after what period of time the process operator was in a low state of readiness to take optimal action if a function goes out of limits. But while this method may give an indication of the length of the actile period it will leave unresolved the question of the quality of the action which is taken at different periods from the commencement of the shift: this can be determined only by comprehensive field work.

Driving

Driving includes any activity in which an individual has to respond continuously to varying signals but may also be required to respond quickly to signals of special importance which may occur at infrequent intervals and in expected or unexpected places. Typical tasks of this kind are driving a motor vehicle, operating a crane or a rolling mill, and controlling a function in a chemical plant by hand. A characteristic failure in driving tasks is often associated not so much with the unexpected location of a signal as with its infrequency or occurrence in circumstances which were not expected. Thus the driver of a motor vehicle, after driving many miles on a wide, fast, almost straight road may miss a halt sign even when the sign is very prominently displayed. This kind of failure is usually considered to be human error but this implies that the driver deliberately missed seeing the relevant signal, or having seen it deliberately decided to ignore it. If it is realized that these mishaps can occur through failure of attention, a much more profitable approach can be made to preventing future incidents. In addition to the possibility of missing infrequent signals, the continuous part of a driving task may also deteriorate with time, but because 'output' in driving can be difficult or impossible to measure, virtually no evidence on this aspect of driving performance seems to exist.

A review of the literature on fatigue in driving road vehicles has been published by Crawford (1961). In it he differentiates between fatigue which results from driving and the effects of fatigue, whatever its cause, on driving performance. In the physical sense, this difference can perhaps be distinguished, but when we consider the perceptual motor factors involved in driving, this distinction becomes less clear. Since we are interested primarily in deterioration in driving performance as a result of having driven for a length of time, decrements in perceptual-motor functions are of importance however they may arise.

According to the present hypothesis, driving performance of any kind would continue at a satisfactory level for a period of time, but at the end of this period it would start to deteriorate. A number of attempts have been made to measure driving performance, and these have ranged from observational studies of drivers' behaviour through attempts to measure alertness in the field, to laboratory experiments on various types of mock-up. In the first class of investigation the critical incident technique has been used by Potts (1951), in which an observer accompanied 17 long-haul truck drivers who made some 20 journeys of 250 miles. The observers reported near accidents in relation to the length of time which had elapsed from the start of the journey. Forty-eight of these were reported and their incidence was higher at the beginning of the journeys than at the end. The location of rests in relation to these incidents is not clear. It is surprising perhaps that this study should make it appear that driving becomes more efficient as time goes on: the finding is, however, supported by other evidence from insurance statistics.

Attempts to make measurements on drivers have taken three main lines. In the first, individuals were given psycho-physiological tests at the end of different periods of time (Ryan and Warner, 1936). This method is difficult to validate, although results suggest that there is a general but slow deterioration in performance at these tests as time goes on. The effect of pauses at different times has been investigated by Lauer and Suhr (1958) who showed that rest pauses at frequent intervals do help to reduce the deterioration of performance due to driving for long periods.

The second type of test attempts to assess the alertness of the driver by means of electrical measurements. The most developed method is the alertness indicator (Kennedy, 1953). Although changes in potential have been found under different driving conditions, such as when driving in concentrated traffic or in towns as compared to the country, this method has not so far proved reliable, and it does not seem to have been used to predict the length of time an individual should drive before taking rest.

A third method of studying the performance of drivers has been to attempt to measure 'spare mental capacity'. Poulton and Brown (1961) gave drivers a supplementary auditory task and estimated the load of driving from performance on the secondary task. So far they have only compared driving performance under different road conditions though the authors say that they propose to measure performance against time. This seems a most promising approach. If actility declines as driving time continues a diminution in spare mental capacity might take place. Thus a time might come when it has diminished to such an extent that the subsidiary task would barely be carried out. This might only be momentary in the first instance, but it might be predicted that occasions on which there was failure to respond to a secondary task would become steadily more frequent.

The remaining type of study of driver performance is that which is conducted in the laboratory. One such experiment is the classic work by Drew (1940) in the Cambridge cockpit. The performance of pilots on instrument flight was studied over a period of 2 hours. Their performance showed a deterioration before the end of the period and it appeared that the instruments which were most remote from the centre of attention were progressively disregarded by the subjects. Although not conducted on a 'driving' task, similar results were shown by Bursill (1958) who exposed his acclimatized subjects to various temperatures and humidities while performing a pursuit-meter task. Supplementary visual signals in peripheral positions were also given. These results suggested that, under high thermal stress, the probability of peripheral signals being missed was related to their displacement from the centre of the visual field. An experiment in a vigilance situation using a driving task has been described by Tarrière and Wisner (1960). The display consisted of a projected picture of a road passing through a forest. Eight lights were located in the trees at the edges of the display which were illuminated in random order and at irregular intervals. Sixteen signals appeared in 1 hour and the test lasted for 1 hour 40 minutes. It was given in a silent room. Twenty per cent of the signals were missed in the first half hour increasing to 35% in the second with a slight improvement towards the end, a result similar to that found in the Clock Test (Mackworth, 1950). This experiment differs from the real life situation in exposing the subjects to the stress of inactivity and isolation so that the results are qualitative rather than quantitative. They confirm that performance does deteriorate but they do not tell us when to expect this deterioration on the road.

It will be seen from this brief review of the rather scanty material that the state of knowledge is not at present such that any useful predictions can be made on the optimum lengths of work period or on the effects of reaching the end of the actile period. There are many factors obviously involved which need explanation; for instance, the hallucinations on American freeways which have been frequently described, or the high incidence of same-lane rear-end collisions on these super highways. Driving behaviour in road vehicles is of great importance, but driving in other situations must not be overlooked, as it can be of equal importance; it would seem that the techniques which are being developed to study road driver performance could with advantage be applied to such jobs as that of crane driver or charger driver in factories.

Machine-Minding

Machine-minding is a job in which an operator is required to watch a number of machines or parts of a machine either to correct faults or to feed a stopped machine with fresh material. The length of time and the rate at which this can be done efficiently depends upon the amount of concentration which the job demands, the speed at which the machines are running and the frequency of stoppages for new materials because of faults. Equally important is the number of machines which an individual may be required to look after, the number and type of signals which the machine will give and the economic consequences of failure to keep the group of machines running at their optimum. An answer to some of these questions is important in manning a group of machines. Up till now this is usually done on the basis of a work study assessment or by experience, but since the true load of this kind of task is perceptual rather than physical, assessments based on physical movement clearly cannot always give the best answer.

The most usual method of deciding how many machines (or components) an operative should mind is based on the process time and the operator's time. It is assumed that the resulting work-load is independent of the number of machines being minded provided that the ratio of the machine number and the process time is approximately constant, that is if, for instance, the rate of output is halved the number of machines can be approximately doubled. It is further assumed that the operator's time is constant and independent of the number of machines being minded.

These assumptions have been questioned by Conrad (1954) who has pointed out that there are factors in the situation which are not taken into account by this simplified approach. In the first place, it is necessary to distinguish between *load* (the number of machines or components) and *speed* (the rate at which events requiring attention will occur) because it seems that when load is increased while speed is kept constant performance will deteriorate but that if load is constant and speed is increased performance can improve. There are thus two factors which must be taken into account which can have differing effects upon output according to the extent to which they are involved. Secondly, the operator's time is not independent of the other two variables but will vary as they are changed in a manner not clearly understood.

Further work on the influence of these factors on efficient machine-minding is clearly needed. But from Conrad's results it seems evident that the relationships which are usually assumed are not as simple as they would seem to be at first sight. The perceptual load which devolves from a small number of signals from a large number of machines is not the same as that from the same number of signals from a smaller number of machines.

A knowledge of factors affecting the perceptual load and their effect on efficiency must obviously be brought to bear when machine-minding is considered. In most instances productivity will depend on rapid attention to a signal which indicates stoppage for a fault or a requirement for fresh material. A lapse of attention may mean that one or more machines are not dealt with as promptly as they should be. Speed of noticing the signals for action may depend on whether the signal is large or small: for example, whether the whole machine stops, or one portion ceases to function, or there is merely a break in the continuity of the product. It may also depend on the opportunity which the operators may have had of learning the probability of occurrence of the events with which they may have to deal, or on spatial distribution of signals for action: one operator, for instance, may be required to look after machines which are arranged in a long line, and he may have come to expect trouble at one end of the line and so be completely oblivious of trouble at the other end which he cannot easily see. It is important that everything possible should be done to reduce the possibility of an event being missed by increasing the attention-getting value of signals which an operator has to spot and by recognizing that there may be periods during the working day when an operator may be much less alert. If this is found to be the case, it may be worthwhile relieving the operators for a short time in order to avoid a pile-up of stopped machines.

'FATIGUE' AND 'BOREDOM'

Many of the biological factors which can influence performance have been discussed in detail and it remains to be seen how these fit in with the commonly accepted ideas of fatigue and boredom.

Fatigue is a word which can mean a multitude of things. Muscio (1921) has suggested that it is too indefinite a word to be used in any scientific discussion because it is in fact meaningless. This is true and it is unfortunate that its indiscriminate use can lead to a good deal of misunderstanding, both in scientific work and in industry, unless the exact sense in which it is being used is very clearly defined.

The term 'fatigue' is commonly used in two main ways: to express a *subjective* feeling or to describe *objective* measurements. In the subjective sense it is used to denote the feeling of tiredness after a period of exercise and because such sensations are experienced to some extent by everyone, the word is attributed a meaningfulness it does not possess and which invariably disappears when an attempt is made to reach a definition.

A working objective definition of fatigue is *the detrimental effect of work upon continued work*, which may manifest itself as a decrement in performance and in this sense it may well be a contributory factor to the shape of long-term performance curves.

The effects of continued work are easier to demonstrate in muscular activity than are the effects of light physical work or 'mental' work. In activity which involves muscular exercise, whether of the whole of the body or of a single limb, methods have already been described for assessing the effects of prolonged work so that objective measurements can be made. Fatigue in the muscular sense has, therefore, a very real meaning under

:ific conditions which can be defined and measured. But once work has
sed to produce 'fatigue' in the muscular sense (and most industrial work
does not) there is no longer any justification for assuming that decrements
of performance are due to the operators becoming physically tired. This
is to confuse cause and effect; a decrement we can measure, but to say
that the cause is fatigue is a circular argument which gets us nowhere.
Rather the cause seems to lie in the central rather than the physical
processes as we have seen.

Heavy Muscular Work

When work takes place at a rate much above 5 Kcal/min, rest pauses have
to be taken if an excessive amount of lactic acid is not to be accumulated
in the body to cause a sensation of muscular fatigue after prolonged
exercise. It was because of this obvious need for rest after muscular activity
that Taylor (1895) in his original experiments developed the concept of
the 'fatigue allowance'. And very properly so, because a well-motivated
workman will perform at a rate of work which should be fairly close to the
physiological optimum and he will take rest pauses when required. A
method has been described of estimating how long a period of work should
be and when rest should be taken. For any particular level of activity there
is an optimum ratio of work and rest, and if work continues beyond this
point (as it can) the rest required will be disproportionately lengthened.
The point at which rest should be taken will not necessarily be very
evident to the man, since subjective 'fatigue' should not by then have been
experienced; so left to himself (as he usually is) he would work for longer
spells than he should, and end the day unduly fatigued, showing a marked
decline in performance. It is therefore likely that current fatigue allowances
are larger than than should be.

It is important therefore when dealing with heavy physical activity to
realize that rest pauses must be taken, *and taken at the right time*, and that
the practice of giving allowances for rest by percentage (as is now customary
in industry) is inappropriate, especially if it is not recognized that the
workmen must stop work to take this rest. Where this has not been recognized, Lehmann (1958b), in his studies of heavy work in German industry,
has shown that individuals will take rest pauses during the working cycle
which are disguised as activity. Lehmann has called these activities 'hidden
rest'. Rest pauses are taken in this way because the workmen are afraid
that if they take a proper rest when they feel it is physiologically necessary
they may be reprimanded by the foreman for being idle.

In industries which are traditionally heavy, this difficulty may not be acute.
For instance, in coalmining which is universally recognized as a strenuous

physical job, men take natural rest pauses as they do also in other h
work. For instance, Vernon and Bedford (1927b) investigated the
pauses taken by roadmen and building labourers in pick and shovel wor
Pauses were taken regularly and amounted to approximately 12 minutes
per hour. They made the suggestion that this may be excessive and the
pauses may have been too frequent and further, that the length of the rest
taken may be due to the men being on time work. When, however, they
carried out an investigation on men in a shoe industry on piece work,
where a certain number of rest pauses were caused by the nature of the
work, they found that the men still took voluntary rests amounting to
about $2\frac{1}{2}$ minutes per hour and that the total rest taken was if anything
greater than the rest taken by men on pick and shovel work.

It is in industries which are not traditionally heavy that some difficulty
with rest pauses is more likely to arise. Lehmann (1958b) quotes results
from certain jobs in the manufacture of motor car engines. He shows that
there was auxiliary work of the type which conceals 'hidden rest' and he
makes the comment that even with very heavy physical work, when the
rest pauses needed can be calculated fairly exactly from the calorie output,
it is always found that the actual rest pauses are shorter than the calculated
ones. Instead of really pausing, the workers prefer to slow down output by
doing 'accessory work' in order to appear occupied.

Lehmann sums up: 'One of the causes of this situation appears to be
the method of supervision of the worker, especially by junior supervisors.
Their principle is that every man must be occupied all the time and a man
who is doing nothing for a certain time is regarded as lazy. Therefore, it
is much better for rest pauses to be organized by management than for the
organization to be left to the discrimination of individual workers.' It is,
of course, important that shop floor supervisors have a clear understanding
of the factors involved in physical work so that they will realize that rest
is physiologically necessary. It may be thought that this will make it
possible to malinger; this is to an extent true with the present methods of
estimating physical work, but with the use of physiological methods
malingering is much less likely to be possible. Furthermore when rest
pauses are properly organized at the correct time during the working cycle
and are of adequate duration, the output is likely to increase rather than
to decrease. Physiological methods of measurement have been developed
now to a point where it is possible to estimate when these rest pauses
should occur. This is a much more profitable approach than adding a
percentage to the allowed time and leaving the workmen to take rest when
they think fit, if they are permitted by supervision to take it at all.

Unfortunately the methods of measuring work involving overall calorie expenditure are not likely to be very profitable when work involves expenditure of energy below 5 Kcal/min; although some increase in pulse rate may be found it is unlikely that it will exceed the limits which have been laid down. Over-exercise of a few groups of muscles can be demonstrated by electromyography, but this method is likely to be too tedious to be of practical application in industry. A more profitable approach is that individuals who are responsible for the organization of work should be made to realize the effects of over-exercise of small groups of muscles and the existence of postural or static activity; work should be planned so as to spread the muscular load as far as possible over different muscle groups. An indication of whether this has been done successfully may come from output curves through the working day, but as the work done becomes lighter other factors obtrude which may have more influence on decrement of performance than any form of muscular fatigue.

Light Physical Work

When light work is undertaken there is unlikely to be any true muscular fatigue and, as a result, rest pauses for recovery from muscular fatigue are usually no longer necessary. This view is supported by several studies in which the actual rate of work has been shewn to increase throughout the day. One study which shows this very clearly, because its objective was to obtain evidence on this point, is by Davis and Josselyn (1953), who carried out an extensive investigation of a semi-skilled light assembly operation, which involved preparing small motor armatures by straightening the wires, pushing the wires into slots and soldering them to the commutator. The pattern of muscular activity in this task was fairly varied and the output showed a decrement towards the end of the day, dropping from a peak of about 13 units/hour, which was achieved towards the middle of the third hour of work, to less than 7 units/hour at the end of the day. A careful analysis showed that the effective operation time varied very little throughout the period, the differences were so small that they were completely insignificant. Further study showed that the decrement in output was due entirely to the cumulative effects of short pauses between each operation, which increased in frequency so that by the end of the working day nonproductive time had amounted to almost 30%. It is quite evident from this study that the operatives were not experiencing physical fatigue in the commonly understood sense, because the rate at which they worked was substantially unaltered. Further evidence on this point can be abstracted from the study of Wyatt and Langdon (1932) of inspecting sporting cartridges and packing them in boxes. They found the usual decrement of performance and attributed this to fatigue. However, there was a very

marked increase in the amount of talking towards the end of the working
period and while this talking was proceeding work would stop. In their
report they give detailed graphs of production time, from which it is
possible to reconstruct the actual rate of work when the operatives were
not talking. This shows that far from slowing down the operatives were
actually working faster, but that it was the time lost through talking which
caused the decrement in output.

Dudley (1958) carried out an extensive study of repetitive work from
which he also concluded that cycle times remained substantially level
throughout the working day; he rejected the hypothesis that an operator
maintains her rate of work until fatigue forces her to slow down; instead he
suggests that an operator will maintain a consistent working pace throughout the day while actually working; changes in output being due to
'personal and operational delays and auxiliary work', which seem to be
grouped at the start and end of the working period. The presence of this
auxiliary work near the end of the work period suggests that operatives
were seeking a change of activity after the end of the actile period had
been reached. Thus in most light jobs actual physical tiredness can be
discounted as a cause of deterioration in performance. This means that if
the true cause of decrement is understood it will be possible to take steps
which will either minimize these decrements or remove them entirely, so
that the prevention of 'fatigue' becomes a matter of organization. Work can
be arranged in appropriate spells of work and rest; operators can change
their jobs or can leave their work places to fetch parts at regular intervals;
the length of the stint can be arranged so that operators can set themselves
targets and operators can know how their output is progressing. It may
even be advisable to ensure that they do not go too long without food. In
this way optimum conditions for output can be established.

Boredom

Like 'fatigue', the term 'boredom' is a rather woolly word which because
it depends on subjective feelings is difficult to define in precise terms.
Actually, if any definition were to be given it would differ very little, if at
all, from that already given for fatigue itself. Perhaps because it is so
difficult to separate fatigue from boredom, it is a subject which has received
relatively little attention. One I.F.R.B. report which deals with boredom
is by Wyatt and Fraser (1929) who investigated monotony both in the
laboratory and on the shop floor. They discussed whether it is fatigue or
boredom which is the cause of variation in output throughout the working
day. They used a subjective assessment of boredom as well as measurements of output and of variability. They claim that when boredom was

said to be experienced the rate of working went down and that this occurs typically after one hour's work in the morning and the first half-hour's work in the afternoon. After this period there is a more irregular rate of working and a general recovery as the end of the period of work is approached.

Relating their observations of the workers' behaviour and the workers' subjective statements to the variability of the output the authors come to the conclusions that the low rate of working in the middle of the spell, which they believe to be symptomatic of the existence of boredom, is accompanied by increased variability in work (how close they come to the ideas proposed here!). Output figures for two of their workers are shown in Fig. 132. Worker B became restless about the middle of the period of work, looking round continually, indulging in talk with her neighbours, scratching and yawning. She said that the work was very monotonous and that she had a job to make herself continue it. The worker whose output is shown in curve A on the other hand said that she liked the work and was

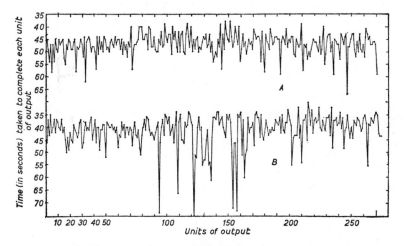

Fig. 132. Variations in rate of work of an operative who was 'fatigued' (A) and 'bored' (B) (from Wyatt and Fraser, 1929; by courtesy of the Controller of Her Majesty's Stationery Office).

never bored and that she found the uniformity of the work very pleasant compared with a more varied type of task on which she had formerly been employed. The difference in their variabilities is very evident.

It would seem that the behaviour patterns which are described as due to 'boredom' may just as readily be explained in terms of the actile period as can 'fatigue'. The means and standard deviations of the above two girls for each 20 units of output are shown in Fig. 133. The output of A who was 'fatigued' improves slightly but gradually throughout the period. Her mean time for the first 140 items was 46·15 sec and for the remainder 45·54 sec and at the same time the standard deviation increases from 3·95 to 4·12 sec. There is a warm-up period covering the first 40 units and if

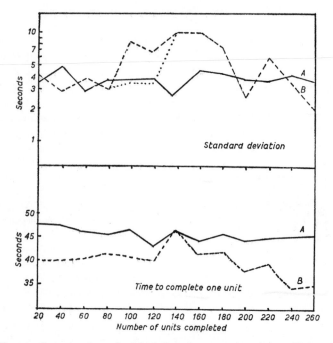

Fig. 133. Output and standard deviation of an operative who was 'fatigued' (A) and 'bored' (B) (from Murrell, 1962; based on data from Wyatt and Fraser, 1929).

this is omitted the standard deviation for 41-140 units is 3·35 sec. There is an increase in variability after 140 are completed; it is here, perhaps, that the actile period might have ended.

Individual B who was 'bored', has a very steady output curve for the first 120 items, averaging 40·11 sec, but at 140 there is a marked break in performance, followed at the end by some very good times indeed, perhaps because the end of the work period is in sight. The standard deviation on the other hand shows a marked change earlier at 100, but inspection of Fig. 132 will show that this is due entirely to one very long time in each of the blocks 81 to 100 and 100 to 120. If these two times are omitted the standard deviation remains unaltered at 3·54 sec until 120 are completed. The long times are given in Table 44.

For B there is a sharp increase in long times after the 120 have been completed: from a fairly regular one-in-twenty the number jumps to nine. For A the incidence of long times starts when 140 have been completed. B, therefore, has a slightly shorter actile period than A but on the evidence

Table 44. Long times for operators subject to 'fatigue' (A) and 'boredom' (B) (after Murrell, 1962b)

Op'tor	Units of Production Completed												
	20	40	60	80	100	120	140	160	180	200	220	240	260
A	1	2	1	1	–	–	–	2	2	1	3	2	3
B	1	–	1	1	1	1	9	3	4	–	3	2	3

of these two individuals alone we would be wrong to ascribe this to any particular cause. What can perhaps be concluded is that up to the end of the actile period there is very little difference between these two individuals; A has a standard deviation of 3·35 sec while that of B is 3·54 sec so that, in this case at any rate, 'boredom' is not characterized by greater variability so long as the actile period lasts. After the actile period, B's performance does go to pieces for a time and it may be tentatively concluded that this is characteristic of someone who finds it an effort to concentrate on a monotonous task.

From the above discussion we can develop further the idea that the length of the actile period will depend on the demands which the task makes on the individual. If a person is unsuited temperamentally to monotonous work (i.e. her actility in relation to repetitive work is rather low) a repetitive job would be more demanding on her than it would be on someone who can tolerate or even likes repetitive work. As a result her actile period may be shorter but, apart from this, performance so long as actility lasts may not differ markedly from one individual to another. If we accept that it is desirable to take some action to renew actility at an appropriate time it seems to be unprofitable to argue whether boredom is present or not. However, it may be that the study of variability of individuals will show whether they are well suited to the work they are given to do.

It might be argued that since output appears to be increasing throughout the work period it would be best to leave well alone. It does not follow necessarily that either girl was doing as well as she could. What is not known is what the effect of introducing a pause at the end of the actile period would have been. Other work which has been quoted suggests that output before a pause will be higher in anticipation of the pause and it may well be that B, who *can* work at a rate of between 35 to 37 sec/unit would work at this rate near the beginning of the work period if she knew that she was to have a pause after 80 minutes, i.e. after about 120 had been completed.

It is sometimes suggested that decrements in output are, under modern conditions, more likely to be due to lack of a desire to work rather than of the ability to work (e.g. Wyatt and Fraser, 1928). It is important that an individual's willingness or otherwise to work at the optimum, which may affect output, should not be confused with the above effects. It is endemic to the idea of actility that performance will start to disintegrate after a period of work however well motivated the worker may be, and that the worker himself will be unaware for a time that this is happening, if he becomes aware of it at all. Under the kind of circumstance we have been discussing a worker who finds it a great effort to carry on a monotonous task may be able to continue longer before disintegration sets in when she is well motivated than when she is not; unfortunately no evidence is available on this point. On the other hand, inadequate motivation may simply have the effect of reducing the general level of performance without influencing the length of the actile period very much.

There are many ways in which management may influence motivation, including applying ergonomics to the work place, giving knowledge of results (which can act both ways if targets for earnings have been set) and setting targets which can be associated with pauses. This is a subject which borders on social psychology, so beyond noting briefly the importance of motivation, it will not be dealt with further.

Summing Up

Most of the factors which contribute to a decrement in output can be manipulated by organization, and of these probably the most important is the length of time during which work should be carried out continuously, which will be determined by the actile period of the operator in relation to the demand of a particular job. It is clearly a mistake to think that the provision of conveyors, service girls and so on, which will make it unnecessary for an operative to leave her work place during the whole of a morning or afternoon's work, will necessarily increase output. On the contrary it is more likely to have the opposite effect. What is often needed is a change of attitude towards the organization of work to achieve the maximum efficiency; hours must not be equated with output, and it must be realized that, in many instances, a far greater output will be achieved from short bursts of work at a high intensity than from longer periods at a lower intensity. It is only by breaking away from the notion of fatigue and thinking instead of decrement of performance, by attempting to determine the true cause of any decrement, and by making such administrative changes as will ensure that the optimum conditions for maximum output are obtained that an advance in work organization can take place.

There follows a check list which sets out some questions which should be in the mind of people involved in work organization. This list is complementary to that already given for equipment design.

(1) If physical work is involved, how long should be allowed for recovery, and how should these recovery periods be spaced? If there is a heat load on the operator, how long should he work in heat and how long should he be allowed for recovery?

(2) How long should light work continue before a pause is given and how long should this be? Should there be more than one pause in a working period? Should pauses be spaced by management or left to the will of the operator? Should some change in activity be introduced during a pause? Should the operator be kept on the same job all the time or should jobs be undertaken in rotation?

(3) To what extent should the job be broken down? How can knowledge of results be given and targets set?

(4) What should be the size of the working group and the physical spacing between members of the group? Should talking be permitted?

(5) Are buffers provided between operator and operator or operator and machine, and are they of adequate size? Does the machine speed in paced work bear the optimum relationship to the performance and variability of the people employed on the job? Are the machines being run for the optimum length of time?

(6) Is the variability of the people it is proposed to employ on repetitive work within limits which can be accommodated by the process and can uniform performance be maintained which is normal for the process?

CHAPTER EIGHTEEN

Organizational Factors III. Inspection

An important part of a firm's activities must be the inspection of its products. The success or otherwise of inspection can profoundly affect the quality of the product which reaches the customer and upon this may depend the customer's willingness to continue to purchase from the firm. Depending as it does on the use of human sensing abilities, inspection is one of the industrial activities which is not likely to become mechanized as automation proceeds; in fact human inspection may become relatively more important. In spite of this, there seem to be a number of firms who appear to have given inspection relatively little thought, and who fail to appreciate many of the human factors involved in maintaining its quality; this is not very surprising since these factors may be quite complex and in spite of the research which has been directed towards solving them in recent years, some are still not very well understood.

Inspection can be divided into two broad types: first, inspection on a moving belt or similar equipment, when the inspector has little or no control over the speed at which he works, so that the task is paced, and secondly, inspection of individual objects at the inspector's own rate of work, so that the task is self-paced. In the first type, inspection can only be visual and consequently this is probably the industrial productive task nearest to the vigilance situation. If this is true, the length of the actile period in paced inspection will be relatively short and may play a more important part in the quality of the inspection than some other factors.

In the second type of inspection, the inspector is able to look at and handle individual objects in his own time; this may range from objective gauging to subjective judgement of surface quality or shape. The consistency of inspectors can be quite variable, even on such objective inspection as 'go'/'no-go' gauging, though this will obviously be more consistent than inspection which depends on a judgement of quality.

Points which must be reviewed in any study of inspection, whether paced or unpaced, will include the personal qualities of the inspectors, their eyesight or ability to make auditory or tactual judgement, the training which they have received and the setting of standards which are clearly understood both in relation to quality and to the number of defects which are likely to occur as the result of the production process which is currently in use. Working conditions which may also affect the success of inspection are quality of lighting, general environment, isolation, interruption and so on. The actual number of defects found may influence the inspectors in setting up norms, which will lead them to expect a particular incidence of faults from which they will not readily depart. Experience of defects may also lead them to expectations of the nature of the defects for which they are looking and they may miss faults which have a low frequency of occurrence. Inspectors who are dealing with an unvarying object will usually be more successful than those who are inspecting a product which shows a wide variety of shape, size, or quality. Most inspectors will be subject to social pressure, from individuals or from groups within an organization. Thus it will be seen that the achievement of a satisfactory standard of inspection can be quite complicated and, as a result, there is often relatively little understanding of the parameters of the process. The basis of this lack of understanding seems to be an unwillingness to calibrate inspectors, with the result that the quality of the inspection may be very much over-estimated, and this may not be realized until a substantial number of complaints have been received. There seems to be a general belief that inspectors are almost infallible and therefore the need for calibration does not arise.

The way in which pre-conceived ideas of inspector accuracy may be rudely shaken is illustrated in an account by Jacobson (1953) of studies carried out on quality control inspectors in an electrical plant. Before these tests were made it had been believed that the accuracy of the inspectors was at least as good as 95% on overall inspection and might have been as good as 98%. On the inspection of soldered joints anything worse than one loose connection in 10,000 was considered to be unsatisfactory. In order to calibrate the inspectors, test pieces were made up containing a number of both wiring and solder defects. It was expected that almost 100% of the latter would be found by the inspectors. The experimenters were rather shocked to find that the average 'catch' was only 80% of the soldered joints, whilst on the overall inspection, the 'catch' ranged between 32 and 65% only. In view of this, a scheme of training was instituted in order to improve the accuracy with which defective soldered joints would be found. A second test was then constructed, consisting of 1,000 connections. Some of the defects were poor soldering, but others consisted simply of wrapping a wire round the terminal without any solder being applied at all. This test was viewed by 39 inspectors, each of whom was allowed 1½ hours. Of these, only four found all the defective joints, whilst five found 95% and nine found 90%. At the other end of the scale, four found 65% and one found only 45%. The over-all efficiency had increased slightly to 83%. What is interesting in the results is the number of solderless connections which were missed, two men missing 50% and three missing 40%. Only nine out of the 39 inspectors found all the solderless joints. There might perhaps be some doubt as to whether a soldered joint was 'dry' or not, but there could be no doubt at all when a joint was completely solderless. It should be noted that inspectors worked for $1\frac{1}{2}$ hours on the test which, for a task of this kind, may have been rather a long period of concentrated work. The operators were expected to finish the task in this time and so they may have been under a degree of stress. It would be interesting to know whether the number of faults found varied with clock time, but this unfortunately is not recorded.

Further investigations were carried out in order to discover some of the possible causes of this variation. The number of times each defect was found was studied and it became clear that, since none of the defects were found by all the inspectors, the nature of the defects themselves cannot have influenced the results. Visual acuity was then investigated, using the Orthorater, which disclosed that of the 18 men who had scored 90% or better, only two were below standard. All four inspectors who scored 100% had exceptional vision. The age of the inspectors was also investigated and it was found that accuracy increased until the age of 35 was reached with an average accuracy of 90% and then there was a gradual decline to the age of 55, when accuracy was about 75%. The one man who scored very low was between the age of 45 and 50. On this particular task, therefore, and from this particular investigation, it would appear that the optimum age for inspection is about 30 to 35 years of age, though the 25 to 29 age group was only about 4% inferior. Psychological tests were also given to the inspectors which were claimed to give good agreement with accuracy of inspection. As a result, later applicants for inspection were given a battery of tests and their visual acuity was also measured.

This study shows some of the problems which will arise when inspection is studied. The first step which must be taken is to carry out a calibration of the inspectors under existing conditions and this will often be done by putting a known number of defective parts in a batch and passing them through the inspection process. There are a number of ways of doing this: the defective parts can be marked with a material which will fluoresce under ultra-violet light or they can be of a different material treated to look the same as the parts being inspected, so that they can be separated magnetically. Tests carried out in this way will enable not only the number of defective parts which are caught to be estimated, but also the kind of defect which is caught or passed can be evaluated. It is not sufficient simply to count the number of parts rejected by inspectors, since it is quite possible for the correct total to be reached by rejecting a certain number of 'good' parts. In one experiment reported by Mackenzie (1958) on an inspection task involving making measurements with a micrometer, there were three defective parts in a batch of 50. Three inspectors took part and each rejected three parts which, had the results not been checked further, might have suggested a very reasonable accuracy, but actually only one of the inspectors found all three defectives on the first test. A second inspector rejected two defectives and one 'good', while the third inspector rejected one defective and two 'good'. Only one of the three defective parts was rejected by all three inspectors. The experiment was replicated twice so that the batch was inspected, altogether, nine times during which, to quote Mackenzie, 'The catch was no better than one defective caught on six out of nine occasions, another was caught five times and the third only two times in the nine inspections'. Mackenzie has carried out a number of other experiments of this kind and in his paper he gives details, not only of these, but also of the work of others. Those who are interested should refer to his paper and also that of Mackenzie and Pugh (1957).

The calibration of inspectors will show the 'catch' which can be expected. Whether this is acceptable or not will depend upon a number of factors including, perhaps, consumer research. It may well be that the standards set by the manufacturer are at variance with those set by the customer. If the customer is a retail purchaser these standards may be difficult to arrive at, but if the customer is another manufacturer some agreement on standards may be desirable, especially if the customer also carries out his own inspection. When agreement has not been reached, all kinds of differences can and will be found. The author was recently told of a manufacturer who inspected articles in accordance with what he believed to be the customer's requirements, but these articles were in turn inspected by the customer and a number were returned as being defective. He decided on one occasion to see how consistent was the inspection by the customer. He returned a batch of rejected products, which had previously been marked for identification, and found that not a single one of them was returned a second time. This story emphasizes the waste of time, money and product which can occur when lack of agreement on quality causes either unnecessary or over-zealous inspection to be carried out.

While there may often be doubt about the standard to be achieved, there will also be many instances where the quality of the product will be unmistakable, making close inspection essential. This is certainly true of the process studied by Jacobson, for a dry joint will render an equipment unsatisfactory and cannot be tolerated. Therefore if the calibration suggests that inspection is not of the standard required, there are a number of factors, which have already been briefly listed, which must be taken into account in any attempt to improve the quality of inspection. These will be considered in detail, but before doing so it must be emphasized that in any study of inspection, only one factor should be changed at a time and the effects of this change should be carefully evaluated before any further change is made.

There will clearly be personal qualities in the inspector which will make him suitable or unsuitable for the job. In many firms inspection is part of the productive process and the inspectors may be older workpeople who have been put on to inspection in the mistaken idea that it is a nice, easy job into which to 'retire'. Furthermore, it is often thought that a knowledge of the process or the skill which has been learned as a journeyman will be of value in carrying out inspection. There are some social reasons, which will be referred to in a moment, why this is not a very good idea, but another reason is that too much knowledge of a part of a process may cause an unbalanced view.

Apart from the obvious tests of vision, hearing and so forth, which may be carried out when selecting inspectors, it would appear that there are no well-established psychological tests which can be used to detect potential inspectors with any certainty. Wyatt and Langdon (1932) found only a very small correlation between inspection efficiency and the intelligence tests which were used, and similar results were obtained by the author's colleagues in results at present unpublished. It would seem, therefore, that successful selection of inspectors should be related to the particular type of inspection which is going to be carried out.

Once people have been chosen for inspection, they should go through a course of training in order that they should clearly understand the factors for which they have to look. Here again, the use of older experienced people may be a disadvantage, because they may resent the idea that anybody can teach them how to detect a good or a poor product. Even after training has been completed, the inspectors should be checked at regular intervals, in order to ensure that their standards have not drifted; the older individual taken from the shop floor may resent the idea that it is necessary for a check to be made upon his work, whereas a younger person, hired especially for the job, may take this as a part of the general condition of employment. It may well be that resistance by older experienced people to having any check made on their work is one of the reasons why so little calibration of inspectors seems to be carried out at present. On balance, therefore, it would seem that it would be best to select young people specially for the job of inspector. In any event, it is vitally important that a proper programme of training should be undertaken and that the inspectors should be calibrated at regular intervals.

The choice of inspectors is clearly closely related to the system of inspection in the organization. There are two main situations: the inspector may be part of the production team responsible to the same supervisor as are the individuals on production, or he may be part of a separate inspection organization responsible to an independent supervisor. In either event, he can be in a difficult position, standing as it were between production and management. A high rate of rejection, even if it is justified, is bound to interfere with production and this can make difficulties, either for the inspector himself if he is part of the production team, or between production and inspection departments if he is a part of the inspection team. On the other hand, if he rejects fewer parts than he ought to and the customer complains, he may incur the displeasure of management. But this is a more remote contingency and there must be a tendency on the part of the inspector to pass parts which, under conditions free from pressure, should be rejected, and to send back for repair parts which should be discarded completely. The extent to which he does this will depend on the personal pressures which are put upon him. As Mackenzie says, 'inspection is always, if implicitly, of people; inspection decisions about a man's work directly reflect upon him'. Therefore, the more closely integrated the inspector is with the production line, the heavier the pressure is likely to be from production to allow doubtful products to pass, and this will be particularly true when the inspection is largely subjective so that whether a product passes or not must largely be a matter of opinion. If the bad parts have to be rectified in the workmen's own time, the pressure to pass doubtful cases will be even greater and a decision whether this system is worthwhile will clearly depend on a recognition of this fact and on just how much it matters that a certain number of defective products are passed. Belbin (1957) has described a knitwear factory where the knitters would lose pay for defective work, more pay being lost for 'seconds' than for work classed as 'mendable'. Belbin found that the inspectors reported very few 'seconds', nearly all the rejects being classified as 'mendable'. It appears that the inspectors were unable to resist pressure from the knitters to give a classification which would cause them to lose the least amount of money. The economic value of 'not paying for bad work' must therefore be studied with the greatest care before such a system is used. It seems so obviously right and yet it can perhaps lose more business than the money saved.

It has already been suggested that inspection is primarily a job for younger people. This in itself can cause social difficulties. Workpeople may resent their products being judged, they would think, by someone young enough to be their child; this may be particularly true with older men, if the inspector is a girl. The author was told of one factory manufacturing a fashion article, where a young woman was put on inspection with the idea that she would be more fashion-conscious than a man. This expectation was only too true and she set her standards so high that she virtually stopped the factory. Her presence appeared to be greatly resented by the older workpeople who, in their own eyes, were far more experienced, and they felt that the quality of their work should not be judged by a 'chit of a girl'.

Another cause of difficulty of a personal nature may be of the opposite kind; an inspector may identify himself with one particular group and so give more favourable treatment to the products of this group, or conversely, may look with disfavour on the product of another group or of a particular individual. The author has had a personal experience of this on a production line with which he was involved. It was found that the output suddenly went down. When this was investigated it was found that an inspector who was at the last but one station in the line, was deliberately putting all the inspected parts the wrong way round, so that the packer had to turn them all before placing them in the cartons. This extra movement, which had not been allowed for, was sufficient to reduce the output. It turned out that the packer and the inspector had had a flaming row over something quite outside the factory. An interesting feature of this example was that although the remainder of the production line, being on piece work, were losing money, they had taken sides in the quarrel, some supporting the inspector and some supporting the packer. The only remedy in this particular case was to move the inspector to another job.

The method of payment may also influence the performance of inspectors and this may be an added complication in the social factors already discussed. If an inspector is put on bonus and is paid for the number of defects found, he may tend to reject all doubtful parts and thus increase the tension which may exist between himself and production. On the other hand, if he is paid on the total number handled, or on the number passed, he may tend to skimp the inspection, or to pass all the borderline cases. It must be realized that inspection is largely a perceptual task and that, provided the inspector is not deliberately making decisions which he knows to be wrong, it seems that any attempt to put an inspector on bonus must cause difficulties since successful perception is not a process which can be readily influenced by an incentive, however great. In an investigation conducted by one of the author's colleagues (Collin, 1960), it was found that even the offer of substantial financial reward did not improve the inspector accuracy to any great extent.

The evidence suggests that inspectors will tend to build up norms based on previous experience. The extent of these norms will depend upon the number of faults which can occur. If this number is small, the norm may be related to the number of rejects which an inspector will expect to find in a particular batch. If, however, the number of possible faults is large, it seems quite likely that there is a limit to the number of faults which an inspector can carry in his mind at one time, so that he will build up norms based on the frequency with which the faults occur. In other words he will build up a repertoire of the most frequently occurring faults and through tending to concentrate on these, may completely fail to see a fault, however glaring, which is outside the current repertoire, because it occurs very infrequently. There is little reported evidence to show how large this repertoire may be; it will clearly vary from product to product, but should it be of great importance to catch all defectives, inspection by more than one person, each having a clearly defined repertoire, is probably necessary. It seems likely that the repertoire of an inspector who is viewing an unvarying product may be somewhat larger than that of an inspector who inspects a varying product. In the former case, it has been suggested by Thomas (1962) that an inspector will build up a mental picture of a perfect article and that he will reject any article which does not match this image. Under these circumstances, an inspector can reject an article and when asked immediately why he has done so, he will have to examine it carefully to see in what way it failed to meet his mental specification. Where the article is varying this kind of picture cannot be built up, and therefore it seems likely that expectations will play a much larger part. Expectations may also be based on a knowledge of the process. In one inspection of printed matter, investigated by a collaborator of the author, the inspectors appeared to have built up a norm based on the knowledge that if a defect develops in a printing plate it will not produce just one faulty impression, but a series. In this experiment, therefore, single defects were missed by everybody, but when eight defects were presented every inspector who found the first, found the lot. Inspection was carried out by flicking and it seemed to the investigator that the rate of flicking was so fast that individual defects would not be seen; only defects which occurred in series could produce an image sufficient to register.

Finally, norms may be set up in relation to the actual number of rejects found which may bear no relationship to the quality of the work coming forward. These norms may have been established on the basis of what a supervisor may consider to be adequate or they may be based on past experience. Where a norm of this kind has been established, it may be more important to an inspector or his supervisor that the right number should be rejected, rather than that the rejects should in fact be defectives. In another experiment carried out by Collin (1960), perfect batches of articles were put through for inspection; it was noted that the inspectors were extremely worried when they did not find the expected number of rejects and they spent a great deal of time trying to discover defects which did not exist. In fact, a certain number of the articles were rejected for reasons which, normally, would have been quite inadequate. In the same way, if an unexpectedly large number of defective parts occur, a larger proportion will be passed than is normal, because once an inspector has achieved his norm he will appear to be satisfied that he has 'caught' everything that there is to catch. Thus norms may not be associated with production rate on a percentage or any other statistical basis. In one instance, production was doubled due to new methods but the reject rate was halved, because the inspectors appeared to relate the number which they would reject not to the number of parts going through their hands, but to the total 'catch' for the day. This may, to some extent, be due to lack of communication between production and inspection. Inspection can pass information back to production about defects which are occurring so that the production processes may be corrected; this is not an uncommon feature. What is uncommon, however, is for information to be passed forward from production to inspection that, due to trouble with plant or material, a larger number of defective parts may pass into the inspector's hands.

An inspector may also find himself under pressure to accept a norm set for him by a supervisor. Mackenzie (1958) describes one experiment in which 25 products were inspected by a team of inspectors in turn in a noise-testing task. One inspector rejected 11 of these, a second only three and a third five. The charge hand found this result unacceptable and claimed that the first inspector was putting his standard much higher than he usually did because he was taking part in an experiment. He was apparently extremely annoyed at this and proceeded to demonstrate how wrong this inspector was. It appeared that he had a norm in mind which was much lower than the number of rejects found by this inspector so that when he inspected the 25 products himself, he rejected only five, but classed three as borderline. Each of these three had been rejected by the first inspector on both trials and the two of them had been rejected by the other two inspectors as well. Moreover of the three rejected by the second inspector, one was passed by the charge hand as good and the other two were classed as borderline. This led Mackenzie to suspect that had he not been trying to prove that the first inspector was wrong, the chargehand would have rejected all eight. The work of the other two inspectors passed without criticism, presumably because their rejection rate was within the chargehand's norm, and their difference in performance was not commented upon. It would appear from this that as long as the right number was rejected, the chargehand was not very concerned whether the rejects were justified or not.

In setting up conditions under which an inspector has to work it is difficult to give any form of general recommendation. The many principles of ergonomics which have been described are just as important for inspection as they are for any other type of job. A task may be assisted by good lighting, by suitable background, by adequate temperature, by lack of interruption and so forth. What is more difficult to decide is whether inspectors should work with the remainder of the production team, or in

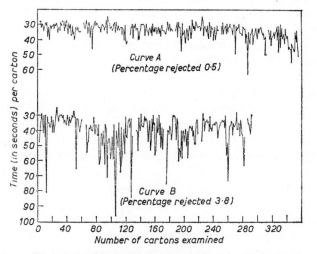

Fig. 134. The relationship between the rate of work and the number of rejects found by the same worker (from Wyatt and Langdon, 1932; by courtesy of the Controller of Her Majesty's Stationery Office).

a room by themselves, isolated from production. If the latter course is adopted, the inspector may be free from distraction, but he may, due to his isolation, find it more difficult to maintain his concentration for any length of time. This brings us to the applicability of the actile period to inspection. As occurs in other types of work, the ability to maintain concentration depends on the variety in the job – the less the variety, the more likely it is that the effectiveness of the inspection will fall off after a period of time. There is not a great deal of industrial evidence to show the effects of continuous inspection on the performance of inspectors. Wyatt and Langdon (1932) showed that when the examination of cartridges was broken up by filling cartons with the examined cartridges, gumming and removing them, the output was substantially higher than when the operators were examining the cartridges continuously. The results are shown in Fig. 134. It will be noted that when the work was uniform there was a very marked decrease in output before $2\frac{1}{4}$ hours had passed; the effect on the rejection rate is not given. A further study was carried out on the inspection of metal cases when a known number of rejects was inserted. As an alternating task, metal discs were examined in half the experiment. When metal cases only were examined, the 'catch' was lower than when the discs were examined in alternate hours. For the three workers who took part in this experiment, the 'catch' under the unvaried conditions was 75, 74 and 72·5% and under the varied conditions was 79·3, 77·8 and 75·2% respectively. Belbin (1957) gives details of a task which involved the viewing of radiographs in order to find very small defects in metal parts. As this was in the aircraft industry it was very important that none of these defects should be missed. The viewers appeared to work for approximately half an hour at a time and then take a break of 15 minutes or so, during which time they did other work. This division of work appears to have risen spontaneously. Belbin also gives an example of inspectors who were viewing sheets of glass and who were not permitted to have a rest period during either the morning or afternoon periods of work. The glass passed the inspectors on a continuous conveyor, so it would seem that they would be subject to some form of speed stress. There is no reason to suppose that the quality of the glass varied greatly from period to period, yet the number of faults found increased steadily until the fourth hour of the shift was reached, (when the 'catch' was more than double that at the beginning of the shift) then decreased again to a figure equal to that at the start, the pattern being similar on both the morning and afternoon shifts. The number of faults found on the morning shift, which started at 6.00 a.m. was substantially lower for the first three hours than the number of faults found in the same period by the afternoon shift which started at 2.00 p.m. This may perhaps be associated with the diurnal body rhythm. Raphael (1942) has suggested that where inspectors work on batches, the size of the batch should be such that it is completed in less than 2 hours and that the inspectors should, themselves, have to leave their place of work in order to fetch each batch. On the other hand, when the nature of the article changes frequently, Collin (1960) found no difference in the performance of inspectors at different times of the day.

These examples, meagre though they are, do suggest that the actile

period will be related to the nature of the inspection task: the more continuous the inspection and the less active the inspector, the shorter the
actile period will be. Where the task is unvaried, a pause is desirable and
the position of the pause may be determined by studying the performance
of inspectors in relation to the passage of time. Changes of occupation
which are introduced either by varying the product inspected or by making
the inspectors leave their place of work to remove finished parts or to fetch
new work, are likely to be of benefit when effectiveness of inspection and
not absolute output are considered. Where inspection is continuous, which
makes this procedure impossible, it would seem that actual breaks in
inspection should be introduced.

Finally it cannot be too strongly emphasized that inspection is largely a
perceptual task and, as such, standards of performance other than those
normally applied to factory work will have to be adopted. The effectiveness
of an inspector will be related to his 'catch' and this can only be determined
by calibration. All the factors which have been discussed throughout this
book which are related to the effectiveness of perceptual tasks will apply
to inspection. In particular, it must be realized that, just because an
inspector is apparently doing nothing but sitting and looking, inspection
must not be thought to be an easy task. Of all the jobs to be found in
production it is probably one of the most arduous, and the organization
of inspection in a firm must be based on this realization.

CHAPTER NINETEEN

Organizational Factors IV.
Shift Work

This chapter will discuss the biological background to the planning of shift work. It must be borne in mind, however, that this is only one of a number of factors which must be taken into consideration when a shift system is adopted (Banks, 1956). Social factors may be of great importance also. It will be remembered that the temperature of the body fluctuates throughout the working day; being lowest in the early hours of the morning and highest in the middle of the afternoon, Fig. 11, p. 34. This varying temperature cycle is just one manifestation of a cycle of bodily activity which goes on throughout the 24 hours. Other functions which fluctuate in a similar way are glandular functions, the secretion of urine, metabolic rate, use of oxygen and rate of excretion of carbon dioxide and so on; all these functions being less active at night and most active in the early afternoon.

The importance of this diurnal rhythm in industry lies in the effect which it may have on an individual's ability to do his work efficiently. When a man transfers from day shift to night shift, he is having to work at a period when his body wants to rest, and when he attempts to rest his body is at its most active. In addition, it would appear from the work of Kleitman and others (Kleitman, 1939) that speed and accuracy as well as reaction time may be related to the diurnal cycle, performance being at its best when the body is most active. Against this biological background night shift work should, in theory, be rather unsatisfactory. Unfortunately the evidence on this point is by no means clear so that it is difficult to assess the exact importance of this biological factor. Such evidence as there is can only be reviewed briefly here. For a fuller review, articles by Teleky (1943) or Walker (1961) should be studied.

There are two aspects of this problem: the effect of going on night shift on the 24-hour biological cycle and on the efficiency of night work. The only readily measurable variable of the biological cycle is body temperature and it is this which will be referred to in the discussion which follows.

The effect of changes of pattern of activity on the temperature cycle has been investigated by a number of workers both in the laboratory and in industry. The results seem to suggest that the temperature cycle can be inverted in a number of individuals but that the rate and extent of inversion depends on the amount of bodily activity undertaken (Teleky, 1943). That is, an individual who has a fairly active job, say in a machine shop, is much more likely to achieve temperature inversion on night shift than is a watchman who may spend most of his time sitting at a desk. This tempera- ture inversion seems to take up to six days to complete in those individuals who do invert. Other individuals appear to be able to achieve only partial inversion whereas a few will not be able to invert at all. de Jong (1960) reported that in an extensive investigation in the Netherlands it was found that most men inverted in one to three days and would revert even on a single day off. Reid (1961), without giving the source of his information states that a British survey of shift workers showed that 27% inverted their temperature in one to three days, 12% took four to six days, 23% inverted in longer than six days and 38% did not invert at all. It is not entirely clear to what extent temperature inversion represents a corres- ponding inversion of other bodily functions such as digestion. An experi- mental change carried out during the period of polar night when all work was carried out by artificial light, showed similar variability in the time needed for inversion although eventually all the experimental group of 26 appeared to have achieved a reversal. However, the most difficult change seemed to be in the time of defecation (Lewis and Lobban, 1957).

From this evidence it appears that some individuals are more suitable for night work than others and that men who are unable to obtain a temperature inversion should be put on to permanent day work. It seems, however, that those who are only able to achieve partial inversion may not be lucky in this respect and they may do night work under a certain degree of biological strain. It has been noted that night work involving physical activity is more likely to cause inversion than night work in which very little activity is involved. This may mean, therefore, that jobs such as process controller are a much greater strain on an individual than more active jobs.

The one point that seems to have been fairly well established is that the majority of individuals who do achieve inversion will, on going back to day shift, revert to the normal cycle in one to three days. Thus the commonly accepted shift system of week-on, week-off nights is biologically, at any rate, likely to be the most unsatisfactory and to subject the individuals involved to the greatest strain: no sooner has the body adapted itself to the new pattern of activity by inverting the temperature cycle than it is forced to revert to a different pattern, and so on from week to week. But what alternative shift system is the most likely to be the best? To help in answering this question the effect of shift work upon output and on social activity must be examined.

We have already noted the suggestion by a number of workers that performance will follow roughly the diurnal rhythm and that it will reach its peak towards the middle of the working period (Kleitman *et al.*, 1938). Much of the work on which this view is based has covered too short a time

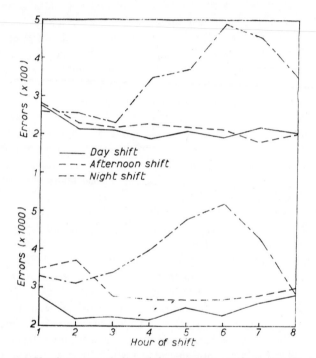

Fig. 135. Total errors made in each hour on three shifts in Gas Works I Preliminary investigation (above) and main investigation (below) (after Bjerner *et al.*, 1955).

to be free of the effects of chance variations. This is not, however, true of the studies carried out by Bjerner *et al.* (1955) who analysed records kept at a gas works between 1901 and 1943. It is not clear which years were covered by their preliminary investigation of 61,296 entries, but their main investigation covered the years 1912-1931 with 175,000 entries. All errors made in logging were recorded and the resulting hourly totals are shown in Fig. 135 plotted against the hours of each shift. Shifts started at

6 a.m., 2 p.m. and 10 p.m. Additional data were obtained from a second gas works and from a paper mill, Fig. 136. The morning and afternoon shifts show no very definite pattern, some improving as the shift proceeds, others deteriorating. What is noticeable is the rather poor performance in the second hour (3 p.m.) in the afternoon shifts. We are not told what

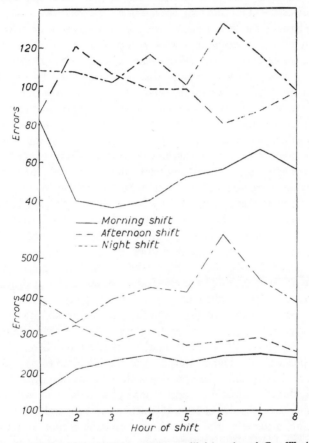

Fig. 136. Total errors made in a paper mill (above) and Gas Works II (below) in three shifts (after Bjerner *et al.*, 1955).

pattern of living was adopted by the afternoon shift, it seems likely that they lived by a fairly normal routine and ate the usual large Swedish mid-day meal after a morning of relatively inactive wakefulness. This might account for the peak of errors about 1 hour after coming to work. The night shifts are much more definite; all show a very marked deterioration up to

3 a.m. which may correspond to the 'low' in the temperature curve since the men involved were on nights for only a week at a time and were in the process of inverting. Unfortunately by its very nature this study lacks the biological data on the men involved which would make it possible to relate their diurnal rhythm to their performance. What is evident is that when shifts are changed weekly the performance at night is characterized by many more errors in logging than are found on the other two shifts.

On the basis of this rather scanty evidence it would seem, *a priori*, that during the inversion, in the first week of a period on nights, production should be lower than during subsequent periods. Some of the earliest evidence on this point was obtained during World War I by the Health of Munitions Workers Committee whose final report was published in 1918. A finding of this Committee was that when one group of individuals was on continuous night work their output was 4·4% less than when they had been on discontinuous night work; while in another group there was an improvement of 11% when continuous night work was changed to discontinuous night work. The discontinuous night work referred to by the Committee was week-on and week-off; with this system the output was slightly inferior to that of individuals who were on continuous day work. There is, however, less lost time on the week-on week-off system than there was on continuous day work. Other output figures quoted show that when two weeks at a time were worked on each shift, output in the first week of night shift was about 2% higher than in the second week, whereas during the day shift the output on the first week was 3% lower than in the second week; the night shift hourly output was slightly higher than that of the day shift. Figures are also given for lost time which bear a direct relation to the figures given for hourly output. The lost time of approximately 6% for a group of women and girls on the first week of a night shift corresponded with the highest output, and the lost time of 8½% on the first week of the day shift corresponded with the lowest hourly output. It is possible, therefore, that these figures may be influenced to some extent by the presence or absence of other distractions at night and may not be a true measure of the effects of the biological cycle on output. It must also be remembered that these measurements were taken during a period of intensive activity during a war when very long hours were being worked and an all-out effort was being made to make good a serious deficiency in munitions. Nevertheless the Health of Munitions Workers Committee emphasized the physiological advantage of making relatively infrequent shift changes, although these are difficult to reconcile with family and other claims.

More recently an extensive investigation was carried out by Wyatt and

Marriott (1953) who got output data from a number of factories operating on weekly, fortnightly and monthly shift systems. Unfortunately, as they were dealing with a wide range of products, the only way they could express their results was in terms of percentage of combined output of comparable night and day shifts which makes it rather difficult to make a comparison between the true outputs of groups of men working on these three systems. Thus, although there is the smallest difference in output between day and night shifts on the fortnightly system, it would be impossible to conclude from this kind of evidence whether production would actually be increased if a fortnightly system was changed to, say, a monthly system. The figures do however show that both on the fortnightly system and on the monthly system the first week of the new shift has a lower output than the subsequent weeks. The output of workers on a four-week system steadily increased towards the end of a month, and on a percentage basis the output in the first week was at a slightly lower level than that of workers on a fortnightly system. Assuming that on a true output basis there is likely to be little difference in output between the first week of a fortnightly and the first week of a monthly shift, then owing to the steady improvement throughout the month, it would seem that by the end of the month, output would be slightly greater than at the end of a fortnight.

It was also found that the percentage absence during the second week of a two-week night shift is greater than the absence during the first week, whereas on the day shift the absence in the second week is less than that in the first week. Absence on the night shift as a whole is less than on the day shift. This same effect is to be found also on the four-week system in that the day shift has a slightly greater absence than the night shift. Absence decreases on the day shift as the month proceeds. On the whole the absence on both shifts of the four-week system is about half the absence on the two-week system.

In the Dutch investigation differences in output between shifts were found generally to be quite small and inconclusive. Where there were significant output differences between night and day shifts they were quite small, being between 1% and 4% (de Jong, 1960).

Wyatt and Marriott discussed at length the preferences of workers on the various shift systems and those interested should refer to their paper for full details, but in general their findings show a dislike of night shift, a difficulty in adapting to the changed meal times, difficulty in sleeping during the day, a disrupting effect on living patterns at week-ends and dislocation of normal home life. It is fairly clear that, although the majority of men dislike night work, their attitude did not appear to greatly affect their performance. Differences in attitude between individuals may be the result of some men finding it much easier to adapt themselves to working at night than others.

A factor in shift work which seems to have received very little attention is the living pattern adopted by shift workers. The normal living pattern is sleep – work – recreation, but it would seem that people on afternoon and night shift will adopt a sleep – recreation – work pattern in an effort to fit into the activities which are going on in the everyday world. That afternoon shift workers would find it difficult to do anything else seems evident, but night workers could sleep in the afternoon and this might prove to be a less stressful living pattern than that usually adopted. Afternoon shift workers might be helped if their shift ended at 9 p.m. (before the pubs close) either by moving the whole system forward an hour or adopting uneven shifts of perhaps 21.00-06.00, 06.00-13.00, 13.00-21.00 hours to give 9, 7, and 8 hours per shift, or some variant of this arrangement. de Jong (1960) suggests (for the Dutch) 23.30-07.30, 07.30-16.30, 16.30-23.30 hours, which also gives uneven shift duration, but differently distributed. He is of the opinion that the present system starting at 6 a.m. causes unnecessary dislocation in social and domestic activity.

It will be seen from the evidence quoted that the organization of night work in circumstances when night work is necessary is by no means easy. Ideally it would seem that periods of about one month on each shift will give the best results, but this is usually unacceptable socially and the recent Dutch work (de Jong, 1960; Wesseldijk, 1961) has indicated that the importance of longer periods on nights may be over-emphasized. In reviewing the various alternative methods of organizing shift work we must distinguish between continuous shifts and weekly shifts. In both types of organization, however, the principle behind the alternative to the standard methods which will be proposed is based on the idea that if there is any validity at all in the biological arguments which have been presented, the logical conclusions must be that short periods on nights are better than intermediate periods, that is two or three days are better than seven, although it is possible that 14 or 28 days are better than either. It is clear that when only a small number of night shifts are worked at a time, extensive inversion will not take place in two or three days and that when returning on the day shift there will be no need to revert from a night rhythm on to a day rhythm. This idea is in part confirmed by a statement made to the author by the medical officer of a company which has been operating short night shifts, who said that complaints of ill health had been reduced since the introduction of this system. This appears to support the idea that the health and/or output of men on shifts which change frequently

Table 45. Arrangement of various shift systems

CONTINUOUS SHIFTS

	Su	M	T	W	Th	F	S	Su	M	T	W	Th	F		
2-2-2															*Shifts*
A	1	1	2	2	3	3	0	0						8 days per	1 = 06-14 hrs
B	2	2	3	3	0	0	1	1						cycle	2 = 14-22 hrs
C	3	3	0	0	1	1	2	2						7 cycles in	3 = 22-06 hrs
D	0	0	1	1	2	2	3	3 etc.						8 weeks	
2-2-3															*Shifts*
A	1	1	2	2	3	3	3	0	0					9-9-10 days	As 2-2-2
B	2	2	3	3	0	0	0	1	1					per cycle	
C	3	3	0	0	1	1	1	2	2					3 cycles in	
D	0	0	1	1	2	2	2	3	3 etc.					4 weeks	
Rotating 9 hour															*Shifts*
A	1	4	7	2	5	8	0	0	0	3	6	–		12 days per	1 = 00-09 hrs
B	2	5	8	0	0	0	3	6	–	1	4	7		cycle	2 = 09-18 hrs
C	0	0	0	3	6	–	1	4	7	2	5	8		7 cycles in	3 = 18-03 hrs
D	3	6	–	1	4	7	2	5	8	0	0	0 etc.		12 weeks	4 = 03-12 hrs
															5 = 12-21 hrs
															6 = 21-06 hrs
															7 = 06-15 hrs
															8 = 15-00 hrs
Rotating 6 day															*Shifts*
A	1	1	1	1	1	1	0	0						24 days per	As 2-2-2
B	2	2	2	2	0	0	1	1						cycle	
C	3	3	0	0	2	2	2	2						7 cycles in	
D	0	0	3	3	3	3	3	3 etc.						24 weeks	

WEEKLY SHIFTS

		Su	M	T	W	Th	F	S	Su	M	T	W	Th	F		
2-3																*Shifts*
A	3	3	0	1	1	1	0	0	2	2					21 days per	As 2-2-2
B	0	1	1	2	2	2	0	3	3	0					cycle	
C	0	2	2	3	3	3	0	0	1	1 etc.						

	Su	M	T	W	Th	F	S		
6-5-4									*Shifts*
A	1	1	1	1	1	1	0	21 days per	As 2-2-2
B	2	2	2	2	2	0	0	cycle	
C	3	3	3	3	0	0	0	etc.	
6-5-2-2									*Shifts*
A	1	1	1	1	1	1	0	21 days per	As 2-2-2
B	2	2	2	2	2	0	0	cycle	
C	3	3	0	3	3	0	0	etc.	
5-5-5									*Shifts*
A	1	1	1	1	1	0	0	21 days per	As 2-2-2
B	2	2	2	2	2	0	0	cycle	
C	3	3	3	3	3	0	0	etc.	

Table 46. Data in relation to various shift systems

Type of Shift	Length of Cycle (days)	Phase (weeks)	Consecutive Night Shifts	Shifts Between Days Off	Average Hours Working per Week	No. of Cycles	No. of Shifts	Periods Off No.	Duration Hours	Shifts	Total Shifts Off
Continuous Shifts					In 24 Weeks						
2-2-2	8	8	2	6	42	21	126	21	48	2	42
2-2-3	9-9-10	4	3	7	42	18	126	12	48	2	42
								6	72	3	
Rotating 9 hour	12	12	4	8	42	14	112	14	90	3	42
Rotating 6 day	24	24	6	6	42	7	126	14	56	2	42
								7	80	2	
Weekly Shifts											
2-3	21	3	3	5-2-3	40	8	120	16	72	2	48
								8	48	1	
								8	24	1	
6-5-4	21	3	4	6-5-4	40	8	120	8	72	3	48
								8	72	2	
								8	48	1	
6-5-2-2	21	3	2	6-5-2-2	40	8	120	8	72	2	48
								8	48	2	
								8	48	1	
5-5-5	21	3	5	5	40	8	120	8	40	1	48
								8	80	2	
								16	56	2	

is better than that of men who work for periods of about one week on one shift, but since data on these shifts have not been published, quantitative estimates of the effect of working short periods on nights cannot be given.

Since periods of one month on the same shift is an arrangement which is clearly impractical for continuous shift work, three alternative systems to the usual rotating six or seven day system are presented. These are the 2-2-2 system, the 2-2-3 system, and the rotating 9 hour system. The arrangement of these systems and the data relating to them are shown in Tables 45 and 46. In the 2-2-2 system, only two shifts are worked at a stretch, which with two days off makes a cycle of eight days, during which each team has two nights off work. There will be six shifts between each off period. The cycle will repeat itself once in every eight weeks. The 2-2-3 system is very similar except that in each cycle of nine days one spell of three shifts is worked including night shifts. Every third cycle will last ten days and the three cycles will be completed in four weeks. Both these systems can be readily organized within the framework of the existing shift starting times. There is no particular reason why the 2-2-3 should be preferred to the 2-2-2; it gives the impression that the shift pattern will fit into the seven day cycle, but this is in fact an illusion since the cycle, with the days off, is nine days and not seven. There seems to be nothing to choose between the two systems unless it is that under the 2-2-3 system there are periods of three days off at a spell amounting to 72 hours which do not occur under the 2-2-2 system when all the periods off work are of two day duration.

A novel and somewhat unusual system is the rotating nine-hour system, this is illustrated in the tables and it will be noted that there are eight different shift times, four of them being night shifts and four being day shifts. The starting time of each third shift will be advanced by three hours due to the total duration of three shifts being 27 hours, this means that from time to time shifts will have to start and end at midnight and at 3 o'clock in the morning. Clearly this can only be done if there is proper transport available. Under this system four night shifts are worked at a time, the first starts at 6 o'clock in the evening and the starting time gets progressively three hours later over the four shifts so that the fourth in the series starts at 3 o'clock in the morning. Obviously there can be no question of inversion taking place under this system; the advantages which have been suggested for this arrangement are that fewer shifts are worked in a given period of time: in 24 weeks for instance, 112 shifts are worked as against 126 shifts which would be worked on the 2-2-3 system. In addition there is a period of 18 hours between shifts instead of 16 and all the off periods consist of three days at a time. There are, however, fewer of these off periods compared with the other systems. It will take 12 days to complete a cycle and there will be eight shifts between the off periods.

We turn now to the weekly or discontinuous shifts. The most common system in use at present is to have one week (4/5/6 shifts) on nights followed by one or two weeks on days, according to the type of shift system used. There are so many of the 'five-day week' arrangements that we cannot consider them all, so the system which will be given for comparison with the alternatives is that in which work is done in three shifts a day, and five days a week. Under these circumstances five consecutive night shifts will be worked and the object of the alternatives is to reduce this number. The first of these is the 2-3 system; this is an adaptation of the 2-2-3 system to weekly working. To obtain the shift change at the end of the second day of the week it is necessary to introduce a day off for the team which has been on the night shift and this is achieved by starting the night shift at 10 o'clock on Sunday night and not working a night shift on Tuesdays. Under this arrangement three consecutive night shifts will be worked in one week followed by two spells of two nights in subsequent weeks. The total cycle lasts seven days.

In the 6-5-4 system the day shift works six shifts per week, the afternoon shift works five shifts and the night shift works four shifts. This reduces the number of shifts at night from five to four and makes up the loss by working the extra morning shift. Whereas under the previous system work starts on Sunday night and skips a night shift on Tuesday, under this system work will start on Monday morning and there will be no night shift on Friday night. The distribution of the time off is not so advantageous under this system as it is under the 2-3 system. The 6-5-2-2 system is a variant of the foregoing with the night shift split into two periods of two nights at a time with a night off in between. It is similar to the 2-3 system in that the night without a shift is taken in the middle of the period of work. Here the work will begin with the morning shift on Mondays and will continue to the end of the morning shift on Saturday. Although the periods of time off are somewhat shorter in duration, the total hours off between shifts are slightly more under this system than they are under the previous system.

The various systems which have been discussed are by no means exhaustive. There are all kinds of ways in which the 24 hours may be split up, especially when the shifts are uneven in length – de Jong in his investigation collected some 200 different shift schemes! Some systems have a longer night shift with perhaps only four shifts a week, whereas some systems have a shorter night shift. But enough has been said, it is hoped, to show the approach which can be adopted if it is thought desirable to depart from systems which leave a man on night shift just long enough to have completed his inversion of bodily function and which have relatively little to recommend them from the physiological point of view. It is probable that these systems will put the maximum strain on the human organism although socially they are probably more acceptable than longer periods on night shift. Wyatt and Marriott in one of their reports quote one interviewee as saying 'a week is quite long enough for the wife to be left alone at night'. The alternative systems which have been outlined will certainly meet this difficulty since the length of period on nights can be reduced and for this reason alone these alternative systems seem to be socially quite acceptable. On the other hand should there be no social difficulties it would seem that there is something to be said for working for a fortnight or a month on one shift. On balance it seems likely that slightly greater output and slightly less physiological strain may result; but it must be admitted that at present there is not sufficient evidence to support a decision on the superiority of either system.

Finally, it must be remembered when discussing shift work that there are some individuals that can adapt themselves very readily to night work and who may even prefer night work. There seems to be a case from the purely biological point of view for attempting to discover which individuals are most suited to night work. In saying this it is clearly realized that difficulties will arise if certain individuals are selected for night work or conversely if others are rejected because they are considered unsuitable. Those who are selected may feel that they are being harshly treated and that the others are getting off too lightly. Since it is not easy openly to demonstrate an individual's unsuitability for work at night this can cause difficulty as was shown by a strike which occurred in 1958 due to a man being transferred to day work out of his turn a week after he came back from his honeymoon! But whatever the difficulties it seems that selection of individuals who are biologically suited for shift work must inevitably result in less strain on the working force as a whole and should result in an increased output though the extent of this is somewhat problematical.

CHAPTER TWENTY

Organizational Factors V.
Age

Throughout this book there have been references to the effects of age on performance. In this chapter various aspects of the changes which take place with age will be considered in greater detail and in particular those which have to be taken into account in work design. There has been and still is an immense amount of research interest in human ageing, covering changes in physical as well as mental functions. Additionally, research and social workers have been studying what happens to elderly people and have concerned themselves with problems of retirement and the care of the older section of the population. On the other hand, industry has shown relatively little interest in the research results. It seems, rather, that industrial interest in ageing has been confined to finding fairly easy jobs for the loyal older men who are approaching or have passed retiring age. This is perhaps not very surprising, because it has only recently been shown that the ageing of the work force may be an important factor which has to be considered when planning how work shall be done. Evidence is now accumulating which suggests that the ageing process, in so far as it may affect industrial efficiency, can begin to show itself in certain circumstances at ages as low as 35, and that unless changes in ability with age are taken into consideration for individuals over the age of about 40-45 years, the optimum use will not be made of the work force. The importance of this problem is shown by the fact that of about 12 million men between the ages of 20 and 64 who were employed in British Manufacturing Industry in 1958, some 52% are over the age of 40, therefore at least half the work force may be showing some effect of ageing.

There is a great body of knowledge from research on the biological changes which take place with age. What is much less well known is how these changes will affect employment of the over-forties in the vast variety of tasks which are carried out in industry. If we were to accept the more commonly held industrial view we would conclude that the effect is very small indeed; but while this view will not stand up to critical scrutiny it is still not very easy to make firm recommendations to industry on the various factors in the working environment which should be planned to ensure the maximum efficiency and health of the older part of the work force.

A first step, therefore, in trying to decide to what extent age is important, can be to find out whether there are particular types of job on which fewer older men seem to be employed, and to see if it is possible to determine the characteristics of these jobs which may be related to age. This approach was undertaken by the author and his colleagues in their studies of the light engineering industry. They found first that in this industry there appears to be a relationship between job and age (Murrell, Griew and Tucker, 1957; Murrell and Griew, 1958), which is fairly consistent throughout the country and seems to have changed relatively little over the

Table 47.   The ranking of jobs by age
in the light engineering industry
(after Murrell, Griew and Tucker, 1957)

Job	Median Age
Pattern makers	31·0
Electricians	34·2
Millers	35·1
Grinders	38·0
Sheet Metal Workers	38·9
Testers	39·2
Borers	39·6
Drivers	39·9
Fitters	40·9
Turners	42·1
Welders	42·9
	Md.
Drillers	43·6
Building Maintenance	44·0
Foundry Workers	44·4
Polishers	44·8
Plating and heat treatment	46·0
Inspectors	49·2
Storekeepers	49·5
Packers	53·0
Labourers	55·0
Factory services	56·5

The Median Age of the Industry is 43·2 years.

post-war years, at any rate in one firm studied. The ranking of jobs by age as found by these studies is shown in Table 47. What is most noticeable about this ranking is that the jobs which are manned by the younger men – the 'young jobs' – are mainly those which may have a high perceptual content, whereas the jobs which are manned by the older men are mainly jobs of a physical nature. This led to the view that the age of men on a job may be related to the complexity of the job, on the grounds that the more complex the job, the greater were the perceptual demands. Confirmatory evidence of this view has come from other workers. Belbin (1955) found that the older workers tended to be on physical jobs and so did Richardson (1953), who pointed out that even although older men remained on physical work, they tended to move to jobs which were done at a slower pace. Heron and Chown (1961) who carried out a survey of Merseyside industry, have found that there were fewer older men 'on jobs making severe demands on attention to fine detail or involving sustained care and concentration'. Griew (1958a), in a study of the relationship between age,

Table 48. Productive and non-productive workers
(after Murrell, Griew and Tucker, 1957)

	Mean Age	% over 44
Fitters		
Productive	45·53	50
Assembly	44·77	51
Non-productive	41·34	41
Maintenance	42·77	40
Outside maintenance	42·89	32
Machinists		
Millers – productive	36·57	22
Millers – *non-productive*	33·50	10
Turners – productive	38·06	25
Turners – *non-productive*	39·84	25
Borers – productive	38·00	23
Borers – *non-productive*	32·00	0

accidents and jobs, has shown that older men on some 'young jobs' tend to have more accidents than would be predicted on the basis of the incidence of accidents in the whole of the employed population in the particular factory studied. Further evidence uncovered by Murrell *et al.* (1957), showed that job for job, men on non-productive work, such as work in the tool room where perceptual demands can be high, tended to be younger than men on production, Table 48. There may be a number of causes for the effects demonstrated. Some of these may be social and will not concern us here, but if it is accepted that age may be related to the demand made by jobs, the application of ergonomic knowledge can go a long way towards improving conditions for older people. Conditions will also often improve for younger people, but it may well be that far better use can be made of the skill, knowledge and experience of older men, if changes of the kind which will be suggested are made.

An immediate question will come to mind: if older people are finding some jobs difficult, what will happen? Will their output fall, will they move to other jobs, or will they separate from their firms and seek other jobs elsewhere? The evidence of decline in output with age is distinctly scanty. McFarland and O'Doherty in a chapter in Birren (1959) quote two studies, neither of which show any output decline. On the other hand, Anderson in his Chapter describes two other studies which show the opposite effect. A recent experiment by the present author (Murrell and Edwards, 1963) on two younger and two older tool room turners using two different measuring devices showed that the cutting time for the older men was less than that of the younger men under both conditions (Fig. 137). Clay (1956) has pointed out that it seems likely that men whose output has fallen below a level which is acceptable either to the management or themselves, due to age or to some other cause, will separate from

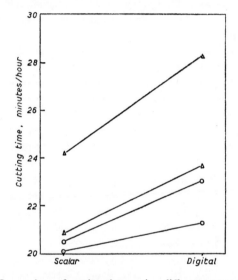

Fig. 137. Comparison of cutting times using different measuring devices with two younger subjects (above) and two older subjects (below) (after Murrell and Edwards, 1963).

their firms, either voluntarily or involuntarily. This will produce a measure of natural selection, in that the output of the older men who do remain in their employment cannot be very much below that of their younger contemporaries. As a result, any study of output in relation to age at a particular point in time, is not liable to show any great effect. What are required are longitudinal studies over many years, and these unfortunately seem to be completely lacking.

If productivity does decline or if an older man finds that job demands are becoming insupportable it seems that he is more likely to leave his firm than to take a less demanding job which might be considered to be 'inferior'. This view is supported by a study of internal migration in one firm over a period of eight years in which Murrell *et al.* (1956) found that of 528 inter-job moves, only 22 could be considered to be inferior jobs with lower perceptual demands. Older men seemed to have been leaving the 'young' jobs before normal retiring age. In this particular firm other work would have been available and so it would appear that the older men who left were unwilling to accept the loss of status which may have been implicit in the easier (and often lower paid) job. We cannot here consider in detail the problems which will face men who have to make their 'first retirement'. They have been reviewed by Murrell (1959); but what is of concern is that the demands of jobs should be studied in relation to age and a suitable progression planned, so that older men are not forced to make this first retirement at all.

A number of studies have suggested that any slowing of output rate which does take place will occur first in the perceptual elements of a task, rather than in the physical movement which a man may make. In a study of a large number of stop-watch times taken in one engineering firm, Murrell and Forsaith (1960), could find no difference in times of bodily movement between older and younger men. Perceptual elements are usually too short to be timed, but cumulatively they may produce a loss of productivity. But for some time before this occurs, experience may enable older men to achieve satisfactory performance by an economy of force and action which will help to compensate for any slowing down. In one study of physical activity in a machine shop, Griew and Tucker (1958) found that older men appeared to be able to achieve the same results with fewer control movements than younger men working on similar machines. In a study of pillar drilling, the performance of older naïve subjects was substantially **worse than** that of young naïve subjects, but the performance of older professional drillers obtained from industry, was if anything, slightly better than that of young drillers (Murrell, Powesland and Forsaith, 1962). This clearly demonstrates the role of experience in compensating for increasing age. But this compensation can occur only up to a point and it should be the object of good management to discover those factors which make the greatest demands upon the older worker and if possible to modify them.

This is not always as easy as it sounds. Most industrial tasks are immensely varied and it may not always be easy to determine which particular task or part of a task may be causing difficulty to the older people. One approach is to look at various jobs in order to see what particular factors in a job situation are age-related, while another approach is to look at the various ways in which age has been demonstrated in the laboratory to influence performance, and to draw conclusions from these changes and to test them in practice. The first approach was followed by Murrell and Tucker (1960) in a pilot study from which they isolated four factors which appear to bear a relationship to age; these were the accuracy with which a task has to be carried out, the size of the detail in a task, the nature of the measuring instruments used and the type of instructions given.

These four factors are not exclusive but are typical of job components which may influence or be influenced by the various biological changes which take place with age. The most important of these would seem to be a decrease in visual acuity and loss of acuteness of hearing (presbycusis); a loss in channel capacity or, as Gregory (1959) has suggested, an increase in 'neural noise'; a decrease in the speed of discrimination and in the speed of making decisions; a loss of short term memory and a tendency for greater variability (Welford, 1958). We will consider these briefly and indicate some of the things which may be done to mitigate their effects.

Loss of visual acuity, accommodative power and pupil size, may affect the size of detail which can be seen and the ability to read fine scales. While to some extent these losses can be offset by increased lighting and by the use of adequate optical aids, it would seem that some very fine work is probably unsuitable for older people. For instance, under an illumination of 8 lm/ft^2, small type on poor paper can be read quickly by men of 25 years of age. Their performance improved only 14% when the illumination was increased to 500 lm/ft^2. Under similar conditions, men of 45 years of age gave an increase in speed of nearly 50% (Weston, 1949). Much can be done to assist in the reading of instruments with the use of digital displays, instead of scalar displays. In the engineering industry, this principle can be applied both to micrometers, Figs. 138 and 139, and to the scales found on machine tools, Figs. 140 and 141. The latter application can also act as a 'built-in memory', since the operator will no longer have to work by differences and by having to remember a series of values; he will instead be able to use a single dimension related to a datum to which he is working. This will reduce the need to use short-term memory which will also play a part in the transfer of information from a drawing to a machine. If drawings are obscure and rather complex, the older person may have some difficulty in remembering the various dimensions involved. He may also find interpretation difficult, particularly if he has to translate one set of dimensions in order to arrive at the figures which he requires. Other situations in which short-term memory may play a part are where the operator may have to learn and remember limits on instruments where these are not marked, or where he may have to remember a fact or a dimension while he is securing a second dimension to go with the first. This means that information on which an older person has to act must be given in the clearest possible manner so that the minimum amount of memory and translation should be required. Good design will help older people to make decisions more quickly, but decision making may also be influenced by other factors of which the most important is a reduction of channel capacity which requires as much redundant information as possible to be eliminated from the task. If only four instruments are required to run a plant, they should be clearly separated from instruments which give information which is used at very rare intervals. Operative controls should be kept to a minimum and the relationship between these controls and the movements they effect should be compatible. Older people find it more difficult to cope with incompatible situations and appear also to find it more difficult to be certain that the decisions they have made are in fact the correct ones. The reduction in channel capacity will also affect the rate at which older people can receive information and this may limit the speed of machines upon which older people have to work. It appears that older men may tend to gravitate towards machines which run more slowly, or towards machines where the cutting time is longer in relation to the setting time.

The effect of increased variability with age is most likely to be found on paced tasks. This has been shown experimentally by Brown (1957) to be the case and studies in industry carried out by Shooter and Belbin (reported in Welford, 1958) show that on jobs with time stress, there was a substantial deficiency of men over the age of 45 and in women over the age of 40. Welford (1958) makes the suggestion that another factor may enter here: if the work is paced in a way that causes signals for action to occur at times which are not of the subject's own choosing, and if responses have to be carried out within a defined period, breakdown may arise because of the limited capacity for information transfer and short-term retention which, at high speeds, may be overloaded. Under these circumstances, responses may lag further and further behind the signals, until 'blocking' takes place (Vince, 1949). It seems therefore that the introduction of 'buffers', which will benefit all workers on this type of work, is likely to be of particular benefit to the older part of the work force. The effect of piece work and bonus schemes on older workmen is an aspect of pressure for speed which is not quite so obvious. It is improbable that the slowing down of older men would be shown up by stop watch studies of element times (Murrell and Forsaith, 1960), so that the normal process of rate-fixing will usually be unable to take age into account. Predetermined time

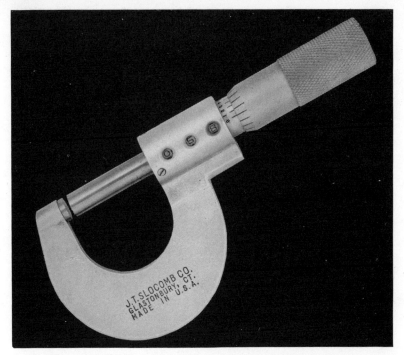

Fig. 138. A digital micrometer (by courtesy of J. T. Slocomb Co.)

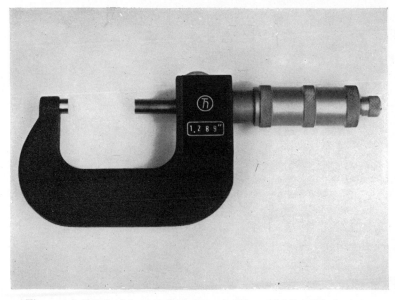

Fig. 139. A digital micrometer (by courtesy of Ernst Thielicke Werkzeug-maschinen).

Facing page 448

Fig. 140. Indicator of machine tool travel (by courtesy of English Numbering Machines Ltd).

Fig. 141. Indicators of machine tool travel fitted to a centre lathe (by courtesy of English Numbering Machines Ltd). ▲

systems also seem to have as a datum the performance of younger people, so that the time values which will be set, since they take no account of the perceptual components of work, will not necessarily be truly representative of the level of performance which can be expected from older men. Unpublished work carried out by the author and his colleagues has suggested that older men dislike individual bonus or working on piece work and prefer some form of group bonus scheme, whereas younger men take the opposite view. Heron and Chown (1960) report that in an assembly department where a bonus target was set by informal leaders of the teams, one or two teams would not carry older men who were thought to be slowing them down. Firms reported that because of this they had had to move several older men each year, usually in their middle forties, and nearly always to lower paid jobs. It is clear, therefore, that the method of payment and the need to maintain a comparable earning level with that of younger people may put time stress on older men which they may find in time to be intolerable. It should be borne in mind that this may be a form of 'added stress' which may have the effect of reducing performance, a reduction which would not occur if this stress were not present. In other words older men may put up a better performance on the same job if they are not on piece work than if they are.

Anyone reading the results of the laboratory experiments could be forgiven for imagining that any person who achieves the age of fifty will have become a slow, forgetful, half-blind, half-deaf, palsied character of little use in industry. In fact, many older men and women hold down jobs with complete satisfaction to their employer. This does not mean that the experimental findings are fallacious. The apparent anomaly seems to derive from the use in the laboratory of subjects who are naïve in the practice of the particular faculty which is being tested. Whether the effect of continual practice of a particular faculty in a job will result in the decrements predicted by the horizontal experimental techniques not occurring, has not been investigated experimentally—but early indications of a laboratory study of latency suggests that with adequate practice age differences may disappear. This practice will be part of the experience which has been built up over the years and will almost certainly be specific to a job situation. Heron and Chown (1961) found that older men tended to be working on older machines; when new machines are bought they appear to be given, not to the most senior and experienced men as one might have expected, but to the young men who are sometimes only just out of their apprenticeship.

If this is correct, there will be a population of older men, each of whom is able to build up 'experience' specific to the job he is doing, which is enabling him to continue to be employed profitably at an age when he would have been expected to show output decrement; and it would seem that so long as he continues to do the job to which his experience applies, he will be able to be employed up to retiring age. But in industry things do not stand still. Techniques change, new machines are bought, new processes are installed and it would seem that, as often as not, people will be transferred without any really adequate study as to whether the new task has features which make use of their accumulated experience, and they will not get adequate training in the new job. It would hardly be surprising then, if men tended to find these new tasks difficult, and to separate from their firm to take jobs which are less demanding. This explanation of the migration of older men is an alternative to the one given earlier, that they were finding their existing jobs increasingly difficult.

In a study carried out in a large engineering firm (Murrell and Edwards, 1963) the conventional graduated scale on the cross slide of a centre lathe was replaced by a device which gave the dimension being cut directly in digits (Fig. 140). It was hypothesized that this would prove to be of great assistance to older turners. One of the turners tested was 63 years of age. Contrary to expectation, his initial performance on the new device was, if anything, inferior to his performance when he was using the standard graduated scale with which he was familiar. It was only after about seven weeks that he began to show the kind of improvement found with younger men who took part in the experiment. As this was only an experiment, he continued to use the new device under protest; but what would have happened if this had been a permanent change? Would he have persevered until he acquired proficiency, or would he have decided after five or six weeks that he could not carry on and have left the job?

To sum up, technical change is bound to make it necessary for operators to move on to jobs which may either be variants of their old job, or may, in some circumstances, be new jobs entirely. This change can be satisfactorily carried out only after a careful study has been made of the skills and experience which can be transferred from one job to another, and when a programme of retraining is undertaken. The need for this retraining will be fairly obvious when men move to completely new jobs; but what is less obvious and just as important, is the need for retraining when quite small changes are made in the nature of the job; changes which may seem to be so small that they might be neglected, but which may have the effect of upsetting the delicate balance between age and experience. The methods which should be adopted if older workers are to be successfully retrained have been summarized by Belbin (1964).

We may turn now to the physical aspect of ageing. It has already been mentioned that it appears that older people tend to gravitate towards physical jobs, which also it seems should be free from any time stress. Provided that physical jobs are not too heavy and that they have been accustomed to them for a good deal of their lives, older workers appear to be able to cope fairly well. Older people appear to be less tolerant of heat, which may put a greater strain on the cardio-vascular function. Belding and Hatch (1955) have suggested that older people should have a lower heat stress index than that required for 'fit acclimatized young men'. We have already noted in discussion on heat that Vernon and Bedford (1931) found that older people had more accidents in hot conditions than did younger people, Fig. 93, p. 259. Industrial studies of older men in relation to shift work have not been traced, but Pearson (1957) has pointed out that one of the important changes to which older men appear to attach great importance is from shift work to day work. But, again, in all these situations involving physical factors, there may be a process of natural selection taking place, which may disguise the effects being looked for if *ad hoc* studies are carried out.

Posture may also be important to older people. Jobs such as miller or turner, which involve an unsatisfactory posture, may not be so well tolerated and older men may, over the years, be showing the effects of poor posture, which may even have caused some form of disability. The author has found in discussion with machinists, that many of the older men express their dislike in no uncertain terms, of jobs which involve bending, and Griew and Tucker (1958) found fewer older men on jobs which involved bending and poor posture. The design of jobs to ensure a good posture is therefore of extra importance where older workers are concerned.

This Chapter has dealt with that very small part of ageing which is relevant to the design of work. Attention to the factors which have been reviewed should help to reduce the initial demands of jobs so that as men get older they will not be working as near to the limit as they might otherwise be. In other words there will be less chance that, if additional demands are made, the older individual may find himself at or approaching a condition of overload. There are a number of other important aspects of ageing in industry; these are largely organizational or social problems and, as such, do not come within the scope of this book. They will include the nature of job changes, the effect of promotion, of status, of training and of re-employment prospects. These aspects are more fully discussed by Murrell (1959 and 1962c). A full discussion of the nature of the changes of skill, as revealed by laboratory studies, is given by Welford (1958) and of the more basic biological changes in Birren (1959).

Appendix

GLOSSARY OF TERMS USED IN CHAPTER NINE

Scale base. The line, actual or implied, running from end to end of the scale which defines, or corresponds with, the index path.

Scale base length. The distance between the centre lines of the terminal scale marks, measured along the scale base.

Maximum scale value. The greatest value of the measured quantity which the scale is graduated to indicate.

Minimum scale value. The smallest value of the measured quantity which the scale is graduated to indicate.

Scale range. The difference between the nominal values of the measured quantities corresponding to the terminal marks.

Scale division. A part of a scale delimited by two adjacent scale marks.

Called division. The part into which a scale division must be divided by eye if the scale is to be read with the required tolerance.

Numbered division. The part of scale between two numbered major marks.

Scale spacing. The distance between the centres of two adjacent scale marks, measured along the scale base.

Called spacing. The distance along the scale base corresponding to a called division.

Scale interval. The increment of the measured quantity corresponding to the scale spacing.

Called interval. The increment of the measured quantity corresponding to the called spacing.

Major marks. The longest and heaviest scale marks which correspond to the numbering system adopted for the scale.

Intermediate marks. Scale marks of intermediate length and weight which may replace major marks when the number of these becomes excessive.

Minor marks. The shortest and lightest scale marks which correspond to the limit of a scale division.

Numbering system. The sequence of numbers corresponding to the major marks.

Basic scale pattern. The sequence of major, intermediate and minor scale marks.

*Denotes definitions given in B.S.3693/1964.

References

AGATE, J. N., and DRUETT, H. A. (1947). 'A study of portable vibrating tools in relation to the clinical effects which they produce.' *Brit. J. industr. Med.*, **4**, 141

ÅKERBLOM, B. (1948). *Standing and sitting posture*. Stockholm, A. B. Nordiska Bokhandeln

ÅKERBLOM, B. (1954). 'Chairs and sitting.' In *Human factors in Equipment Design* (eds. Floyd, W. F., and Welford, A. T.), pp. 29-35. London, H. K. Lewis

ALCOCK, P. H. (1961). *Report of work done on a $3\frac{1}{2}$ in. centre lathe with an inclined bed*. London, Rural Industries Bureau

ALPHIN, W. (1951). 'Muscle action potentials – as related to visual tasks in industry.' *Illum. Eng.*, **46**, 188

THE AMERICAN ACADEMY OF OPTHALMOLOGY AND OTOLARYNGOLOGY (1957). *Guide for conservation of hearing in noise*. Los Angeles

AMERICAN STANDARDS ASSOCIATION (1954). *Relations of hearing loss to noise exposure*. Report Z24-X-2. New York

ANDREAS, B. G. (1953). *Bibliography of perceptual motor performance under varied display control relationships*. Scientific Rept. No. 1, University of Rochester, N.Y.

ANDREAS, B. G., MURPHY, D. P., and SPRAGG, S. D. S. (1955). *Speed of target acquisition as functions of knob v. stick control, positioning v. velocity relationship, and scoring tolerance*. Scientific Rept. No. 3, University of Rochester, N.Y.

ANON (1963). *Noise*. Final Report of Committee on the problem of Noise. *Cmnd.* 2056. London, H.M.S.O.

ARKIN, A. M. (1941). 'Absolute muscle power.' *Arch. Surg.*, **42**, 395

ARMY SERVICES DATA (WORLD WAR II). *Height and weight data for men inducted into the Army and for rejected men*. Report No. 1-BM, Office of the Surgeon General, Medical Statistics Division

ASHE, W. F., BODENMAN, P., and ROBERTS, L. B. (1943). *Anthropometric measurements*. Project No. 9, File No. 741-3, Armored Force Medical Research Lab., Fort Knox, Kentucky

ÅSTRAND, P. O. (1952). *Experimental studies of physical working capacity in relation to sex and age*. Copenhagen, Ejnar Munksgaard

ÅSTRAND, P. O., and RYHMING, I. (1954). 'A nomogram for calculation of aerobic capacity (physical fitness) from pulse rate during submaximal work.' *J. app. Physiol.*, **7**, 218

ATKINSON, W. H. L. M., CRUMLEY, L. M., and WILLIS, M. P. (1952) *A study of the requirement for letters, numbers and markings to be used on trans-illuminated aircraft control panels. Part 5. The comparative legibility of three fonts for numerals.* Naval Air Material Center. Aero Med. Equipment Lab. Rept. TED NAM EL-609

BACH, L. M. V., SPERRY, C. J., and RAY, J. T. (1956). *Studies of the effects of flickering light on human subjects.* New Orleans, Touland University

BAHRICK, H. P., NOBLE, M., and FITTS, P. M. (1954). 'Extra-task performance as a measure of learning a primary task.' *J. exp. Psychol.*, **48**, 298

BANKS, O. (1956). 'Continuous shift work: the attitude of wives.' *Occup. Psychol.*, **30**, 69

BARBOUR, J. L., and GARNER, W. R. (1951). 'The effect of scale numbering on scale reading accuracy and speed.' *J. exp. Psychol.*, **41**, 298

BARKLA, D. (1961). 'The estimation of body measurements of British population in relation to seat design.' *Ergonomics*, **4**, 123

BARNES, R. M. (1936). *An investigation of some hand motions used in factory work.* University of Iowa. Studies in Engineering, Bulletin 6

BART, E. E. (1946). 'Effects of high speed vibrating tools.' *Occup. Medicine*, **6**, 515

BARTLETT, F. C. (1953). 'Psychological criteria of fatigue.' In *Symposium on Fatigue* (eds. Floyd, W. F., and Welford, A. T.). London, H. K. Lewis

BARTLETT, F. C., and MACKWORTH, N. H. (1950). *Planned Seeing.* London, H.M.S.O.

BAXTER, B. (1942). 'A study of reaction time using a factorial design.' *J. exp. Psychol.*, **31**, 430

BAXTER, B., and TRAVIS, R. C. (1938). 'The reaction time to vestibular stimuli.' *J. exp. Psychol.*, **22**, 277

BEDALE, E. M. (1924). *The effects of posture and rest in muscular work.* I.F.R.B. Rept. No. 29. London, H.M.S.O.

BEDFORD, T. (1936). *The warmth factor in comfort at work.* I.H.R.B. Rept. No. 76. London, H.M.S.O.

BEDFORD, T. (1940). *Environmental warmth and its measurement.* M.R.C. War Memo. No. 17. London, H.M.S.O.

BEDFORD, T., and WARNER, C. G. (1934). 'The globe thermometer in studies of heating and ventilation.' *J. Hyg.*, **34**, 458.

VAN BEEK, H. G. (1961). *Working on assembly lines.* Eindhoven, Philips.

BELBIN, R. M. (1955). 'Older people and heavy work.' *Brit. J. industr. Med.*, **12**, 309

BELBIN, R. M. (1957). 'New fields for quality control.' *Brit. Mgmt. Rev.*, **15**, 79

BELBIN, R. M. (1964). *Methods of training for older Workers.* Paris, O.E.C.D.

BELDING, H. S., and HATCH, T. F. (1955). 'Index for evaluating heat stress in terms of resulting physiological strains.' *Heat. Pipe and Air Cond.*, **27**, 129

BELDING, H. S., HERTIG, B. A., and RIEDESEL, M. L. (1960). 'Laboratory simulation of a hot industrial job to find effective heat stress and resulting physiologic strain.' *Amer. industr. Hyg. Ass. J.*, **21**, 25

BELLIS, C. J. (1933). 'Reaction time and chronological age.' *Proc. Soc. exp. Med. Biol.*, **30**, 801

BERGER, C. (1944). 'Stroke width, form and horizontal spacing of numerals as determinants of the threshold of recognition.' *J. app. Psychol.*, **28**, 208-231 and 336-346

BERRY, P. C. (1961). 'Effect of colored illumination upon perceived temperature.' *J. app. Psychol.*, **45**, 248

BEVAN, W. (1958). 'Perception: evolution of a concept.' *Psychol. Rev.*, **65**, 34

BIESELE, R. L. (1950). 'Effect of brightness ratios on visual performance.' *Illum. Eng.*, **45**, 733

BIRMINGHAM, H. E., and TAYLOR, F. V. (1954). 'A design philosophy for man/machine control systems.' *Proc. Inst. Radio Engineers*, **42**, 1748

BIRREN, J. (1959). *Handbook of ageing and the individual.* Chicago, Univ. of Chicago Press

BJERNER, B., HOLM, Å., and SWENSSON, Å. (1955). 'Diurnal variation in mental performance.' *Brit. J. industr. Med.*, **12**, 103

BLACKWELL, H. R. (1952). 'Brightness discrimination data for the specification of quantity of illumination.' *Illum. Eng.*, **47**, 602

BLACKWELL, H. R. (1954). 'The problem of specifying the quantity and quality of illumination.' *Illum. Eng.*, **49**, 93

BLACKWELL, H. R. (1959). 'Specifications of interior illumination.' *Illum. Eng.*, **54**, 317

BONJER, F. H. (1959). 'The effect of aptitude, fitness, physical working capacity, skill and motivation on the amount and quality of work.' *Ergonomics*, **2**, 254

BONJER, F. H., and VAN ZUILEN, D. (1958). 'Research on industrial noise in the Netherlands,' quoted by Murrell, K. F. H., *Fitting the Job to the Worker*, p. 98. Paris, European Productivity Agency

BORNEMANN, E. (1942). 'Untersuchungen über der grad der geistigen beanspruchung.' *Arbeitsphysiol.*, **42**, I. Aus arbeitung der methode, 142. II. Praktisch ergebrisse, 173

BRADLEY, J. V. (1954). 'Desirable control display relationships for moving scale instruments.' *U.S.A.F. W.A.D.C. Tech. Rept.* 54-423

BRIGGS, S. J. (1955). *A study in the design of the work area.* Unpublished doctorial dissertation. Purdue University, Lafayette

BROADBENT, D. E. (1953). 'Noise, paced performance and vigilance tasks.' *Brit. J. Psychol.*, **44**, 295

BROADBENT, D. E. (1956). 'Listening between and during practical auditory distraction.' *Brit. J. Psychol.*, **47**, 51

BROADBENT, D. E. (1957). 'Effects of noise on behavior.' In *Handbook of noise control* (ed. Harris, C. M.), Ch. 10. New York, McGraw-Hill

BROADBENT, D. E. (1958). *Perception and communication.* London, Pergamon Press

BROADBENT, D. E., and LITTLE, E. A. J. (1960). 'Effects of noise reduction in a work situation.' *Occup. Psychol.*, **34**, 133

BROUHA, L. (1960). *Physiology in Industry.* Oxford, Pergamon Press

BROWN, F. R. (1953). *A study of the requirements for letters, numbers and markings to be used on trans-illuminated aircraft control panels. Part 4.* Naval Air Material Center. Aero Med. Equipment Lab. Rept. TED NAM EL-609

BROWN, J. S., and SLATER-HAMMEL, H. T. (1949). 'Discrete movements in the horizontal plane as a function of their length and direction.' *J. exp. Psychol.*, **39**, 84

BROWN, R. A. (1957). 'Age and paced work.' *Occup. Psychol.*, **31**, 11

BROWNE, R. G. (1954). 'Figure and ground in a two-dimensional display.' *J. app. Psychol.*, **38**, 462

BROZEK, J., and KEYS, A. (1944). 'Flicker fusion frequency as a test of fatigue.' *J. industr. Hyg.*, **26**, 169

BURG, A., and HULBERT, F. (1961). 'Dynamic visual acuity as related to age, sex and static acuity.' *J. app. Psychol.*, **45**, 111

BURGER, G. C. E., and DE JONG, J. R. (1962). 'Aspects of Ergonomic Job Analysis.' *Ergonomics*, **5**, 185

BURNETT, I. (1925). *An experimental investigation into repetitive work.* I.H.R.B. Rept. No. 30. London, H.M.S.O.

BURNS, W., and LITTLER, T. S. (1960). 'Noise.' In *Modern trends in occupational health* (ed. Schilling, R. S. F.), Ch. 17. London, Butterworth

BURSILL, A. E. (1958). 'The restriction of peripheral vision during exposure to hot and humid conditions.' *Quart. J. exp. Psychol.*, **10**, 113

BURTON, A. C., and EDHOLM, O. G. (1955). *Man in a cold environment.* London, Arnold

CALDWELL, L. S. (1961). *The relationship between the maximum force exertable by the hand in a horizontal pull and the endurance of a sub-maximal holding response.* U.S. Army Med. Res. Lab. Rept. 470 Proj. 6X95-25-001 Fort Knox, Ky

CARLSÖÖ, S. (1961). 'Muscle load in working postures.' *Ergonomics*, **4**, 193

CATHCART, E. P., BEDALE, E. M., BLAIR, C., MACLEOD, K., and WEATHERHEAD, E. (1927). *The physique of women in industry.* I.F.R.B. Rept. No. 44. London, H.M.S.O.

CHANNEL, R. C., and TOLCOTT, M. A. (1948). *The use of human engineering data in equipment design problems.* Special Devices Center Rept. No. 151-1-16

CHAPANIS, A., GARNER, W. R., and MORGAN, C. T. (1949). *Applied Experimental Psychology.* New York, Wiley.

CHAPANIS, A., and LEYZOREK, M. (1950). 'Accuracy of visual interpolation.' *J. exp. Psychol.*, **40**, 655

CHRISTENSEN, E. H. (1953). 'Physiological valuation of work in Nykroppa Iron Works.' In *Symposium on Fatigue* (eds. Floyd, W. F., and Welford, A. T.). London, H. K. Lewis.

CHRISTENSEN, J. M. (1952). *Quantitative instrument reading as a function of dial design, exposure time, preparatory fixation and practice.* U.S.A.F. Aero-Medical Lab. Rept. No. AF52/116

CHURCHILL, A. V. (1956). 'The effect of scale interval length and pointer clearance on speed and accuracy of interpolation.' *J. app. Psychol.*, **40**, 358

CHURCHILL, A. V. (1959). 'Optimal interval length for visual interpolation. The effect of viewing distances.' *J. app. Psychol.*, **43**, 125

CHURCHILL, A. V., and ALLEN, D. G. (1955). *Experimental dial design.* Def. Res. Med. Lab. (Canada) DRML, Proj. No. 164

CLARK, K. S. (1929). *Music in industry.* New York, National Bureau for the Advancement of Music

CLAY, H. M. (1956). 'A study of performance in relation to age in two printing works.' *J. Gerontol.*, **11**, 417

CLEMENTS, E. M. B. (1954). 'Body measurement of the working population.' In *Human Factors in Equipment Design* (eds. Floyd, W. F., and Welford, A. T.). London, H. K. Lewis

CLEMENTS, E. M. B., and PICKETT, K. G. (1952). 'Stature of Scotsmen aged 18 to 40 years, in 1941.' *Brit. J. soc. Med.*, **6**, 245

COBB, P. W., and MOSS, F. K. (1927). 'Four fundamental factors in vision.' In Luckiesh and Moss, *Interpreting the science of seeing into lighting practice*, **1**. 1927-1932. Cleveland, General Electric Co.

COERMANN, R. (1938). 'Untersuchungen über die Einwirkung von Schwingungen auf den Menschlichen Organismus Ausgabe: Ausrüstung.' *Jahrb, deut, Luftfahrtforschung*, **3**, 111

COLLIN, D. A. (1960). Personal communication

COLLINS, J. B., and HOPKINSON, R. G. (1954). 'Flicker discomfort in relation to the lighting of buildings.' *Trans. Illum. Eng. Soc. (Lond.)*, **19**, 135

CONNELL, S. C. (1950). *The variables affecting instrument check reading.* U.S.A.F. Aero-Medical Lab. Rept. No. 6024

CONRAD, R. (1954). 'Speed Stress.' Ch. 13 in *Human Factors in Equipment Design* (eds. Floyd, W. F., and Welford, A. T.). London, H. K. Lewis

CONRAD, R. (1955). *Setting the Pace.* M.R.C. App. Psychol. Res. Unit, Rept. No. 232/55. Cambridge

CONRAD, R., and HILLE, B. A. (1953). 'A comparison of paced and un-paced performance at a packing task.' *Occup. Psychol.*, **29**, 15

COOK, T. W., and SHEPHARD, A. H. (1958). 'Performance on several control display relationships as a function of age and sex.' *Percept. and Motor Skills*, **8**, 339

COPE, F. W. (1960). 'Problems in human vibration engineering.' *Ergonomics*, **3**, 35

CRAIK, K. W. (1941). *Instrument lighting for night use.* Flying Personnel Research Committee Rept. No. 342

CRAIK, K. W. (1947). 'Theory of the human operator in control systems. I. The operator as an engineering system.' *Brit. J. Psychol.*, **38**, 56

CRAIK, K. W. (1948). 'Theory of the human operator in control systems. 2. Man as an element in a control system.' *Brit. J. Psychol.*, **38**, 142

CRAWFORD, A. (1961). 'Fatigue and Driving.' *Ergonomics*, **4**, 143

CRAWFORD, B. H., and STILES, W. S. (1935). 'Brightness difference threshold data for the evaluation of glare from light sources.' *J. Sci. Instrum.*, **12**, 177

CREELMAN, J. A., and MILLER, E. E. (1956). 'Evaluation of a "moving airplane" attitude indicator.' U.S. Naval School of Aviation Medicine Research Project NM 001.109.107 Rept. No. 3

CROCKFORD, G. W., and HELLON, R. F. (1962). 'Protective clothing for very hot conditions.' *New Scientist*, **13**, 146

CROCKFORD, G. W., HELLON, R. F., HUMPHREYS, P. W., and LIND, A. R. (1961). 'An air-ventilated suit for wear in very hot environments.' *Ergonomics*, **4**, 63

CROSSMAN, E. R. F. W. (1956). 'Perception study. A complement to motion study.' *The Manager*, **24**, 141

CROUCH, C. L. (1945a). 'Illumination and vision.' *Trans. Illum. Eng. Soc.*, **11**, 747

CROUCH, C. L. (1945b). 'Brightness limitations of luminaires.' *Illum. Eng.*, **40**, 427

CROUCH, C. L. (1958). 'A new method of determining illumination required for a task.' *Illum. Eng.*, **53**, 416

CROWDEN, G. P. (1928). *The physiological cost of muscular movement involved in barrow work.* I.F.R.B. Rept. No. 50. London, H.M.S.O.

DALE, H. C. A. (1958). 'Fault finding in electronic equipment.' *Ergonomics*, **1**, 356

DANIELS, G. S., MEYERS, H. C., and SHERYL, W. H. (1953a). *Anthropometry of WAF basic trainees.* W.A.D.C. Training Report 53-12, U.S.A.F., Wright Air Development Center, Wright-Patterson Air Force Base, Ohio

DANIELS, G. S., MEYERS, H. C., and CHURCHILL, E. (1953b). *Anthropometry of male basic trainees.* W.A.D.C. Technical Report 53-49, U.S.A.F. Wright Air Development Center, Wright-Patterson Air Force Base, Ohio

DARCUS, H. D. (1948). *The anatomical principles related to sighting.* Rept. No. R.N.P. 48/474. London, Med. Res. Council

DARCUS, H. D. (1954). 'Range and strength of joint movement.' In *Human Factors in Equipment Design* (eds. Floyd, W. F., and Welford, A. T.). London, H. K. Lewis

DAVIS, L. E., and JOSSELYN, P. D. (1953). 'How fatigue affects productivity.' *Personnel*, **30**, 54

DAVIS, S. W. (1955). 'Auditory and visual flicker fusion as measures of fatigue.' *Amer. J. Psychol.*, **68**, 654

DAVY, J. (1961). 'Arctic Wind in Acton.' *Observer*, June 18th, p. 3.

DEMPSEY, C. A. (1953). *Development of a workspace measuring device.* W.A.D.C. Wright-Patterson A.F. Base, Ohio. Tech. Rept. No. 53-53

DEMPSTER, W. T., GABEL, W. C., and FELTS, W. J. L. (1959). 'The anthropometry of the manual work space for the seated subject.' *Amer. J. phys. Anthrop.*, **17**, 289

DIEKMANN, D. (1958). 'A study of the influence of vibration on man.' *Ergonomics*, **1**, 347

DIEHL, H. S. (1933a). 'Height and weight of American College men.' *Human Biol.*, **5**, 445

DIEHL, H. S. (1933b). 'The height and weight of American College women.' *Human Biol.*, **5**, 660

DREW, G. C. (1940). *An experimental study of mental fatigue.* Air Ministry F.P.R.C., Rept. 227

DUCROS, E. (1955). 'Statistique de biometrie medicale elementaire relatives au personnel navigant de l'Armée de l'Air Française.' In *Anthropometry and Human Engineering.* London, Butterworth

DUDLEY, N. A. (1958). 'Output pattern in repetitive tasks.' *Inst. Prod. Eng. J.*, **37**, 187

DUNBAR, C. (1938). 'Necessary values of brightness contrast in artificially lighted streets.' *Trans. Illum. Eng. Soc. (Lond.)*, **3**, 187

EDHOLM, O. G. (1957). 'Energy expenditure.' *Advance Sci.*, **13**, 486

EDHOLM, O. G. (1962). An experiment in acclimatization. *New Scientist*, **16**, (315), 500

EDHOLM, O. G., FLETCHER, J. G., WIDDOWSON, E. M., and MCCANCE, R. A. (1955). 'The energy expenditure and food intake of individual men.' *Brit. J. Nutr.*, **9**, 286

EDWARDS, E. (1964). *Information transmission.* London, Chapman & Hall

ELLIOTT, D. N., and HOWARD, E. M. (1956). 'The effect of position on warning light effectiveness.' *Percep. Motor Skills*, **6**, 69

FARMER, E. (1921). *Motion study in metal polishing.* I.F.R.B. Rept. No. 15. London, H.M.S.O.

FERREE, E. C., and RAND, G. (1922). 'The effect of variation of visual angle, intensity and composition of light on important ocular functions.' *Trans. Illum. Eng. Soc.*, **17**, 69-102

FISHER, O. (1906). *Theoretisch Grundlagen für eine Mechanik der Lebenden Körper.* Leipzig

FITTS, P. M. (1947). 'A study of location discrimination ability.' In *Psychological research on equipment design* (ed. Fitts, P. M.). Army Air Force, Aviation Psychol Programme, Res. Rept. No. 9.

FITTS, P. M., and JONES, R. E. (1947a). *Analysis of factors contributing to 460 'pilot error' experiences in operating aircraft controls.* U.S.A.F. Air Materiel Command Memo Rept. No. TSEAA-694-12

FITTS, P. M., and JONES, R. E. (1947b). *Psychological aspects of instrument display.* I. *Analysis of 270 'pilot error' experiences in reading and interpreting aircraft instruments.* U.S.A.F., Air Materiel Command, Wright-Patterson A.F. Base. Memo Rept. No. TSEAA-694-12A

FITTS, P. M., and SIMON, C. W. (1952). *The arrangement of instruments, the distance between instruments, and the position of instrument pointers as determinants of performance in an eye-hand co-ordination task.* U.S.A.F., Wright Air Dev. Center. Tech. Rept. 5832.

FLANAGAN, J. C. (1954). 'The critical incidents technique.' *Psychol Bull.,* **51,** 327

FLETCHER, H., and MUNSON, W. A. (1933). 'Loudness, its definition, measurement and calculation.' *J. acoust. Soc. Amer.,* **5,** 82

FLOYD, W. F., and ROBERTS, D. F. (1958). *Anatomical, physiological and anthropometric principles in the design of office chairs and tables.* B.S. 3044. London, British Standards Institution

FLOYD, W. F., and SILVER, P. H. S. (1955). 'The function of the Erectores Spinae muscles in certain movements and postures in man.' *J. Physiol.,* **129,** 184

FLOYD, W. F., and WELFORD, A. T. (1953). Eds. *Fatigue.* London, H. K. Lewis

FOLEY, P. J. (1956). 'Effect of background on the critical flicker frequency.' *Canad. J. Psychol.,* **10,** 200

FOOD AND AGRICULTURAL ORGANIZATION OF THE UNITED NATIONS (1949). *Dietary surveys. Their technique and interpretation.* Nutritional studies No. 4. Washington, U.N.

FORBES, O. G. (1945). 'The effect of certain variables on visual and auditory reaction times.' *J. exp. Psychol.,* **35,** 153

FORBES, P. W. (1946). 'Auditory signals for instrument flying.' *J. of Aeronautical Sciences,* **13,** 255

FORBES, T. W., and HOLMES, R. S. (1939). 'Legibility distances of highways destination signs in relation to letter height and width.' *Proc. Highway Res. Bd.,* **19,** 321

FRASER, D. C. (1950). 'The relation between the angle of display and performance in a prolonged visual task.' *Quart. J. exp. Psychol.,* **2,** 176

FRASER, D. C. (1957a). 'Environmental stress and its effect on performance.' *Occup. Psychol.,* **31,** 248

FRASER, D. C. (1957b). 'Discussion II in Symposium on vigilance.' *Advanc. Sci.,* **13,** 409

FRITZE, C., and SIMONSON, E. (1951). 'A new electronic apparatus for the measurement of the fusion frequency of flicker.' *Science*, **113**, 547

FRY, G. A. (1954). 'A revaluation of the scattering theory of glare.' *Illum. Eng.*, **49**, 98

GAGGE, A. P., BAZETT, H. C., and BURTON, A. C. (1941). 'A practical system of units for the description of the heat exchange of man with his environment.' *Science*, **94**, 428

GARRY, R. C., PASSMORE, R., WARNOCK, G. M., and DURNIN, J. V. G. A. (1955). *Studies on expenditure of energy and consumption of food by miners and clerks, Fife, Scotland.* Med. Res. Council Special Rept. Series No. 289. London, H.M.S.O.

GARDNER, J. F. (1954). *Speed and accuracy of response to five different attitude indicators.* U.S.A.F. Wright Air Development Center, Tech. Rep. 54-236

GARVEY, W. D., and TAYLOR, F. V. (1959). 'Interactions among operator variables, system dynamics and task induced stress.' *J. app. Psychol.*, **43**, 79

GEOGHEGAN, B. (1953). *The height of R.N. male personnel.* Operational Efficiency Sub-committee of the Royal Naval Personnel Research Committee, Report No. I.E.S. 227 R.N.P. 53/733

GIBBS, C. B. (1953). 'The continuous regulation of skilled response by kinaesthetic feedback.' M.R.C. App. Psychol. Res. Unit Report No. 190/53. Cambridge

GLOAG, H. L. (1951). 'The development of the use of colour in British factories.' *Proc. Building Res. Cong. Div.*, **3**, Part II, 181

GLOAG, H. L., and KEYTE, M. J. (1957). 'Rational aspects of colouring in building interiors.' *Arch. J.*, **125**, 399 and 443

GLOVER, J. R. (1960). 'Back Pain and Hyperaesthesia.' *Lancet*, **1**, 1165.

GOLDFARB, W. (1941). *An investigation of reaction time in older adults and its relationship to certain observed mental test patterns.* Teachers Coll. Contribution to education No. 831. New York, Columbia University

GOLDMAN, D. E. (1948). *A review of subjective responses to vibrating motion of the human body in the frequency range 1-70 cycles per second.* U.S. Naval Medical Research Institute. Project NM 004/01 Rept. No. 1

GOLDMAN, D. E. (1957). 'Effects of vibration on man.' Ch. 11 in *Handbook of noise control* (ed. Harris, C. M.). New York, McGraw-Hill

GOLDMAN, D. E. (1961). *The biological effects of vibration.* Armed Forces, N.R.C. Committee on Hearing and Bio-acoustics Rept. 89

GORE, W. C. (1960). 'Information theory.' In *Operations research and system engineering* (ed. Flagle, C. H. et al.). Baltimore, John Hopkins Press

GRAHAM, C. H. (1958). 'Sensation and perception in an objective psychology.' *Psychol. Rev.*, **65**, 65

GRAHAM, N. E. (1952). 'Manual tracking on a horizontal scale and in the four quadrants of a circular scale.' *Brit. J. Psychol. (Gen. Sec.)*, **43**, 70

GRAHAM, N. E. (1956). 'The speed and accuracy of reading horizontal, vertical and circular scales.' *J. app. Psychol.*, **40**, 228

GRAHAM, N. E., BAXTER, I. G., and BROWNE, R. C. L. (1951). 'Manual tracking in response to the display of horizontal, vertical and circular scales.' *Brit. J. Psychol. (Gen. Sec.)*, **42**, 155

GRANDJEAN, E., and PERRET, E. (1961). 'Effects of pupil aperture on variations of flicker fusion frequency.' *Ergonomics*, **4**, 17

GREGORY, R. L. (1959). 'Increase in "neurological noise" as a factor in Ageing.' *Proc. 4th Cong. internat. A. Gerontol. Merano*, 1957, **1**, 314

GRETHER, W. F. (1948). 'Habit interference as a factor in the design of clock dials to be read in 24 hr. (military) time.' *J. app. Psychol.*, **32**, 159

GRETHER, W. F. (1949). 'Instrument reading. (1) The design of long-scale indicators for speed and accuracy of quantitative reading.' *J. app. Psychol.*, **33**, 363

GRETHER, W. F., and CONNELL, S. C. (1948). *Psychological factors in check reading single instruments.* U.S.A.F. Air Materiel Command, Rept. MCREXD 694-17A

GRETHER, W. F., and WILLIAMS, A. C. (1947). 'Speed and accuracy of dial readings as a function of dial diameter and angular separation of scale divisions.' In *Psychol. Res. in equipment design* (ed. Fitt, P. M.). Army Air Force, Aviation Psychol. Prog. Res. Rept. No. 9

GRIEVE, J. I. (1960). 'Thermal stress in a single storey factory.' *Ergonomics*, **3**, 297

GRIEW, S. (1958a). 'A study of accidents in relation to occupation and age.' *Ergonomics*, **2**, 17

GRIEW, S. (1958b). 'Age changes and information loss in performance of a pursuit tracking task involving interrupted preview.' *J. exp. Psychol.*, **55**, 486

GRIEW, S. (1959). 'Complexity of response and time of initiating responses in relation to age.' *Amer. J. Psychol.*, **72**, 83

GRIEW, S., and TUCKER, W. A. (1958). 'The identification of job activities associated with age differences in the engineering industry.' *J. app. Psychol.*, **42**, 278

GUIGNARD, J. C. (1959). *The physiological response of seated men to low-frequency vertical vibration.* Air Ministry Flying Personnel Research Committee, Rept. No. 1062

GUILLEMIN, V., and WECHSBERG, P. (1953). 'Physiological effects of long term repetitive exposure to mechanical vibration.' *J. Aviat. Med.*, **24**, 208

HAINES, G. F., and HATCH, T. F. (1952). 'Industrial heat exposures evaluation and control.' *Heat. and Vent.*, **49**, 93

HANSON, R., and CORNOG, D. Y. (1958). *Annotated bibliography of Physical Anthropology in human engineering.* U.S.A.F. Air Research and Development Command, W.A.D.C. Tech. Rept. No. 56-30

HARRIS, C. M. (1957). Ed. *Handbook of Noise Control*, New York, McGraw-Hill

HARRIS, S. J., and SMITH, K. U. (1954). 'Dimensional analysis of motion. VII. Extent and direction of manipulative movements as factors in defining movements.' *J. app. Psychol.*, **38**, 126

HARRISON, W. (1945). 'Glare Ratings.' *Illum. Eng.*, **40**, 525

HARRISON, W., and MEAKER, P. (1947). 'Further data on glare rating.' *Illum. Eng.*, **41**, 153

HATCH, T. F. (1958). 'Heat control in the hot industries.' Ch. 21 in *Industrial Hygiene and Toxicology*, **1**, 2nd Edn. (ed. Patty, F. A.). New York, Interscience.

HAWLEY, M. E., and KRYTER, K. D. (1957). 'Effects of noise on speech.' Ch. 9 in *Handbook of Noise Control* (ed. Harris, C. M.). New York, McGraw-Hill

HEALTH OF MUNITION WORKERS COMMITTEE (1918). *Industrial health and efficiency*. Final Rept. Cd. 9065. London, H.M.S.O.

HECHT, S., and MINTZ, E. U. (1939). 'The visibility of single lines at various illuminations and the retinal basis of visual resolution.' *J. Gen. Physiol.*, **22**, 593

HELLON, R. F., and CROCKFORD, G. W. (1959). 'Improvements in the globe thermometer.' *J. app. Physiol.*, **14**, 649

HERON, A., and CHOWN, S. (1961). *Ageing and the semi-skilled*. M.R.C. Memo No. 40. London, H.M.S.O.

HERTZBERG, H. T. E., DANIELS, G. S., and CHURCHILL, E. (1954). *Anthropometry of flying personnel* – 1950. W.A.D.C. Technical Report 52-321, U.S.A.F. Wright Air Development Center, Wright-Patterson Air Force Base

HICK, W. E. (1952). 'On rate of gain of information.' *Quart. J. exp. Psychol.*, **4**, 11

HICK, W. E., and BATES, J. A. (1950). *The human operator of control mechanisms*. Permanent record of research and development. Monograph No. 17.204. London, H.M.S.O.

HINCHCLIFFE, R. (1958). 'The pattern of the threshold of perception for hearing and other special senses as a function of age.' *Gerontologia*, **2**, 311

HOLLADAY, L. L. (1926). 'The fundamentals of glare and visibility.' *J. Opt. Soc. Amer.*, **12**, 271

HOLMES, J. G. (1941). 'The recognition of coloured light signals.' *Trans. Illum. Eng. Soc.*, **VI**, 71

HOOTON, E. A. (1945). *A survey in seating*. Gardner, Mass., Heywood-Wakefield Co.

HOOTON, E. A., and DUPERTUIS, C. W. (1951). *Age changes and selective survival in Irish males*. Studies in Physical Anthropology No. 2, Amer. Assn. of Phys. Anthropol.

HOPKINSON, R. G. (1940). 'Discomfort glare in lighted streets.' *Trans. Illum. Eng. Soc. (Lond.)*, **5,** 1

HOPKINSON, R. G. (1950). 'The multiple criterion technique of subjective appraisal.' *Quart. J. exp. Psychol.*, **2,** 124

HOPKINSON, R. G., and LONGMORE, J. (1959). 'Attention and distraction in the lighting of working-places.' *Ergonomics*, **2,** 321

HOPKINSON, R. G., STEVENS, W. R., and WALDRAM, J. M. (1941). 'Brightness and contrast in illuminating engineering.' *Trans. Illum. Eng. Soc. (Lond.)*, **6,** 37

HOUGHTEN, F. C., and YAGLOGLOU, C. P. (1923). 'Determining lines of equal comfort.' *Trans. Amer. Soc. Heat. Vent. Engrs.*, **29,** 163

HOUSTON, R. C., and WALKER, R. Y. (1949). *The evaluation of auditory warning signals for aircraft.* U.S.A.F. Air Materiel Command Tech. Rept. No. 5762

HUGH-JONES, P. (1947). 'The effect of limb position in seated subjects on their ability to utilize the maximum contractile force of the limb muscles.' *J. Physiol.*, **105,** 332

HUNSICKER, P. A. (1955). *Arm strength at selected degrees of elbow flexion.* U.S.A.F. Wright Air Development Center, Tech. Rept. WADC-TR-154-548

HUNT, D. P. (1953). *The coding of aircraft controls.* U.S.A.F. Wright Air Development Center, Tech. Rept. 53-221

HUNT, G. C. (1956). 'Sequential arrays of waiting lines.' *Operations Res.*, **4,** 674

HUTCHINSON, R. C. (1954). 'Effect of gastric contents on mental concentration and production rate.' *J. app. Physiol.*, **7,** 143

ILLUMINATING ENGINEERING SOCIETY LONDON (1961). *The I.E.S. Code, Recommendations for good interior lighting.* London, Illuminating Engineering Society

ILLUMINATING ENGINEERING SOCIETY (U.S.A.) (1959). *Currently recommended levels of illumination.* New York, Illuminating Engineering Society

INTERNATIONAL COMMISSION ON ILLUMINATION (1959). *Colours of Light Signals.* Pub. C.I.E. No. 2 (W-1.3.3.). Paris, International Commission on Illumination

JACKLIN, H. M. (1936). 'Human reaction to vibration.' *S.A.E. Jnl.*, **39,** 401

JACOBSON, H. J. (1953). 'A study of inspector accuracy.' *Engng. Insp.*, **17,** 2

JENKINS, W. O. (1947). 'The tactual discrimination of shapes for coding aircraft-type controls.' In *Psychological research on equipment design* (ed. Fitts, P. M.). Army Air Force, Aviation Psychology Program, Research Report, **19**

JERISON, H. J. (1959). *Experiments on Vigilance.* U.S.A.F. W.A.D.C. Tech. Rept. 58/526

JOHNSGARD, K. W. (1953). 'Check reading as a function of pointer symmetry and uniform alignment.' *J. app. Psychol.*, **37**, 407

JONES, E. D. (1919). *The administration of industrial enterprises.* New York, Longman

JONES, J. C., WARD, A. J., and HEYWOOD, P. M. (1965). 'Reading dials at short distances.' *A.E.I. Engineering*, Jan./Feb., 28.

JONES, R. E., MILTON, J. L., and FITTS, P. M. (1949). *Eye fixations of aircraft pilots. IV. Frequency, duration and sequence of fixations during routine instrument flight.* U.S.A.F. Aero Med. Lab. A.F. Tech. Rept. 5957

DE JONG, J. R. (1960). *An investigation in respect of working in shifts.* Personal communication

JOSEPH, T., and NIGHTINGALE, A. (1956). 'Electromyography of muscles of posture: leg and thigh muscles in women, including the effect of high heels.' *J. Physiol.*, **132**, 465

KAPPAUF, W. E. (1949). *Studies pertaining to the design of visual displays for aircraft instruments, computers, maps, charts, tables and graphs: a review of the literature.* U.S.A.F. Air Materiel Command, Wright-Patterson A.F.B. Tech. Rept. No. 5765

KAPPAUF, W. E., SMITH, W. M., and BRAY, C. W. (1947). *Design of instrument dials for maximum legibility.* U.S.A.F. Aero Medical Lab. Memo Rept. No. TSEAA-694

KAPPAUF, W. E., and SMITH, W. M. (1950). *Designing of instrument dials for maximum legibility. III. Some data on the difficulty of quantitative readings in different parts of a dial.* U.S.A.F. Air Materiel Command Tech. Rept. 5914

KARPINOS, B. D. (1958). 'Height and weight of selective service registrants processed for military service during World War II.' *Human Biol.*, **30**, 292

KAUFMAN, E. L., LORD, M. W., REESE, T. W., and VOLKMANN, J. (1949). 'Discrimination of visual number.' *Amer. J. Psychol.*, **62**, 498

KAY, H. (1954). 'The effects of position in a display upon problem solving.' *Quart. J. exp. Psychol.*, **6**, 155

KAY, R. (1958). *Determination of a population stereotype for turning a vertically mounted door handle.* Report privately circulated.

KEATING, G. F., and LANER, S. (1958). 'Some notes on the effects of excessive noise on the hearing of a group of workers.' *Brit. J. industr. Med.*, **15**, 237

KELLERMANN, F. TH., VAN WELY, P. A., and WILLEMS, P. J. (1963). *Vademecum-Ergonomics in industry.* N. V. Philips' Gloeilampenfabrieken, Eindhoven (Netherlands).

KEMSLEY, W. F. F. (1950). 'Weight and height of a population in 1943.' *Ann. Eugenics*, **15**, 161

KEMSLEY, W. F. F. (1957). *Women's measurements and sizes.* London, H.M.S.O. Published by the Board of Trade, for Joint Clothing Council Ltd.

KENNEDY, J. L. (1953). 'Some practical problems of the alertness indicator.' In *A Symposium on Fatigue* (eds. Floyd, W. F., and Welford, A. T.). London, H. K. Lewis

KLEITMAN, N. (1939). *Sleep and wakefulness.* Chicago, Univ. of Chicago Press

KLEITMAN, N., TITELBAUM, S., and FEIVESON, P. (1938). 'The effect of body temperature on reaction time.' *Amer. J. Physiol.*, **121**, 495

KOFRANYI, E., and MICHAËLIS, H. F. (1941). 'Ein tragbarer Apparat zur Bestimmung des Gasstofwechsels.' *Arbeitsphysiologie*, **11**, 148

KRYTER, K. D. (1950). *The effects of noise on man.* J. Speech Disorders. Mon. Sup. No. 1

KUNTZ, J. E., and SLEIGHT, R. B. (1949). 'The effect of target brightness on normal and subnormal visual acuity.' *J. app. Psychol.*, **33**, 83

KUNTZ, T. E., and SLEIGHT, R. B. (1950). 'Legibility of numerals: the optimum ratio of height to width of strokes.' *Amer. J. Psychol.*, **63**, 567

KURKE, M. I. (1956). 'Evaluation of a display incorporating quantitative and check reading characteristics.' *J. app. Psychol.*, **40**, 233

LAIRD, D. A., and COYE, K. (1929). 'Psychological measurements of annoyance as related to pitch and loudness.' *J. acoust. Soc. Amer.*, **1**, 158

LAMMERS, B. (1960). 'Possibilities and limitations of different working postures.' Paper presented to the XIIIth International Congress on Occup. Hlth. New York. July, 1960

LANDIS, C., and HUNT, W. A. (1939). *The startle pattern.* New York, McFarrar and Rinehard

LANER, S. (1961). 'Ergonomics in the steel industry.' In *Ergonomics in industry*, p. 103. London, H.M.S.O.

LANER, S. (1962). *Fitting the job to the worker–Seminar on Ergonomics for Engineers.* Paris, O.E.C.D.

LATHAM, F. (1957). 'A study in body ballistics: seat ejection.' *Proc. Roy. Soc. Sect. B*, **147**, 121

LAUER, A. K., and SUHR, V. W. (1958). *Road adaptation of a laboratory technique for studying driving efficiency with or without a refreshment pause.* Chicago, National Safety Council

LAURU, L. (1957). 'Physiological studies of motions' (trans. L. Brouha). *Advanc. Manag.*, **22**, (3), 17

LAY, W. E., and FISHER, L. C. (1940). 'Riding comfort and cushions.' *J. Soc. Auto. Engin.* (*Trans*), **47**, 482

LEHMANN, G. (1958a). 'Physiological basis of tractor design.' *Ergonomics*, **1**, 197

LEHMANN, G. (1958b). 'Physiological measurement as a basis of work organization in industry.' *Ergonomics*, **1**, 328

LEMMON, P. W. (1927). *The relation of reaction time to measures of intelligence, memory and learning.* Arch. Psychol. N.Y., **15**, (94), 38

LEWIS, C. E., SHERBERGER, R. F., and MILLER, F. A. (1960). 'The study of heat stress in extremely hot environments and the infrared reflectance of some potential shielding materials.' *Brit. J. industr. Med.*, **17**, 52

LEWIS, P. R., and LOBBAN, M. C. (1957). 'Dissociation of diurnal rhythms in human subjects, living in abnormal time routines.' *Quart. J. exp. Physiol.*, **42**, 371

LEYZOREK, M. (1949). 'Accuracy of visual interpolation between circular scale markers as a function of the separation between markers.' *J. exp. Psychol.*, **39**, 270

LIND, A. R. (1960). 'The effect of heat on the industrial worker.' *Ann. Occup. Hyg.*, **2**, 190

LIND, A. R., and HELLON, R. F. (1957). 'Assessment of physiological severity of hot climates.' *J. app. Physiol.*, **11**, 35

LIND, A. R., HELLON, R. F., WEINER, J. S., JONES, R. M., and FRASER, D. C. (1957). *Reactions of mines-rescue personnel to work in hot environments.* National Coal Board. Medical Research Memo. 1

LIPPOLD, O. C. J., REDFEARN, J. W. T., and VUČO, J. (1960). 'The electromyography of fatigue.' *Ergonomics*, **3**, 121

LITTLER, T. S. (1958). 'Noise measurement, analysis and evaluation of harmful effects.' *Ann. of Occup. Hyg.*, **1**, 11

VAN LOON, J. H. (1963). 'Diurnal body temperature curves in shift workers.' *Ergonomics*, **6**, 267

LOUCKS, R. B. (1945). *Evaluation of aircraft attitude indicators on the basis of Link instrument ground trainer performance.* U.S.A.F. Sch. Aviat. Med. Randolph Field. Proj. 341 Rept. 1

LOVELESS, N. E. (1956). 'Display control relationships on circular and linear scales.' *Brit. J. Psychol.*, **47**, 271

LOVELESS, N. E. (1962). 'Direction of Motion Stereotypes: A Review.' *Ergonomics*, **5**, 357

LOVELL, G. (1954). *Design and development of the R.A.F. dummy of the standard airman.* Royal Aircraft Establishment, Farnborough. Tech. Note No. Mech. Eng. 176

LOWSON, J. C., DRESLER, A., and HOLMAN, S. (1954). 'A practical investigation on discomfort glare.' *Illum. Eng.*, **49**, 497

LUCKIESH, M., and GUTH, S. K. (1946). 'Discomfort glare and angular distance of glare source.' *Illum. Eng.*, **42**, 485

LUCKIESH, M., and GUTH, S. K. (1949). 'Brightness in the visual field, at the border line between comfort and discomfort. B.C.D.' *Illum. Eng.*, **44**, 650

LUCKIESH, M., and MOSS, F. K. (1932). 'The new science of seeing.' In *Interpreting the science of seeing into lighting practice*, **1**, 1927-1932. Cleveland, General Electric Co.

LUNDERVOLD, A. (1958). 'Electromyographic investigation during typewriting.' *Ergonomics*, **1**, 226

LYTHGOE, R. J. (1932). *The measurement of visual acuity.* M.R.C. Special Report Series No. 173. London, H.M.S.O.

MCAVOY, W. H. (1937). *Maximum forces applied by pilots to wheel type controls.* N.A.C.A. Tech. Note No. 623

MCCARTHY, C. (1952). *The use of dials in H.M. Ships.* Unpublished Report. (Quoted in Murrell, 1952b)

MCCORMICK, E. (1957). *Human Engineering.* New York, McGraw-Hill

MCCORMICK, E. J., and NIVEN, J. R. (1952). 'The effect of varying intensities of illumination upon performance on a motor task.' *J. app. Psychol.*, **36,** 193

MCFARLAND, R. A., DAMON, A., STOUDT, H. W., MOSELEY, A. L., DUNLAP, J. W., and HALL, W. A. (1954). *Human body size and capabilities in the design and operation of vehicular equipment.* Boston, Harvard School of Public Health

MCFARLAND, R. A., and DOMEY, R. G. (1960). 'Human factors in the design of passenger cars; an evaluation study of models produced in 1957.' *Proc. Highway Res. Bd.*, **39,** 565

MCFARLAND, R. A., DUNLAP, J. W., HALL, W. A., and MOSELEY, A. L. (1953). *Human factors in the design of highway transport equipment: a summary report of vehicle evaluation.* Boston, Harvard Sch. of Pub. Hlth.

MCFARLAND, R. A., WARREN, A. B., and KARIS, C. (1958). 'Alterations in critical flicker frequency as a function of age and light : dark ratio.' *J. exp. Psychol.*, **56,** 529

MCGHEE, W., and GARDNER, J. (1949). 'Music in a complex industrial task.' *Personnel Psychol.*, **2,** 405

MCGRATH, J. J., HARABEDIAN, A., and BUCKNER, D. N. (1959). *Review and critique of the literature on vigilance performance.* Prepared by Human Factors Res. Inc. Contract No. nr 2649(00). NR 153-199 Tech. Rept. No. 1 for Psychol. Sci. Div., U.S. Office of Naval Research.

MACKENZIE, R. M. (1958). 'On the accuracy of inspectors' *Ergonomics*, **1,** 258

MACKENZIE, R. M., and PUGH, D. S. (1957). 'Some human aspects of inspection.' *J. Inst. Prod. Engrs.*, **36,** 378

MACKWORTH, N. H. (1944). *Notes on the clock test. A new approach to the study of prolonged visual perception to find the optimum length of watch for radar operators.* M.R.C. Rept. No. 46/348. APU Rept. No. 1

MACKWORTH, N. H. (1950). *Researches on the measurement of human performance.* M.R.C. Special Report Series No. 268. London, H.M.S.O.

MACPHERSON, R. K. (1960). *Physiological Responses to hot environments.* M.R.C. Special Report Series No. 298. London, H.M.S.O.

MANN, A. J. (1961). Contribution to Conference on *Ergonomics in Industry*, p. 42. London, H.M.S.O.

MAY, J., and WRIGHT, H. B. (1961). 'Heights and weights of business men.' *Trans. Ass. Ind. Med. Off.*, **11,** 143

MEAKER, P. (1949). 'Brightness v. area in the glare factor formula.' *Illum. Eng.*, **44,** 401

MECH, E. V. (1953). 'Factors influencing routine performance under noise. I. The influence of "set".' *J. Psychol.*, **35,** 283

METZ, B. (1960). *Fitting the job to the Worker: International conference of Zurich.* Paris, Organisation for European Economic Co-operation (E.P.A.)

METZ, B., LAMBERT, G., and SCHIEBER, J. P. (1961). 'Enregistrement continu, analogique et digital, de temps court (0,25 à 10s).' *J. de Physiologie*, **53,** 426

MILES, W. R. (1931). 'Measures of certain human abilities throughout the life span.' *Proc. Nat. Acad. Sci.*, **17,** 627

MILES, G. H., and SKILBECK, O. (1923). 'Experiments on change of work.' *J. Nat. Inst. Ind. Psychol.*, **1,** 236

MILLER, G. A. (1947). 'The masking of speech.' *Psychol. Bull.*, **44,** 105

MILLER, G. A., HEISE, G. A., and LICHTEN, W. (1951). 'The intelligibility of speech as a function of the context of the test materials.' *J. exp. Psychol.*, **41,** 329

MILLER, G. A., and TAYLOR, W. G. (1948). 'The perception of repeated bursts of noise.' *J. acoust. Soc. Amer.*, **20,** 171

MISIAK, H. (1951). 'Decrease of critical flicker frequency with age.' *Science*, **113,** 551

MITCHELL, M. J. H., and VINCE, M. A. (1951). 'The direction of movement of machine controls.' *Quart. J. exp. Psychol.*, **3,** 21

MOWBRAY, G. H., and RHOADES, M. V. (1959). 'On the reduction of choice reaction times with practice.' *Quart. J. exp. Psychol.*, **11,** 16

MORGAN, C. T., and STELLAR, E. (1950). *Physiological Psychology.* New York, McGraw-Hill

MORLEY, N. J., and SUFFIELD, N. G. (1951). *The effect of pointer position on the speed and accuracy of check reading groups of dials.* Admiralty Naval Motion Study Unit Rept. No. 46

MOZELLE, M. M., and WHITE, D. C. (1958). 'Behavioural effects of whole body vibration.' *J. Aviat. Med.*, **29,** 719

MÜLLER, E. A. (1939). 'Die Wirkung sinusformiger Vertikalschwingungen auf den sitzenden und stehenden Menschen.' *Arbeitsphysiologie*, **10,** 464

MÜLLER, E. A. (1953). 'The physiological basis of rest pauses in heavy work.' *Quart. J. exp. Physiol.*, **38,** 205

MÜLLER, E. A. (1959-60). 'Muscular Fatigue.' *Annual Vol. Physiol. and Expt. Med. Sc. India*, **3,** 67

MÜLLER, E. A., and HIMMELMANN, W. (1957). 'Geräte zur kontinuirlichen fotoelektrischen Pulszählung.' *Arbeitsphysiologie*, **16,** 400

MÜLLER, E. A., VETTER, K., and BLÜMEL, E. (1958). 'Transport by muscle power over short distances.' *Ergonomics*, **1,** 222

MURRAY, H. D., Ed. (1952). *Colour in theory and practice.* London, Chapman and Hall

MURRELL, K. F. H. (1951a). *The display of bearing and range information.* Admiralty. Naval Motion Study Unit Report No. 43

MURRELL, K. F. H. (1951b). *The design of dials and indicators.* Admiralty. Naval Motion Study Unit Report No. 48

MURRELL, K. F. H. (1952a). 'The design of instrument scales.' *Instrument Practice*, **6**, 225

MURRELL, K. F. H. (1952b). 'The use and arrangement of dials.' *Instrument Practice*, **6**, 520

MURRELL, K. F. H. (1957). *Data on Human Performance for Engineering Designers.* London, Engineering

MURRELL, K. F. H. (1958). *Fitting the job to the Worker: a study of American and European research into working conditions in industry.* Paris, Organisation for European Economic Co-operation (E.P.A.)

MURRELL, K. F. H. (1959). 'Major problems of industrial gerontology.' *J. Gerontol.*, **14**, 216

MURRELL, K. F. H. (1960a). 'A comparison of three dial shapes for check reading instrument Panels.' *Ergonomics*, **3**, 231

MURRELL, K. F. H. (1960b). *Ergonomics – Fitting the job to the Worker.* London, British Productivity Council

MURRELL, K. F. H. (1962a). 'Note on the optimal length for visual interpolation.' *J. app. Psychol.*, **46**, 41

MURRELL, K. F. H. (1962b). 'Operator variability and its industrial consequence.' *International J. Prod. Res.*, **1**, (3), 39

MURRELL, K. F. H. (1962c). 'Industrial Aspects of Ageing.' *Ergonomics*, **5**, 147

MURRELL, K. F. H. (1963a). 'Controls and instruments – design procedure.' *Auto. Design Eng.*, **2**, (14), 70

MURRELL, K. F. H. (1963b). 'Laboratory studies of repetitive work. I: Paced work and its relationship to unpaced work.' *Int. J. Prod. Res.*, **2**, (3), 169

MURRELL, K. F. H., and EDWARDS, E. (1963). 'Field studies of an indicator of machine tool travel with special reference to the ageing worker.' *Occup. Psychol.*, **37**, 267

MURRELL, K. F. H., and ENTWISLE, D. G. (1960). 'Age differences in movement pattern.' *Nature*, **185**, 948

MURRELL, K. F. H., and FORSAITH, B. (1960). 'Age and the timing of movement.' *Occup. Psychol.*, **34**, 275

MURRELL, K. F. H., and FORSAITH, B. (1963). 'Laboratory studies of repetitive work. II: Results from two subjects.' *Int. J. Prod. Res.*, **2**, (4), 247

MURRELL, K. F. H., and GRIEW, S. (1958). 'Age structure in the engineering industry – A study of regional effects.' *Occup. Psychol.*, **32**, 86

MURRELL, K. F. H., GRIEW, S., and TUCKER, W. A. (1956). *The age structure of firms in the Engineering Industry.* Unit for Research on Employment of Older Workers. Univ. of Bristol. (A fuller version of Murrell *et al.*, 1957)

MURRELL, K. F. H., GRIEW, S., and TUCKER, W. A. (1957). 'Age structure in the engineering industry. A preliminary study.' *Occup. Psychol.*, **31**, 150

MURRELL, K. F. H., LAURIE, W. D., and MCCARTHY, C. (1958). 'The relationship between dial size, reading distance and reading accuracy.' *Ergonomics*, **1**, 182

MURRELL, K. F. H., and MCCARTHY, C. (1951). *Direction of motion relationships between valves and gauges when remounted in panels – Results from a printed test.* Naval Motion Study Unit. Admiralty Report No. 51

MURRELL, K. F. H., POWESLAND, P. F., and FORSAITH, B. (1962). 'A study of pillar-drilling in relation to age.' *Occup. Psychol.*, **36**, 45

MURRELL, K. F. H., and TUCKER, W. A. (1960). 'A pilot job-study of age-related causes of difficulty in light engineering.' *Ergonomics*, **3**, 74

MUSCIO, B. (1921). 'Is a fatigue test possible ?' *Brit. J. Psychol.*, **12**, 31

NADLER, G., and GOLDMAN, J. (1958). 'The Unopar.' *J. Indust. Eng.*, **9**, 58

NAFE, J. P. (1944). *The number, percentage, disposition and places of occurance of errors in 9440 rounds of artillery fire.* U.S. App. Psychol. Panel OSRD-NRC Proj. SoS-11 Informal Memo No. 4

NAYLOR, G. F. K. (1954). 'An approach to the study of dial reading.' *Occup. Psychol.*, **28**, 90

NELSON, H. A. (1957). 'Legal liability for loss of hearing.' Ch. 38 in *Handbook of noise control* (ed. Harris, C. M.). New York, McGraw-Hill

NORTH, J. D. (1956). 'The application of communications theory to the human operator.' *Information Theory; Third London Symposium*, 1955 (ed. C. Cherry). London, Methuen

O'BRIEN, RUTH, and SHELTON, W. C. (1941). *Women's measurements for garment and pattern construction.* Misc. pub. No. 454, U.S. Department of Agriculture in co-operation with the Work Projects Administration, Textiles and Clothing Division, Bureau of Home Economics

ORLANSKY, J. (1949). 'Psychological aspects of stick and rudder controls in aircraft.' *Aero Engineering Rev.*, **8**, 22

OSBORNE, E. E., and VERNON, H. M. (1922). *Two contributions to the study of accident causation.* I.F.R.B. Rept. No. 19. London, H.M.S.O.

PAPALOÏZOS, A. (1961). 'Some characteristics of instrument measuring dials.' *Ergonomics*, **4**, 169

PASSMORE, R. (1956). 'Daily energy expenditure by man.' *J. Nutrition Soc.*, **15**, 83

PEARSON, M. (1957). 'The transition from work to retirement.' *Occup. Psychol.*, **31**, 139

PEPLER, R. D. (1958). 'Warmth and performance: an investigation in the tropics.' *Ergonomics*, **2**, 63

PETHEBRIDGE, P., and HOPKINSON, R. G. (1950). 'Discomfort glare and the lighting of buildings.' *Trans. Illum. Eng. Soc. (Lond.)*, **15**, 39

PICKENS, J. L. (1961). 'How to investigate accidents before they happen.' *Safety Maintenance* (July), 5

PLUTCHIK, R. (1959). 'The effects of high intensity intermittent sound on performance, feeling and physiology.' *Psychol. Bull.*, **56**, 133

POLLACK, I. (1949). 'Loudness as a discriminable aspect of noise.' *Amer. J. Psychol.*, **62**, 285

POLLACK, I. (1952). 'The information of elementary auditory displays.' *J. acoust. Soc. Amer.*, **24**, 745

POLLOCK, K. G., and BARTLETT, F. C. (1932). *Two studies of the psychological effects of noise. I. Psychological experiments on the effects of noise.* I.H.R.B. Rept. No. 65. London, H.M.S.O.

POLYAK, S. L. (1941). *The Retina.* Chicago, University of Chicago Press

POTTS, C. R. (1951). 'A study of long haul truck operations' – quoted in McFarland, R. A., and Moseley, A. L. (1954), *Human Factors in Highway Transport Safety.* Boston, Harvard School of Public Health

POULTON, E. C. (1958). 'Measuring the order of difficulty of visual motor tasks.' *Ergonomics*, **1**, 234

POULTON, E. C., and BROWN, I. D. (1961). 'Measuring the "spare mental capacity" of car drivers.' *Ergonomics*, **4**, 35

PROVINS, K. A. (1955a). 'Effect of limb position on the forces exerted about the elbow and shoulder joints on the two sides simultaneously.' *J. app. Physiol.*, **7**, 387

PROVINS, K. A. (1955b). 'Maximum forces exerted about the elbow and shoulder joint on each side separately and simultaneously.' *J. app. Physiol.*, **7**, 390

PROVINS, K. A. (1958). 'Environmental conditions and driving efficiency: a review.' *Ergonomics*, **2**, 97

PROVINS, K. A., and CLARKE, R. S. J. (1960). 'The effect of cold on manual performance.' *J. Occup. Med.*, **2**, 169

PROVINS, K. A., and SALTER, N. (1955). 'Maximum torque exerted about the elbow joint.' *J. app. Physiol.*, **7**, 393

QUARTERMASTER CORPS. U.S. ARMY (1946). *Survey of body size of army personnel, male and female.* Project No. E-59-46. Unpublished data quoted in Hansen and Cornog (1958)

RAPHAEL, W. S. (1942). 'Some problems of inspection.' *Occup. Psychol.*, **16**, 157

REID, P. C. (1961). 'The human problems of shift operation.' *The Manager*, **29**, 289

RICHARDSON, I. M. (1953). 'Age and work: a study of 489 men in heavy industry.' *Brit. J. industr. Med.*, **10**, 269

RIOPELLE, A. J., HINES, M., and LAWRENCE, M. (1958). *The effects of intense vibration.* U.S. Army Med. Res. Lab. Rept. No. 358

ROBERTSON, J. M., and REID, D. D. (1952). 'Standards for basal metabolism of normal people in Britain.' *Lancet,* **1,** 940

RUSHTON, W. A. H. (1961). 'The chemical basis of colour vision.' *New Scientist,* **10,** 374

RYAN, A. H., and WARNER, M. (1936). 'The effect of automobile driving on the reactions of the drivers.' *Amer. J. Psychol.,* **48,** 403

RYAN, T. A. (1953). 'Muscular potentials as indicators of effort in visual tasks.' In *Fatigue* (eds. Floyd and Welford *q.v.*), p. 109

RYAN, T. A., and BITTERMAN, M. E. (1951). *Investigations of Critical Flicker Fusion Frequency.* Report to the Research Committee of the Illum. Eng. Soc. Reported in Simonson and Brozek (1952).

RYAN, T. A., COTTRELL, C. L., and BITTERMAN, M. E. (1950). 'Muscular tension as an index of effort: the effect of glare and other disturbances in visual work.' *Amer. J. Psychol.,* **63,** 317

SALTER, N., and DARCUS, H. D. (1952). 'The effect of degree of elbow flexion on maximum torques developed in pronation and supination of the right hand.' *J. Anatomy,* **86,** 197

SCHOUTEN, J. F., KALSBEEK, J. W. H., and LEOPOLD, F. F. (1962). 'On evaluation of perceptual and mental load.' *Ergonomics,* **5,** 251

SCHROEDER, F. DE N. (1948). *Anatomy for interior designers.* New York, Whitney

SEASHORE, S. H., and SEASHORE, R. V. (1941). 'Individual differences in simple auditory reaction times of hands, feet and jaws.' *J. exp. Psychol.,* **29,** 342

SENDERS, V. L. (1952). *The effect of number of dials on qualitative reading of a multiple dial panel.* U.S.A.F. Wright Air Development Center Tech. Rept. 52-182

SENDROY, J. JR., and CECCHINI, L. P. (1954). 'Determination of human body surface area from height and weight.' *J. app. Physiol.,* **7,** 1

SEWIG, R. (1936). 'Gestaltung von Instrumenken zum Zwecke der Vereinfachung Misstechnischer Aufgaben.' *z. Instrum. Kde.,* **56,** 349

SHACKEL, B. (1960). 'Electro-oculography: The electrical recording of eye position.' *Proc. 3rd Int. Conf. Med. Electronics,* 323

SHARP, H. M., and PARSONS, J. F. (1951). 'Loss of visibility due to reflections of bright areas.' *Illum. Eng.,* **46,** 581

SIMON, E. W. (1951). 'Effect of stress on performance with preferred and non-preferred instrument design.' *Amer. Psychol.,* **6,** 387

SIMON, J. R., DECROW, T. W., LINCOLN, R. S., and SMITH, K. U. (1954). 'The effect of handedness on tracking accuracy.' *Percept. and Motor Skills,* **4,** 53

SIMONSON, E., and BROZEK, J. (1952). 'Flicker fusion frequency. Background and application.' *Physiol. Rev.,* **32,** 349

SIMONSON, E., and ENZER, E. (1941). 'Measurement of fusion frequency of flicker as a test of fatigue of the central nervous system: observations on laboratory technicians and office workers.' *J. Industr. Hyg.*, **23**, 83

SINCLAIR, D. (1957). *An introduction to functional anatomy*. Oxford, Blackwell

SIPLE, P. A., and PASSEL, C. F. (1945). 'Dry atmospheric cooling in sub-freezing temperatures.' *Proc. Amer. Phil. Soc.*, **89**, 177

SITTIG, J., and FREUDENTHAL, M. (1951). *De Juiste Maat*. Leyden, L. Stafleu

SLEIGHT, R. B. (1948). 'The effect of instrument dial shape on legibility.' *J. app. Psychol.*, **32**, 170

SMITH, A. (1961). 'Seaweed smell earns fish tasters top marks.' *Daily Telegraph*, No. 33142, p. 15, Nov. 7

SMITH, E. L., and LAIRD, D. A. (1930). 'The loudness of auditory stimuli which affect stomach contractions in healthy human beings.' *J. acoust. Soc. Amer.*, **2**, 94

SMITH, F. E. (1955). *Indices of Heat Stress*. M.R.C. Memo No. 29. London, H.M.S.O.

SMITH, J. L. (1952). *Legibility of lettering*. Dept. of Chief Research Officer, British Transport Commission

SMITH, K. U., and WEHRKAMP, R. F. (1951). 'A universal motion analyser applied to psychomotor performance.' *Science*, **113**, 242

SMITH, W. A. S. (1961). 'Effects of industrial music in a work situation requiring complex mental activity.' *Psychol. Rept.*, **8**, 159

SOAR, R. S. (1955). 'Height, width, proportion and stroke width in numeral visibility.' *J. app. Psychol.*, **39**, 43

SOAR, R. S. (1958). 'Numeral form as a variable in numeral visibility.' *J. app. Psychol.*, **42**, 158

SPENCER, J. (1961). 'Observation accuracy in relation to the B.S.I. draft recommendations on industrial instrument scale design.' *Brit. J. app. Physics.*, **12**, 712

SPENCER, J. (1962). Personal communication

SPENCER, J. (1963). 'Pointers for general purpose indicators.' *Ergonomics*, **6**, 35

STEVENS, S. S. (1961). 'To honor Fechner and repeal his law.' *Science*, **80**, 133

STEVENS, S. S. (1962). 'The surprising simplicity of sensory metrics.' *Amer. Psychol.*, **17**, 29

STILES, W. S., BENNETT, M. G., and GREEN, H. N. (1937). *Visibility of light signals with special reference to aviation lights*. Aero Research Committee Rept. No. 1793. London, H.M.S.O.

STOLUROW, L. M., BERGUM, B., HODGSON, T., and SILVA, J. (1955). 'The efficient course of action in "trouble-shooting" as a joint function of probability and cost.' *Educ. psychol. Meas.*, **15**, 462

SUDDON, F. H., and LINK, J. D. (1959). 'Handedness, body orientation and performance on a complex motor task.' *Percept. and Motor Skills*, **9**, 165

SWEENEY, J. H., BAILEY, A. W., and DOWD, J. F. (1957). *Comparative evaluation of three approaches to helicopter instrumentation while in flight.* U.S. Naval Research Lab. Rept. 4952

TANNER, J. M., and WEINER, J. S. (1949). 'The reliability of the photogrammetric method of anthropometry, with a description of a miniature camera technique.' *Amer. J. Phys. Anthrop. N.S.*, **7**, 145

TARRIÈRE, C., and WISNER, A. (1960). 'L'épreuve de vigilance.' *Psychol. Français*, **5**, 261

TAYLOR, F. V. (1957). 'Simplifying the controller's task through display quickening.' *Occup. Psychol.*, **31**, 120

TAYLOR, F. V. (1959). 'The human as an engineering component.' *A.M.A. Arch. Industr. Health*, **19**, 278

TAYLOR, F. V., and GARVEY, W. D. (1959). 'The limitations of a Procrustean approach to the optimisation of man/machine systems.' *Ergonomics*, **2**, 187

TAYLOR, F. W. (1895). 'A piece rate system.' *Trans. A.S.M.E.*, **16**, 861

TAYLOR, F. W. (1915). *The principles of scientific management.* New York, Harper

TEICHNER, W. H. (1954). 'Recent studies of reaction time.' *Psychol. Bull.*, **51**, 128

TEICHNER, W. H., and KOBRICK, J. L. (1955). 'Effects of prolonged exposure to low temperature on visual motor performance.' *J. exp. Psychol.*, **49**, 122

TEICHNER, W. H., and WEHRKAMP, R. F. (1954). 'Visual-motor performance as a function of short duration ambient temperature.' *J. exp. Psychol.*, **47**, 447

TELEKY, L. (1943). 'Problems of night work: influence on health and efficiency.' *Industr. Med.*, **12**, 758

TELFORD, C. W. (1931). 'The refractory phase of voluntary and associative responses.' *J. exp. Psychol.*, **14**, 1

THOMAS, L. F. (1962). 'Perceptual organization in industrial inspectors.' *Ergonomics*, **5**, 429

TINKER, M. A. (1949). 'Trends in illumination standards.' *Trans. Amer. Acad. Opthal. and Otolaryng.*, March-April, 382

TODD, J. W. (1912). *Reaction to multiple stimuli.* Arch. Psychol. N.Y., **3** (25), 65

TUFTS UNIVERSITY (1951). *Handbook of Human Engineering Data.* (2nd edn.). Special Devices Centre Tech. Rept. No. SDC 199-1-2

TURNER, D. (1955). 'The energy cost of some industrial operations.' *Brit. J. industr. Med.*, **12**, 237

TURNER, D. (1958). 'Heat stress in non-ferrous foundries.' *Brit. J. industr. Med.*, **15**, 38

TUSTIN, A. (1947). 'The nature of the operator's response in manual control and its implications for controller design.' *J. Inst. Elect. Eng. (Lond.)*, **94**, 190

UHRBROCK, R. S. (1961). 'Music on the job: its influence on the worker, morale and production.' *Personnel Psychol.*, **14**, 9

VENABLES, P. H., and O'CONNOR, N. (1959). 'Reaction time to auditory and visual stimulation in schizophrenic and normal subjects.' *Quart. J. exp. Psychol.*, **11**, 175

VERNON, H. M. (1924). *On the extent and effect of variety in repetitive work. Pt. A. The degree of variety in repetitive industrial work.* I.F.R.B. Rept. No. 26. London, H.M.S.O.

VERNON, H. M. (1932). 'The measurement of radiant heat in relation to human comfort.' *J. industr. Hyg.*, **14**, 95

VERNON, H. M., and BEDFORD, T. (1927a). *The relation of atmospheric conditions to the working capacity and the accident rate of coal miners.* I.F.R.B. Rept. No. 39. London, H.M.S.O.

VERNON, H. M., and BEDFORD, T. (1927b). *Rest pauses in heavy and moderately heavy industrial work.* I.F.R.B. Rept. No. 41. London, H.M.S.O.

VERNON, H. M., and BEDFORD, T. (1931). *Two studies of absenteeism in coal mines. 1. The absenteeism of miners in relation to short time and other conditions.* I.H.R.B. Rept. No. 62. London, H.M.S.O.

VERNON, M. D. (1946). *Scale and dial reading.* M.R.C. App. Psychol. Res. Unit Rept. No. 59. Cambridge

VINCE, M. A. (1949). 'Rapid response sequences and the psychological refractory period.' *Brit. J. Psychol.*, **40**, 23

VINCE, M. A. (1950). *Learning and retention of an 'unexpected' control display relationship under stress conditions.* M.R.C. App. Psychol. Res. Unit Rept. 125/50. Cambridge

WALKER, J. (1961). 'Shift changes and hours of work.' *Occup. Psychol.*, **35**, 1

WARWICK, M. J. (1947a). 'Direction of movement in the use of control knobs to position visual indicators.' In *Psychol. Research on Equipment Design* (ed. Fitts, P. M.), 137-146. Army Air Forces Aviation Psychol. Program Research Rept. No. 19

WARWICK, M. J. (1947b). 'Direction of motion preferences in positioning indicators by means of control knobs.' *Amer. Psychol.*, **2**, 345

WEHRKAMP, R., and SMITH, K. U. (1952). 'Dimensional Analysis of Motions. II. Travel, distance, effect.' *J. app. Psychol.*, **36**, 201

WEINER, J. S., and LIND, A. R. (1955). 'Working capacity in hot and humid conditions.' *Manager*, **23**, 853

WELFORD, A. T. (1952). 'The "psychological refractory period" and the timing of high-speed performance – a review and a theory.' *Brit. J. Psychol.*, **43**, 2

WELFORD, A. T. (1958). *Ageing and human skill.* London, Oxford University Press

WELFORD, A. T. (1959a). 'Evidence of a single-channel decision mechanism limiting performance in a serial reaction task.' *Quart. J. exp. Psychol.*, **11**, 193

WELFORD, A. T. (1959b). 'Psychomotor performance.' In *Handbook of Ageing and the Individual* (ed. J. Birren), p. 566. Chicago, Univ. Chicago Press

WELFORD, A. T. (1960). 'The measurement of sensory motor performance survey and re-appraisal of twelve years' progress.' *Ergonomics*, **3**, 189

WESSELDIJK, A. TH. G. (1961). 'The influence of shift work on health.' *Ergonomics*, **4**, 281

WESTON, H. C. (1935). *The relation between illumination and industrial efficiency. I. The effect of size of work.* Joint Report Industrial Health Research Board and Illumination Research Committee. London, H.M.S.O.

WESTON, H. C. (1945). *The relation between illumination and visual efficiency – the effect of brightness contrast.* I.H.R.B. Rept. No. 87. London, H.M.S.O.

WESTON, H. C. (1949). *Sight, light and efficiency.* London, H. K. Lewis

WESTON, H. C. (1953). 'Visual fatigue.' *Trans. Illum. Eng. Soc. (Lond.)*, **18**, 39

WESTON, H. C. (1961). 'Rationally recommended illumination levels.' *Trans. Illum. Eng. Soc. (Lond.)*, **26**, 1

WHITE, W. J. (1949). *The effect of dial diameter on occular movements and speed and accuracy of check reading groups of simulated engine instruments.* U.S.A.F. Tech. Rept. No. 5826

WHITE, W. J., WARWICK, M. J., and GRETHER, W. F. (1953). 'Instrument reading. III. Check reading of instrument groups.' *J. app. Psychol.*, **37**, 302

WHITNEY, R. J. (1958). 'The strength of the lifting action in man.' *Ergonomics*, **1**, 101

WHITTENBURG, J. A., ROSS, S., and ANDREWS, T. G. (1956). 'Sustained perceptual efficiency as measured by the Mackworth Clock Test.' *Percept. and Motor Skills*, **6**, 109

WISNER, A. Y. (1961). 'Ergonomics in the Renault Works.' In *Ergonomics in Industry.* London, H.M.S.O.

WISNER, A. Y., and REBIFFÈ, R. (1963). 'Methods of improving work-place layout.' *Int. J. Prod. Res.*, **2**, 145

WOLFF, H. S. (1958). 'The Integrating motor Pneumotachograph.' *Quart. J. exp. Physiol.*, **43**, 270

WOLFF, H. S. (1961). 'The Radio Pill.' *New Scientist,* **12,** 419

WOODHEAD, M. M. (1958). 'The effect of bursts of loud noise on a continuous visual task.' *Brit. J. industr. Med.,* **15,** 120

WOODSON, W. E. (1954). *Human engineering guide for equipment designers.* Berkeley, Univ. California Press

WYATT, S. (1924a). *Two studies on rest pauses in industry. B. Notes on an experiment on rest pauses.* I.F.R.B. Rept. No. 25. London, H.M.S.O.

WYATT, S. (1924b). *On the extent and effect of variety in repetitive work. Pt. B. The effect of changes in activity.* I.F.R.B. Rept. No. 26. London, H.M.S.O.

WYATT, S. (1927). *Rest pauses in industry.* I.F.R.B. Rept. No. 42. London, H.M.S.O.

WYATT, S., and FRASER, J. A. (1925). *Studies in repetitive work with special reference to rest pauses.* I.F.R.B. Rept. No. 32. London, H.M.S.O.

WYATT, S., and FRASER, J. A. (1928). *The comparative effect of variety and uniformity in work.* I.F.R.B. Rept. No. 52. London, H.M.S.O.

WYATT, S., and FRASER, J. A. (1929). *The effect of monotony in work.* I.F.R.B. Rept. No. 56, London, H.M.S.O.

WYATT, S., FRASER, J. A., and STOCK, F. G. L. (1926). *Fan ventilation in a humid weaving shed.* I.F.R.B. Rept. No. 37. London, H.M.S.O.

WYATT, S., and LANGDON, J. N. (1932). *Inspection processes in industry.* I.H.R.B. Report No. 63. London, H.M.S.O.

WYATT, S., and LANGDON, J. N. (1937). *Fatigue and boredom in repetitive work.* I.H.R.B. Rept. No. 77. London, H.M.S.O.

WYATT, S., and LANGDON, J. N. (1938). *The machine and the worker. A study of machine feeding processes.* I.H.R.B. Rept. No. 82. London, H.M.S.O.

WYATT, S., and MARRIOTT, R. (1953). 'Night work and shift changes.' *Brit. J. industr. Med.,* **10,** 164

WYNDHAM, C. H., STRYDOM, N. B., MARITZ, J. S., MORRISON, J. F., PETER, J., and POTGIETER, Z. U. (1959). 'Maximum oxygen intake and maximum heart rate, during strenuous work.' *J. app. Physiol.,* **14,** 927

YAGLOGLOU, C. O., and MILLER, W. E. (1925). 'Effective temperature with clothing.' *Trans. Amer. Soc. Heat. Vent. Engrs.,* **31,** 89

YLLÓ, A. (1962). 'The bio-technology of card punching.' *Ergonomics,* **5,** 75

Author Index

Subject Index